Record Label Marketing

Record Label Marketing

Thomas W. Hutchison
Middle Tennessee State University

Amy Macy
Middle Tennessee State University

Paul Allen
Middle Tennessee State University

ELSEVIER

AMSTERDAM • BOSTON • HEIDELBERG • LONDON
NEW YORK • OXFORD • PARIS • SAN DIEGO
SAN FRANCISCO • SINGAPORE • SYDNEY • TOKYO

Focal Press is an Imprint of Elsevier

Focal Press

Acquisitions Editor:	Catharine Steers
Project Manager:	Andrew Therriault
Assistant Editor:	Stephanie Barrett
Marketing Manager:	Christine Degon
Cover Design:	Eric DeCicco

Focal Press is an imprint of Elsevier
30 Corporate Drive, Suite 400, Burlington, MA 01803, USA
Linacre House, Jordan Hill, Oxford OX2 8DP, UK

 Recognizing the importance of preserving what has been written, Elsevier prints its books on acid-free paper whenever possible.

Library of Congress Cataloging-in-Publication Data
Application Submitted.

British Library Cataloguing-in-Publication Data
A catalogue record for this book is available from the British Library.

ISBN 13: 978-0-240-80787-4
ISBN 10: 0-240-80787-1

For information on all Focal Press publications visit our website at
www.books.elsevier.com

08 09 10   10 9 8 7 6 5 4 3

Printed in the United States of America

Working together to grow
libraries in developing countries

www.elsevier.com | www.bookaid.org | www.sabre.org

ELSEVIER   BOOK AID International   Sabre Foundation

Contents

Acknowledgements

Paul Allen:

Thanks to David Eleazar with the Eleazar Group for his design concept for this book cover; thanks to Tom Hutchison for inviting me to be part of this project; also, thanks to Dr. Marc Singer, Dr. Troy Festervand, Dr. Anantha Babbili, Christian Haseleu, Geoffrey Hull, and Dr. Richard Barnet for their encouragement and support; J. William Denny for opening the door to the music business for me and for his continued friendship; and to Cindy for her love and patience.

Tom Hutchison:

I would like to thank my friends who contributed to this book, including Rob Jacobs, Dexter Story and Edgar Richmond of Universal Music, my dear friend Linda Ury Greenberg of Sony BMG, Vincent Peppe of SESAC, Jed Hilly, Christopher Knab, and Storm Gloor, Dave Yeskel of V2 Records, Jim Donio, Pat Daly and the entire staff of NARM. I would like to thank my former students David Haley and Thad Keim of Compass Records, and Amy Willis of Sony, for their input. Thanks to our colleagues Beverly Keel and Mike Alleyne for reading chapters, thanks to Geoffrey Hull for his advice-legal and personal; to fellow musician Bill Wharton for allowing me to use his website as an example, and to jazz vocalist Inga Swearengen for allowing me a glimpse inside her career. Thanks to my co-writers Amy Macy and Paul Allen, who made this all possible. Praises go out to Chris Palmer for reading the entire manuscript and spending a day with us going over the fine points. And most of all, to my wife Lynne who not only had to put up with this project for a year, but who has also taught me the process of writing and editing books.

Amy Macy:

Tom Hutchison has been such a great colleague and mentor. It was his idea to wrangle Paul Allen and me into writing this book and I thank him for his leadership and guidance through this process and so many others as I have navigated the murky waters of academia. Let me also voice an appreciation to all the great faculty and staff of the Recording Industry Department of Middle Tennessee State University. Its great to work where people are passionate about what they do, are leaders at what they do, and truly enjoy the process of learning. I am honored to be a part of the team.

I have had the great fortune to work with many mentors in my tour of duty within the industry. Thanks to Mike Dungan, Ron Howie, Randy Goodman, Mike Shallett, and Joe Galante. And special thanks to Jim Weatherson, Storm Gloor, and John Conway for their insight into the writing of this book and to Trudy Lartz, VP of Sales and Service for Nielsen SoundScan, who made Chapter 6 possible.

Without Tricia Gentry, our caretaker and dear friend, our family would have been lost – you are appreciated. The writing of this text has inspired a new course offering: "Juggling 101." Thank you Doug Tennant, my gracious and giving husband, who is just as agile, if not more entertaining, at juggling. And to our little girls, Millie and Emma Jo, who are the reason why we learned to juggle.

1 Marketing Concepts and Definitions

Thomas Hutchison

Selling recorded music

The concept of selling recorded music has been around for more than a century. While the actual storage medium for music has evolved, from cylinders to vinyl discs, to magnetic tape, digital discs and now downloads, the basic notion has remained the same: a musical performance is captured to be played back at a later time, at the convenience of the consumer.

In today's marketplace, the consumer is showered with an array of products from which to choose, making the process of marketing more important than ever. Before explaining how records (recorded music) are marketed to consumers, it is first necessary to gain a basic understanding of marketing.

What is marketing?

Marketing is simply defined as the performance of business activities that direct the flow of goods and services from the producer to the consumer (Committee on Definitions, 1960). Marketing involves satisfying customer needs or desires. To study marketing, one must first understand the notions of *product* and consumer (or *market*). The first questions a marketer should answer are, "What markets are we trying to serve?" and, "What are their needs?" Marketers must understand these consumer needs and develop products to satisfy those needs. Then, they must price the products effectively, make the products available

1

in the marketplace, and inform, motivate and remind the customer. In the music business, this involves supplying consumers with the recorded music they desire.

The market is defined as consumers who want or need your product and who have the willingness and ability to buy. This definition emphasizes that the consumer wants or needs something. A product is defined as something that will satisfy the customer's want or need. You may want a candy bar, but not necessarily need one. You may need surgery, but not necessarily want it.

The marketing mix

The marketing mix refers to a blend of product, distribution, promotion and pricing strategies designed to produce mutually satisfying exchanges with the target market. This is often referred to by the *four P's,* which are:

1. Product – goods or services designed to satisfy a customer's need
2. Price – what customers will exchange for the product
3. Promotion – informing and motivating the customer
4. Place – how to deliver and distribute the product

Product

The marketing mix begins with the product. It would be difficult to create a strategy for the other components without a clear understanding of the product to be marketed. New products are developed by identifying a market that is underserved, meaning there is a demand for products that is not being adequately met.

The product aspect of marketing refers to all activities relating to the product development, ensuring that:

- There is a market for the product
- It has appeal
- It sufficiently differs from other products already in the marketplace
- It can be produced at an affordable and competitive price

An array of products may be considered to supply a particular market. Then the field of potential products is narrowed to those most likely to perform well in the marketplace. In the music business, the Artist and Repertoire (A&R) department performs this task by searching for new talent and helping decide which songs will have the most consumer appeal. In other industries, this function is performed by the Research and Development (R&D) arm of the company.

New products introduced into the marketplace must somehow identify themselves as different from those that currently exist. Marketers go to great lengths to position their products to ensure that their customers understand why their product is more suitable for them than the competitor's product. *Product positioning* is defined as the customer's perception of a product in comparison with the competition.

Consumer tastes change over time. As a result, new products must constantly be introduced into the marketplace. New technologies render products obsolete and encourage growth in the marketplace. For example, the introduction of the compact disc (CD) in 1983 created opportunities for the record industry to sell older catalog product to customers who were converting their music collections from the LP (long playing record) to CD. Similarly, when a recording artist releases a new recording, marketing efforts are geared toward selling the new release, rather than older recordings (although the new release may create some consumer interest in earlier works and they may be featured alongside the newer release at retail).

The product life cycle

The product life cycle (PLC) is a concept used to describe the course that a product's sales and profits take over what is referred to as the *lifetime of the product*—the sales window and market for a particular product, from its inception to its demise.

It is characterized by four distinct stages: introduction, growth, maturity and decline. Preceding this is the *product development* stage, before the product is introduced into the marketplace. The *introduction* stage is a period of slow growth as the product is introduced into the marketplace. Profits are nonexistent because of heavy marketing expenses. The *growth* stage is a period of rapid acceptance into the marketplace and profits increase. *Maturity* is a period of leveling in sales mainly because the market is saturated—most consumers have already purchased the product. Marketing is more expensive (to the point of diminishing returns) as efforts are made to reach resistant customers and to stave off competition. *Decline* is the period when sales fall off and profits are reduced. At this point, prices are cut to maintain market share (Kotler, P. and Armstrong, G., 1996).

The PLC can apply to products (a particular album), product forms (artists and music genres), and even product classes (cassettes, CDs and vinyl). Product classes have the longest life cycle—the compact disc has been around since the early 1980s. However, the life cycle of an average album release is 12–18 months.

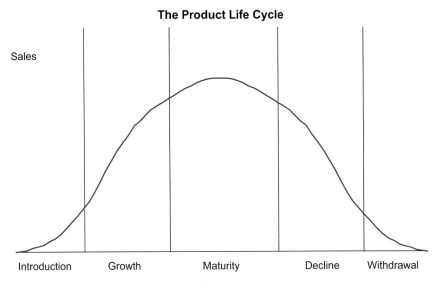

▲ *Figure 1.1 The product life cycle*

Diffusion of innovations

When a product is introduced into the marketplace, its consumption is expected to follow a pattern of diffusion. *Diffusion of innovations* is defined as the process by which the use of an innovation is spread within a market group, over time and over various categories of adopters (*Dictionary of Marketing Terms*, American Marketing Association, 2004). The concept of diffusion of innovations describes how a product typically is adopted by the marketplace and what factors can influence the rate (how fast) or level (how widespread) of adoption. The rate of adoption is dependent on *consumer traits*, the *product,* and the company's marketing efforts.

Consumers are considered *adopters* if they have purchased and used the product. Potential adopters go through distinct stages when deciding whether to adopt (purchase) or reject a new product. These stages are referred to as *AIDA*, which under one model is represented as attention, interest, desire and action. Another model uses awareness, information, decision and action. These stages describe the psychological progress a buyer must go through in order to get to the actual purchase. First, a consumer becomes *aware* that they need to make a purchase in this product category. Perhaps the music consumer has grown tired of his or her collection and directs his/her *attention* toward buying more music. The consumer then seeks out *information* on new releases and begins to gain an *interest* in something in particular, perhaps after hearing a song on the radio or attending a concert. The consumer then makes the *decision* (and *desires*) to purchase a particular recording. The call to *action* is the actual purchase.

This process may take only a few minutes or may take several weeks, depending on the importance of the decision and the risk involved in making the wrong decision. *Involvement* refers to the amount of time and effort a buyer invests in the search, evaluation and decision processes of consumer behavior. If the consumer is not discriminating and dissatisfaction with the decision is not a setback, then the process may take only moments and is thus a low-involvement situation. On the other hand, if the item is expensive, such as a new car, and if consequences of a wrong decision are severe, then the process may take much longer. Several factors influence the consumers' level of involvement in the purchase process, including previous experience, ease of purchase, interest, and perceived risk of negative consequences. As price increases, so does the level of involvement.

Consumers who adopt a new product are divided into five categories:

1. Innovators
2. Early adopters
3. Early majority
4. Late majority
5. Laggards

Innovators are the first 2.5% of the market (Rodgers, 1995) and are eager to try new products. Innovators are above average in income and thus the cost of the product is not of much concern. *Early adopters* are the next 13.5% of the market and adopt once the innovators have demonstrated that the new product is viable. Early adopters are more socially involved and are considered opinion leaders. Their enthusiasm for the new product will do much to assist its diffusion to the majority. The *early majority* is the next 34% and will weigh the merits before deciding to adopt. They rely on the opinions of the early adopters. The late majority represents the next 34%, and these consumers adopt when most of their friends have. The *laggards* are the last 16% of the market and generally adopt only when they feel they have no choice. Laggards adopt a product when it has reached the maturity stage and is being "deep discounted" or is widely available at discount stores. When introducing a new product, marketers target the innovators and early adopters. They will help promote the product through word of mouth.

Products (or innovations) also possess characteristics that influence the rate and level of adoption. Those include:

"Relative Advantage: the degree to which an innovation is perceived as better than what it supersedes. Compatibility: the degree to which an innovation is perceived as consistent with existing values, past experiences, and needs. Complexity: the degree to which an innovation is perceived as difficult to understand and use. Trialability: the degree to which an innovation may be experimented with on a limited basis. Observability: the degree to which the results of an innovation are visible to the receiver and others." (Rogers, E. M., 1995).

One way marketers can increase the potential for success is by allowing customers to "try before they buy." Listening stations in retail stores and online music samples have increased the level of trialability of new music. Marketers can improve sales numbers by ensuring the product has a relative advantage, that it is compatible, that it is not complex, and that consumers can observe and try it before they purchase.

The Boston Consulting Group growth-share matrix

Most companies produce a glut of products, some of which perform better than others. The Boston Consulting Group has come up with a method of classifying a company's products based on market growth and market share. The resulting growth-share matrix can quickly indicate the potential for future sales of each product. Market growth is indicated by the vertical axis, with high growth at the top. Relative market share (relative to the competition) is indicated by the horizontal axis, with high market share located on the left side. Then products are placed in quadrants based on their relative market share and growth potential in that market.

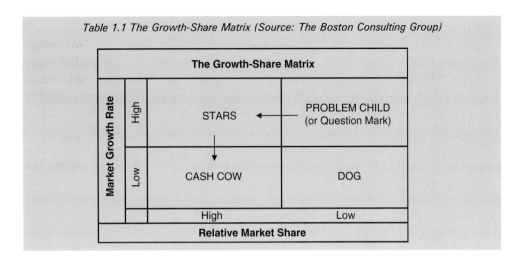

Table 1.1 The Growth-Share Matrix (Source: The Boston Consulting Group)

Stars: Stars are high-growth, high-share entities or products. Since the market for such products is still growing, they consume heavy investment to finance their rapid growth. In the recording industry, a genre of music showing growth (such as rap/hip-hop in the past decade) would be considered in the high market-growth category. An artist with a high relative market share (such as OutKast) would be considered a star.

Cash Cow: These business entities generate a high rate of return for little ongoing investment, and therefore, are very profitable for the company. These are established products with brand recognition and require less marketing dollars per unit sold. In the recording industry, established artists such as Elton John fall in to this category.

Problem Child: The problem child or question mark is a business unit that requires lots of cash and generates little return. They are low-share products in a high-growth market. The goal is to build these into stars by increasing market share while the market is still growing.

Dogs: These are low-growth, low-market share units with no promise of success. For record labels, this may represent artists who are no longer successful (or never had much success) in producing music in a genre that has stopped growing. The dogs should be relegated to niche markets at best.

The overall goal for a company after reviewing the portfolio of products is to attempt to move the problem children into the star category and to groom the star products to become tomorrow's cash cows, while discontinuing the dogs.

Price

Pricing is more complex than, "How much should we charge?" Pricing structure must be based on maximizing marketing success. Once the wholesale and retail pricing is determined, price-based incentives must then be considered. For example, should the product be put on sale? Should coupons be issued? Should the retailers receive wholesale price breaks for quantity orders or other considerations?

There are generally three methods for deciding the retail price of a product: cost-based pricing, competition-based pricing and consumer-based (value-based) pricing.

Cost-based pricing

Cost-based pricing is achieved by determining the cost of product development and manufacturing, marketing and distribution, and company overhead. Then an amount is added to cover the company's profit goal. The weakness with this method is that it does not take into account competition and consumer demand.

$$\text{Break-even point (units)} = \frac{\text{Total fixed costs}}{\text{Price} - \text{Variable costs per unit}}$$

When determining cost-based pricing, consideration must be given to the fixed-costs of running a business, the variable costs of manufacturing products, and the semi-variable costs related to marketing and product development. Fixed-costs include items such as overhead, salaries, utilities, mortgages, and other costs that are not related to the quantity of goods produced and sold. In most business situations, variable costs are usually associated with manufacturing and vary depending on the number of units produced. For recorded music, that would include discs, CD booklets, jewel cases, and other packaging. **However, within the marketing budget of a record label, the costs of manufacturing and mechanical royalties are considered fixed once the number of units to be manufactured is determined and the recording costs have been computed.** Then there are other (semi) variable costs, which can vary widely, but are not directly related to the quantity of product manufactured. They would include recording costs (R&D) and marketing costs. For purposes of the formula below, the costs of marketing are considered variable costs, and the recording and manufacturing costs, having already been determined for the project, are considered fixed at this point.

$$\text{Break-even point (units)} = \frac{\text{Total fixed costs}}{\text{Price} - (\text{total variable costs/units})}$$

Table 1.2 Standard Business model vs. Record Label Business model

	Variable costs	Fixed costs	Semi-variable
Standard business model	Manufacturing cost per unit, royalties, licensing, or patent fees per unit, packaging per unit, shipping and handling per unit	Overhead, salaries, utilities, mortgages, insurance	Marketing, advertising, research and development (R&D)
Record label business model	All marketing aspects, including advertising, discounts, promotion, publicity expenses, sales promotions, etc.	Overhead, salaries, utilities, mortgages, insurance, manufacturing costs, production costs (A&R)	

Competition-based pricing

Competition-based pricing attempts to set prices based on those charged by the company's competitors—rather than demand or cost considerations. The company may charge more, the same, or less than its competitors, depending on its customers, product image, consumer loyalty and other factors.

Companies may use an attractive price on one product to lure consumers away from the competition to stimulate demand for their other products. This is known as *leader pricing.* The item is priced below the optimum price in hopes that sales of other products with higher margins will make up the difference. The most extreme case of this is *loss leader pricing*. Under loss leader pricing, the company actually loses money on one product in an attempt to bring in customers who will purchase other products that are more profitable. Compact discs are commonly used by retailers as *loss leaders* because it will generate traffic to the store and increase sales of other items.

Companies are prohibited from coordinating pricing policies to maximize profits and reduce competitive pricing. Such *price fixing* is illegal under the

9

Sherman Act (1890), which prevents businesses from conspiring to set prices (Finch, J. E., 1996). The Robinson-Patman Act (1936) prohibits any form of price discrimination, making it illegal to sell products to competing buyers at different prices unless those price differentials can be justified.

Value-based (consumer-based) pricing

Compact discs are among those items priced based upon the consumer's perceived value, rather than on any cost-based or competition-based factors. Consumer-based pricing uses the buyer's perceptions of value, a reversal from the cost-plus approach. The music business has at times sought to maintain a high-perceived value for compact discs by implementing controversial pricing programs such as the *minimum advertised price (MAP)*. In the 1990s, labels sought to discourage retailers from using CDs as loss leaders over concerns that, a) consumers would balk at paying full retail price for CDs that were not offered at the reduced price, and b) traditional retail stores would be driven out of business.

Promotion

Promotion includes the activities of advertising, personal selling, sales promotion and public relations. It involves informing, motivating and reminding the consumer to purchase the product. In the recording industry, the four traditional methods of promotion include: radio promotion (getting airplay), advertising, sales promotion (working with retailers), and publicity. More recently, record labels have become more aggressive in marketing through street teams, Internet marketing, digital distribution, tour support, and tie-ins with other products.

Push vs. pull strategy

There are two basic promotion strategies: *push promotion* and *pull promotion*. A *push* strategy involves "pushing the product through the distribution channel to its final destination in the hands of consumers." Marketing activities are directed at motivating channel members (wholesalers, distributors and retailers) to carry the product and promote it to the next level in the channel. In other words, wholesalers would be motivated to inspire retailers to order and sell more product. This can be achieved through offering monetary incentives, discounts, free goods, advertising allowances, contests, and display allowances.

All marketing activities are directed toward these channel members and are regarded as *trade* promotion and *trade* advertising. With a push strategy, channel members are motivated to "push" the product through the channel and ultimately on to the consumer.

Under the pull strategy, the company directs its marketing activities toward the final consumer, creating a demand for the product that will ultimately be fulfilled as requests for product are made from the consumer to the retailer, and then from the retailer to the wholesaler. This is achieved by targeting consumers through advertising in consumer publications and creating "consumer promotions." With a pull strategy, consumer demand pulls the product through the channels.

Most large companies employ both strategies in a combination of consumer advertising and consumer promotions (coupons and sale items), and trade promotion (incentives).

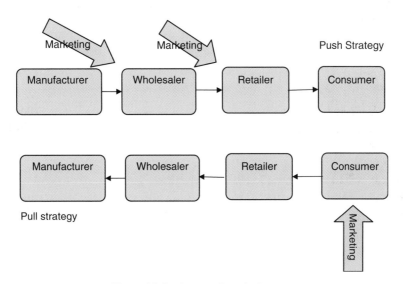

▲ *Figure 1.2 Push vs. pull marketing strategy*

Place

This aspect of marketing involves the process of distributing and delivering the products to the consumer. Distribution strategies entail making products available to consumers when they want them, at their convenience. The various methods of delivery are referred to as *channels of distribution*. The process of distribution in the record business is currently in a state of evolution, as digital distribution becomes a reality. Questions remain as to how rapidly this new method of delivery will become adopted in the marketplace and what effect that will have on traditional distribution channels that deal with physical product.

Distribution systems

Most distribution systems are made up of *channel intermediaries* such as wholesalers and retailers. These channel members are responsible for processing large quantities of merchandise and dispersing smaller quantities to the next level in the channel (such as from manufacturer to wholesaler to retailer). Manufacturers engage the services of distributors or wholesalers because of their superior efficiency in making goods widely accessible to target markets (Kotler, P., 1980).

The effectiveness of intermediaries can be demonstrated in the following chart. In the first example, no intermediary exists and each manufacturer must engage with each retailer on an ongoing basis. Thus, the number of contacts equals the number of manufacturers multiplied by the number of customers or retailers (M, X, C). In the second example, a distributor is included with each manufacturer and each customer contacting only the distributor. The total number of contacts is the number of manufacturers plus the number of customers (M + C).

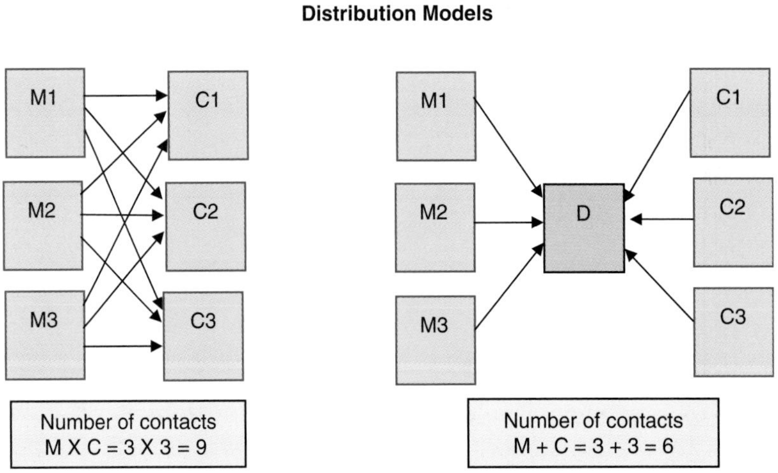

Distribution Models

Number of contacts
M X C = 3 X 3 = 9

Number of contacts
M + C = 3 + 3 = 6

▲ *Figure 1.3 Distribution models*

Types of distribution systems

The three basic types of distribution systems are corporate, contractual, and administered. All three are employed in the record business.

Corporate systems involve having one company own all distribution members at the next level in the channel. A record label would use a distribution system owned by the parent company, such as the relationship between Mercury Records and Universal Music and Video Distribution. Both companies are owned by Universal. This type of ownership is often referred to as *vertical integration*.

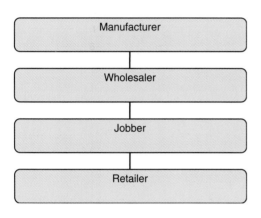

Distribution Chain

▲ *Figure 1.4 Distribution chain*

When a manufacturer owns its distributors, it is called *forward vertical integration*. When a retailer such as Wal-Mart develops its own distribution or manufacturing firms, it is called *backward vertical integration*.

Contractual distribution systems are formed by independent members who contract with each other, setting up an agreement for one company to distribute goods made by the other company. Independent record labels commonly set up such agreements with independent record distributors. Before the major record labels developed in-house distribution systems in the 1960s and 70s, nearly all recorded music was handled through this type of arrangement.

In *administered distribution systems*, arrangements are made for a dominant channel member to distribute products developed by an independent manufacturer. This type of arrangement is common for independent record labels that have agreements to be handled by the distribution branch of one of the major labels.

Retail

Retailing consists of all the activities related to the sale of the product to the final consumer for personal use. Retailers are the final link in the channel of distribution. There are numerous types of retailers, each serving a special niche in the retail environment. Independent stores focus on specialty products and customer service, while mass merchandisers such as Wal-Mart concentrate on low pricing. Online retailing of physical product through Amazon.com, CD Baby, and others has been growing in the past few years and the potential for selling music through downloads is creating both hope and fear among industry insiders.

In recent years, retailers have developed marketing tools to measure the effectiveness of product location in their stores. Universal product codes (UPC), or bar codes, and computerized scanners have helped retailers to determine the optimal arrangement of products within the store. They can move products around to various locations and note its effects on sales. As a result, retailers are charging manufacturers for prime space in the store, including counter space, end caps and special displays. Retail stores sometimes charge *slotting fees*—a flat fee charged to the manufacturer for placement of (mostly introductory) products on the shelves for a limited period of time. If the product fails to sell, the retail store has reduced its risk. A similar concept has developed in the record business, but it is also tied to promotion and favorable location for the product. It is referred to as *price and position* and will be discussed in the chapter on retailing.

Marketing strategy

The marketing strategy is the fundamental marketing plan by which the company intends to achieve its marketing objectives. It consists of a coordinated set of decisions based on target markets, the marketing mix, and the marketing budget. The marketing strategy is the "road map" or guide to growth and success.

The marketing plan for a company should start with an assessment of the company's internal strengths and weaknesses, followed by an examination of the market environment. This situation analysis is sometimes referred to as a *SWOT analysis* (Strengths, Weaknesses, Opportunities and Threats). It is a review of the company's present state, and an evaluation of the external and internal factors that can affect future success.

As a part of the SWOT analysis, the *internal marketing audit* is designed to examine the company's areas of strength (the company's competitive advantages) and weaknesses (the company's vulnerable areas and underperforming units). Then, an assessment is conducted on the external opportunities (areas for growth) and threats (roadblocks and challenges) that exist for the company. The resulting investigation yields the SWOT analysis—strengths, weaknesses,

opportunities and threats. A SWOT assessment allows the marketing department to determine:

- Where the company is right now
- Where the company needs to go
- How to best get there

A SWOT analysis for a record label may yield some of the following findings.

Table 1.3 The SWOT analysis

Strengths	Weaknesses
• Strong roster	• Underperforming artists on roster
• Strong catalog	• Expenses are too high
• Skilled label personnel	• Unhappy artists leaving label (or suing)
• Effective distribution	• Artist legal problems
• Abundant finances	• Loss of label personnel
• Reputation	
Opportunities	**Threats**
• New technologies	• Piracy
• New markets (international or domestic)	• Censorship
• Piracy controls	• Competition
• Trends swing your direction	• Economic downturn
• International trade agreements	• Loss of retail opportunities
	• New technology

The SWOT helps the marketing department determine marketing strategy,
including the identification of suitable markets and appropriate products to meet
the demand. The company can then engage in sales forecasting, budgeting, and
projection of future profits.

In the book *Marketing: Relationships, Quality, Value*, Nickels and Wood use
Motown Records in their example of a SWOT analysis (Nickels, W. G. and Wood,
M. B., 1997). Motown's main strength is its strong catalog of 1960s and 70s
music, including the Supremes, Four Tops, The Jackson Five, Michael Jackson
and Boyz II Men. A weakness is their lack of start-up urban and rap artists
compared to other labels such as Interscope. Opportunities exist in the nostalgia
fads that periodically emerge, creating a demand for the old recordings.
A threat they mention is "a shift in popular interest away from the label's music."

The SWOT analysis helps the company discover its core competencies, which
enables the company to surpass the competition. However, it is the product

marketing plan that helps to coordinate the strategic efforts to promote a particular product. It is a carefully planned strategy "with specialists in the areas of artist development, sales, distribution, advertising, promotion, and publicity joining forces in a coordinated effort to break an artist and generate sales" (Lathrop, T. and Pettigrew, J., 1999).

Glossary

Channels of distribution – The various methods of distributing and delivering products ultimately to the consumer.

Competition-based pricing – Attempts to set prices based on those charged by the company's competitors.

Consumer-based pricing – Using the buyer's perceptions of value to determine the retail price, a reversal from the cost-plus approach.

Cost-based pricing – Determining retail price based on the cost of product development and manufacturing, marketing and distribution, and company overhead and then adding the desired profit.

Diffusion of innovations – The process by which the use of an innovation (or product) is spread within a market group, over time and over various categories of adopters.

EAN – The European Article Numbering system. Foreign interest in UPC led to the adoption of the EAN code format, similar to UPC but allows extra digits for a country identification, in December 1976.

Involvement – The amount of time and effort a buyer invests in the search, evaluation and decision processes of consumer behavior.

Loss leader pricing – The featuring of items priced below cost or at relatively low prices to attract customers to the retail store.

Marketing – The performance of business activities that direct the flow of goods and services from the producer to the consumer (American Marketing Association, 1960). Marketing involves satisfying customer needs or desires.

Marketing mix – A blend of product, distribution, promotion and pricing strategies designed to produce mutually satisfying exchanges with the target market. Often referred to as the four P's.

Price fixing – The practice of two or more sellers agreeing on the price to charge for similar products.

Product class – A group of products that are homogeneous or generally considered as substitutes for each other.

Product form – Products of the same form make up a group within a product class.

Product life cycle – The course that a product's sales and profits take over what is referred to as the lifetime of the product.

Product positioning – The customer's perception of a product in comparison with the competition.

Pull strategy – The company directs its marketing activities toward the final consumer, creating a demand for the product that will ultimately be fulfilled as requests for product are made from the consumer.

Push strategy – Pushing the product through the distribution channel to its final destination through incentives aimed at retail and distribution.

Situation analysis – The systematic collection and study of past and present data to identify trends, forces, and conditions with the potential to influence the performance of the business and the choice of appropriate strategies.

SWOT analysis – An examination of both the internal factors (to identify strengths and weaknesses) and external factors (to identify opportunities and threats).

Universal Product Code (UPC) – An American and Canadian coordinated system of product identification by which a ten-digit number is assigned to products. The UPC is designed so that at the checkout counter an electronic scanner will read the symbol on the product and automatically transmit the information to a computer that controls the sales register.

Bibliography

American Marketing Association (1960). *Marketing Definitions: A Glossary of Marketing Terms, Committee on Definitions,* Chicago: American Marketing Association.
Committee on Definitions (1960). *Marketing Definitions: A Glossary of Marketing Terms,* Chicago: American Marketing Association.

Dictionary of marketing terms: American Marketing Association (2004). www.marketing power.com.

Finch, J. E. (1996). *The Essentials of Marketing Principles,* Piscataway, NJ: REA.

Kotler, P. and Armstrong G. (1996). *Marketing: An Introduction,* Upper Saddle River, NJ: Prentice Hall.

Kotler, P. (1980). *Principles of Marketing,* Englewood Cliffs: Prentice-Hall.

Lathrop, T. and Pettigrew, J. (1999). *This Business of Music Marketing and Promotion,* New York: Billboard Books.

Nickels, W. G. and Wood, M. B. (1997). *Marketing: Relationships, Quality, Value,* New York: Worth Publishers.

Rogers, E. M. (1995). *Diffusion of Innovations (4th edition),* New York: The Free Press.

2 Markets, Market Segmentation and Consumer Behavior

Thomas Hutchison

Before a marketing plan can be designed, it is necessary to fully understand the market. It is not possible to make decisions about where to advertise, where to distribute the product, how to position the product, and so forth, until a thorough examination of the market is conducted. This chapter will explain how markets are identified, segmented, and how marketers learn to understand groups of consumers and their shopping behavior. A market is defined as a set of all actual and potential buyers of a product (Kotler, P., 1980). The market includes anyone who has an interest in the product and has the ability and willingness to buy. Markets are identified by characteristics that are measurable. The basic goal of market segmentation (subdividing a market) is to determine the target market.

Target markets

Simply stated, a *target market* consists of a set of buyers who share common needs or characteristics that the company decides to serve (www.marketing-teacher.com, 2004). Once market segments have been identified, the next step is choosing the target markets within those segments. Target marketing is choosing the segment or segments that will allow the organization to most effectively achieve its marketing goals. To get a product or service to the right person or company, a marketer would initially segment the market, then target a single segment or series of segments, and finally position the product within the segment (Finch, J. E., 1996).

Market segmentation

Because some markets are so complex and composed of people with different needs and preferences, markets are typically subdivided so that promotional efforts can be customized—tailored to fit the particular submarket. For most products, the total potential market is too diverse or heterogeneous to be treated as a single market. To solve this problem, markets are divided into submarkets called *market segments*. Market segmentation is defined as the process of dividing a large market into smaller segments of consumers that are similar in characteristics, behavior, wants or needs. The resulting segments are homogenous with respect to characteristics that are most vital to the marketing efforts. This segmentation may be made based on gender, age group, purchase occasion, or benefits sought. Or they may be segmented strictly according to their needs or preferences for particular products.

For example, the fast food chain McDonalds caters to a wide variety of customers, from kids to grandparents. But McDonalds does not use the same marketing messages and techniques to reach children as they do to reach adults. For children, McDonalds has developed special meal packages, recreational areas and their signature clown. To reach young adults with healthy appetites and meager finances, they developed the "supersize" menu and the 99-cent menu. They reach working adults who need fast food fast by providing packaged breakfast items and hot coffee through their drive-through service windows.

Marketers segment markets for several reasons:

1. It enables marketers to identify groups of consumers with similar needs and interests and get to know the characteristics and buying behavior of the group members.
2. It provides marketers with information to help design custom marketing mixes to speak to the particular market segment.
3. It is consistent with the marketing concept of satisfying customer wants and needs.

In order to be successful, segmentation must meet these criteria:

1. **Substantiality** – the segments must be large enough to justify the costs of marketing to that particular segment.
2. **Measurable** – marketers must be able to conduct an analysis of the segment and develop an understanding of their characteristics. The result is that marketing decisions are made based on knowledge gained from analyzing the segment.

3. **Accessible** – the segment must be reachable through existing channels of communication and distribution.
4. **Responsiveness** – the segment must have the potential to respond to the marketing efforts in a positive way, by purchasing the product.

The process of segmenting markets is done in stages. In the first step, segmentation variables are selected and the market is separated along those partitions. The most appropriate variables for segmentation will vary from product to product. The appropriateness of each segmentation factor is determined by its relevance to the situation. After this is determined and the market is segmented, each segment is then profiled to determine its distinctive demographic and behavioral characteristics. Then the segment is analyzed to determine its potential for sales. The company's target markets are chosen from among the segments determined at this stage.

Market segments

There is no single correct way to segment markets. Segmentation must be done in a way that maximizes marketing potential. This is done by successfully targeting each market segment with a uniquely tailored plan—one that addresses the particular needs of the segment. Markets are most commonly segmented based upon a combination of geographics, demographics, personality or psychographics, and actual purchase behavior. The bases for determining geographic and demographic characteristics are quite standardized in the field of marketing. Psychographics, lifestyle, personality, behavioristic, and purchase characteristics are not as standardized, and the categorization of these variables differs from textbook to textbook. Evans and Berman (Evans, J. R. and Berman, B., 1992) combine these into one category of consumer life-styles. Kotler (1980) breaks these down into psychographics and behavioristic. Psychographics includes personality, beliefs, lifestyle and social class. Behavioristic includes both attitudes toward the product and actual purchase behavior. Hall and Taylor segment according to demographics, geographics, psychographics, buyer thoughts and feelings, and purchase behavior (Hall, C. and Taylor, F., 2000).

Geographics

Geographic segmentation involves dividing the market into different geographical units such as towns, cities, states, regions and countries. Markets also may be segmented depending upon population density, such as urban, suburban or rural. Location may reflect a difference in income, culture, social values, and types of media outlets or other consumer factors (Evans, J. R. and Berman, B., 1992). Media research companies such as Arbitron and Nielsen use geographic units called the *area of dominant influence (ADI)* or *designated market area (DMA)*. DMA is defined by Nielsen as an exclusive geographic area of counties in which the home market television stations hold a dominance of total hours viewed. The American Marketing Association describes both ADI and DMA as the geographic area surrounding a city in which the broadcasting stations based in that city account for a greater share of the listening or viewing households than do broadcasting stations based in other nearby cities (The American Marketing Association, 2004). Following is an index chart for contemporary hit radio (CHR) listening by geographic region.

This chart shows the American national listenership to CHR radio by geographic region. With the national index, or average, being 100, it is easy to see which regions of the country tend to prefer CHR and which ones do not listen to the format as much as the average.

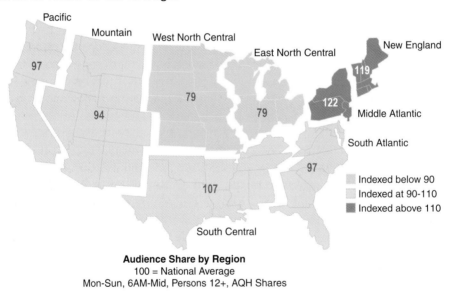

Audience Share by Region
100 = National Average
Mon-Sun, 6AM-Mid, Persons 12+, AQH Shares

▲ *Figure 2.1 Audience share by region – CHR (Source: Arbitron)*

When the marketer of recorded music and live performances uses geographic segmentation in this manner, it can help develop the logic for the use of resources to support the marketing plan. Marketing strategies are then easily tailored to particular geographic segments. For example, tour support relies on geographics to determine which cities to include in the promotion.

As the label's act tours from city to city, the marketing department geographically shifts the focus of their efforts ahead of the actual performance date in each market.

Personal demographics

Personal demographics are basic measurable characteristics of individual consumers and groups such as age, gender, ethnic background, education, income, occupation and marital status. Demographics are the most popular method for segmenting markets because groups of people with similar demographics have similar needs and interests that are distinct from other demographic segments. Also, demographics are easier and cheaper to measure than more complex segmentation variables such as personality and consumer behavior.

Age is probably the demographic most associated with changing needs and interests. Consumers can be divided into several age categories such as children, teens, young adults, adults and older adults. Recently, the segment of "tweens" has been added to the mix to account for the enormous spending power of this group of 23 million preadolescents between the ages of 8 and 13. This concept of segmenting based on age has even been extended to dog food, with companies positioning their various products to appeal to different stages of canine life.

Gender is also a popular variable for segmentation, as the preferences and needs of males are perceived as differing from those of females. Differential needs based upon gender are obvious for product categories such as clothing, cosmetics, hairdressing and even magazines. But even in the area of music preferences, differences in taste exist for males and females. Arbitron reports that males are more likely to listen to alternative, rock or news/talk radio, while women prefer top 40, country, adult contemporary and urban radio (Arbitron Radio Today: How America Listens to Radio, 2004).

Income segmentation is popular among certain product categories such as automobiles, clothing, cosmetics and travel but is not as useful in the recording industry. *Educational* level is sometimes used to segment markets. Well-educated consumers are likely to spend more time shopping, read more, and are more willing to experiment with new brands and products (Kotler, P., 1980). Segmentation by *race* is also important as certain product classes are targeted toward one race over another, including specific genres of music.

Multivariable segmentation

The process of combining two or more demographic or other variables to further segment the market has proven effective in accurately targeting consumers. By combining age, gender and income, marketers can better tailor the marketing messages to reach each group. For example, older males may prefer news/talk radio, whereas younger males prefer modern rock radio.

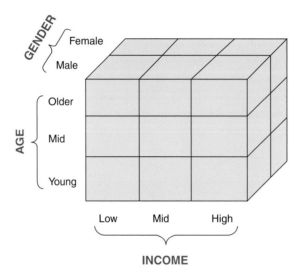

▲ *Figure 2.2 Multivariable segmentation*

Psychographics

Psychographic segmentation involves dividing consumers into groups based upon lifestyles, personality, opinions, motives or interests. Psychographic segmentation is designed to provide additional information about what goes on in minds. While demographics may paint a picture of what consumers are like, psychographics adds vivid detail, enabling marketers to shape very specific marketing messages to appeal to the target market.

By understanding the motive for making purchases, marketers can emphasize the product attributes that attract buyers. For example, if consumers are driven by price, pricing factors such as coupons can be emphasized. If another segment is driven by convenience, this issue can be addressed through widespread product distribution. Lifestyle segmentation divides consumers into groups according to the way they spend their time and the relative importance of things in their life.

Psychographics are more difficult to measure and are a moving target. Marketers are constantly updating information to stay abreast of changes in the marketplace. One such system is the VALS2 (values and lifestyles) segmentation system, developed by the Stanford Research Institute. VALS™ is a marketing and consulting tool that identifies current and future marketing opportunities by segmenting the consumer markets on the basis of the personality traits that drive consumer behavior. The system places consumers into three self-orientation categories of people primarily inspired either by ideals, achievement or self-expression. The market is further subdivided into eight segments by adding the variables of resources and innovation. Those consumers who possess abundant resources are placed in the upper categories. For example, *thinkers* are conservative and motivated by ideals; *experiencers* are young and enthusiastic and motivated by self-expression, excitement and innovation. More information is available at http://www.sric-bi.com/VALS/.

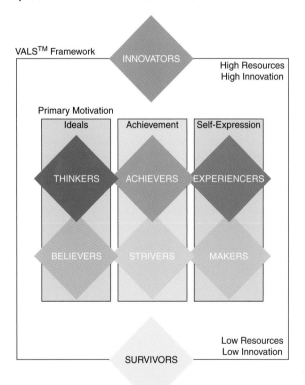

▲ *Figure 2.3 Values and lifestyle segments (Source: SRI Consulting Business Intelligence)*

Researchers at the University of Texas have found that personal music preferences can be linked to personality traits. Rentfrow & Gosling (Rentfrow, P.J.

& Gosling, S.D., 2003) found that people's music preferences typically classify them into one of four basic dimensions: 1) reflective and complex, 2) intense and rebellious, 3) upbeat and conventional, or 4) energetic and rhythmic. Preference for each of the following music dimensions is differentially related to one of these basic personality traits.

Genres	Personality
Classical, jazz, blues, folk	Reflective and complex
Alternative, rock, heavy metal	Intense and rebellious
Country, pop, religious, soundtracks	Upbeat and conventional
Rap/hip-hop, soul/funk, electronica/dance	Energetic and rhythmic

The quest for cool

Marketers use a variety of research techniques to spot lifestyle trends as they develop, so that their products will be in the marketplace ahead of the demand and the competition. Known in vernacular terms as "cool hunting" or the "quest for cool," these marketers are hired by well-known brands to track and understand youth culture. Then the principles of diffusion theory are applied to target opinion leaders in the hopes of penetrating the market (Grossman, L., August 31, 2003). Cool hunting involves observing alpha consumers (cool people) talking, eating, dressing or shopping and then predicting what the rest of the market will be doing a year from now. One such company, Look-Look^{TM}, uses a variety of research techniques including surveys, field observations, ethnographies, mall intercepts and focus groups. They document their findings with photographs and video recordings and report their conclusions to a variety of major companies who are targeting the youth market (Look-Look, TM company website, 2004).

The millennial generation

Neil Howe and William Strauss discuss the emergence of a millennial generation, which they define as the generation of young adults, the first of whom have "come of age" around the time of the millennium (Howe, N. and Strauss, W., 2000). This segment includes people born between 1982 and 1995. The estimates range between 60 and 74 million, and they are expected to be a huge economic force. This segment is characterized by a sharp break from Generation X, and hold values that are at odds with the baby boomers. They are described as optimists—team players who follow the rules and accept authority. For this market segment, technology is a part of life, and staying connected to others is important (Bock, W., 2002).

Behavioristic

Behavioral segmentation is based on actual customer behavior toward products. Behavioral segmentation has the advantage of using variables that are closely related to the product and its attributes. Some of the more common behavioral segmentation variables are: benefits sought, user rate, brand loyalty, user status, readiness to buy and purchase occasion.

Product usage

Product usage involves dividing the market based on those who use a product, potential users of the product, and nonusers—those who have no usage or need for the product. How a product is used (and for what purposes) is also of importance. For example, Kodak found that disposable cameras were being placed on tables for wedding guests to help themselves to. As a result, Kodak modified the product by offering five-pack sets of cameras in festive packaging (Nickels, W. G, and Wood, M., 1997). The recording industry has adapted to usage situations by repackaging music that is customized for usage occasion, such as wedding music compilations, party mixes, romantic mood music, and so forth.

Benefits sought

It is said that people do not buy products, they buy benefits. They buy aspirin to alleviate pain; toothpaste to whiten or protect teeth; and music to elevate mood, combat boredom, provide companionship, or create atmosphere. Benefit segmentation divides the market according to benefits sought by consumers. In the toothpaste market, some consumers are seeking a product that whitens teeth, while others may consider fresh breath most important. Still others may seek fluoride protection against cavities or protection against gum disease. To address these segments, toothpaste companies have all created product variations to accommodate each segment.

User status

User status defines the consumer's relationship with the product and brand. It involves level of loyalty and propensity to become a repeat buyer. Stephan and Tannenholz identify six categories of consumers based on user status: sole users, semi-sole users, discount users, aware non-triers, trial/rejectors, and repertoire users.

User Status

Sole users are the most brand loyal and require the least amount of advertising and promotion.

Semi-sole users typically use brand A, but have an alternate selection if it is not available or if the alter nate is promoted with a discount.

Discount users are the semisole users of competing brand B. They don't buy brand A at full price, but perceive it well enough to buy it at a discount.

Aware non-triers are category users, but haven't bought into brand A's message.

Trial/rejecters bought brand A's advertising message, but didn't like the product.

Repertoire users perceive two or more brands to have superior attributes and will buy at full price. These are the primary brand switchers; therefore, the primary target for brand advertising.

Stephan and Tannenholz

i Stephan and Tannenholz in Arens, William F. (2002). Market Segmentation and the Marketing Mix: Determinants of Advertising Strategy. McGraw-Hill Online Learning Center. http://highered.mcgraw-hill.com/sites/0072415444/student_view0/chapter5/els.html

▲ *Figure 2.4 User status (Source: Stephan and Tannenholz)*

Consumer behavior

Consumer behavior is defined as the buying habits and patterns of consumers in the acquisition and usage of goods and services. According to the American Marketing Association, consumer behavior is defined as "the dynamic interaction of affect and cognition, behavior, and environmental events by which human beings conduct the exchange aspects of their lives." One consumer behavior construct is the hierarchy of effects model that combines involvement and left-right brain processing to determine the process a consumer goes through in making a purchase decision.

High and low-involvement decision making

Involvement refers to the amount of time and effort a consumer invests in the search, evaluation and decision process of consumer behavior (See AIDA models in chapter one). The level of involvement in the purchase depends on the risk of making the wrong decision and the product's economic and social importance to the consumer. Consumers who search for information about products and brands in order to evaluate them thoroughly are engaging in *high involvement decision making*. They want to know as much as possible about all choices before making their decision.

In low *involvement decision-making*, the consumer perceives very little risk, low identification with the product and little or no personal relevance. Low involvement decision-making is often habitual, associated with consumer package goods (prone to repeat purchases) and characterized by brand familiarity and loyalty.

Several factors can influence a consumer's level of involvement in a purchase process:

1. **Previous experience** – If consumers have previous experience with the product or brand, they are more likely to have a low level of involvement.
2. **Interest** – The more personal interest consumers have in the product, product class, or benefits associated with the product, the more likely they are to engage in high involvement behavior.
3. **Perceived risk of negative consequences** – If the stakes are high and the consequences of making the wrong decision are dire, consumers engage in high involvement behavior. Price is a variable associated with risk. High-priced purchases bear more risk. As price increases, so does the level of involvement.
4. **Situation** – Circumstance may play a role in temporarily raising the level of involvement. A food or beverage purchase casually made under most circumstances may call for more scrutiny if guests are coming to dinner. A CD bought for personal use is a rather low involvement purchase; when given as a gift it has higher involvement because of the risk of a bad decision.
5. **Social visibility** – Involvement increases as the social visibility of a product increases. Designer clothing falls into this situation.

Decision-making process

Chapter 1 mentioned the stages involved in the consumer's decision-making process. They were identified as awareness, information, decision and action (AIDA). This is part of the hierarchy of effects theory, which involves a series of steps by which consumers receive and use information in reaching decisions about what actions they will take (e.g., whether or not to buy a product).

A more elaborate model involves the following six steps:

1. **Problem recognition** – The consumer realizes that a purchase must be made. Perhaps a new need has emerged, or perhaps the consumer's supply of a commonly used product has run low.
2. **Information search** – The consumer will seek out information about the various options available for satisfying the want, need, or desire.
3. **Alternative evaluation** – The consumer will look for alternate opinions of the product under consideration.
4. **Purchase or acquisition** – The consumer makes the purchase.
5. **Use** – The consumer uses the product, thereby gaining personal experience.
6. **Post-purchase evaluation** – Experience with the product leads to evaluation to determine if the product meets expectations and should be purchased again next time the need arises.

Cognitive vs. emotional decisions

Much research has been done in the field of advertising on the relative roles of affective factors vs. cognitive factors in motivating consumers. Cognition-based attitudes are thoughts and beliefs about an attitude object. Affect-based attitudes are emotional reactions to the attitude object. Attitudes can have both of these components, and can be based more or less on either cognitions or affect. While there is probably an element of both in each purchase decision, the relative contribution of each as well as the process involved, may differ depending upon the type of product and the level of involvement.

The model of the hierarchy of effects has been subsequently modified to include three constructs: *cognition* (awareness or learning), *affect* (feeling, interest or desire), and *conation* (action or behavior) – CAB. The order in which these steps are taken is subject to the type of product and the level of involvement. According to Richard Vaughn, product category and level of involvement may determine the order of effects as well as the strength of each effect

(Vaughn, R., Feb/Mar, 1986). His model uses involvement (high/low) and think/ feel (cognitive or affective components) as the two dimensions for classifying product categories and ordering these three steps.

To begin with, high-involvement situations suggest that the *cognitive* stage and the *affective* stage usually appear first, and these two stages are followed by the *conation* (behavioral) stage. In low-involvement situations, advertising may create awareness first, but attitudes or brand preferences are formed after product trial or experience. Thus, conation or action occurs before opinions are formed about the product. In other words, the thinking and/or feeling occur before the action in high-involvement situations, whereas opinions are formed only after trial of the product in low-involvement situations. Low involvement products are more subject to impulse purchases.

The resulting model is a four-quadrant grid with quadrant one (informative) containing high involvement cognitive products, quadrant two (affective) for high involvement emotional purchases, quadrant three (habitual) for low involvement rational purchases, and quadrant four (satisfaction) for low-involvement emotional purchases. Music purchases generally fall into the satisfaction quadrant, although as prices increase, so does the level of involvement. Some music purchases can be considered impulse buys, if correctly positioned at retail. Other purchases may require creating the emotional (affective) response in the customer. Since music by its very nature evokes emotional responses, exposure to the product is known to increase purchase rate.

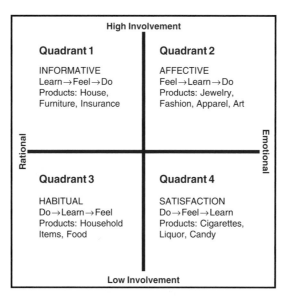

▲ *Figure 2.5 Hierarchy of effects (Source: Foote, Cone and Belding)*

Needs and motives

When a product is purchased, it is usually to fill some sort of need or desire. There is some discrepancy between the consumer's actual state and their desired state. In response to a need, consumers are motivated to make purchases. Motives are internal factors that activate and direct behavior toward some goal. In the case of shopping, the consumer is motivated to satisfy the want or need. Understanding motives is a critical step in creating effective marketing programs. Psychologist Abraham Maslow developed a systematic approach of looking at needs and motives.

Maslow's hierarchy of needs

Maslow states that there is a hierarchy of human needs. More advanced needs are not evident until basic needs are met first. Maslow arranges these needs into five categories of physiological, safety, love, esteem and self-actualization. Marketers believe that it is important to understand where in the hierarchy the consumer is before designing an effective marketing program.

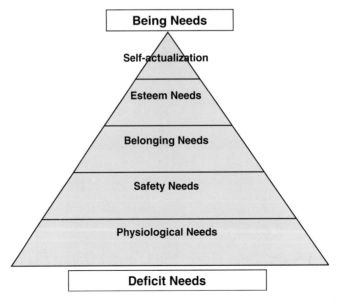

▲ *Figure 2.6 Maslow's hierarchy of needs (Source: Abraham Maslow)*

Physiological needs are the basic survival needs that include air, food, water, sleep, and sex. Until these needs are met, there is little or no interest in fulfilling higher needs. Survivors in the VALs system fit into this category. Safety needs deal with establishing a sense of security. Safety needs motivate consumers to seek political and religious solutions. Until these needs are met, the need for love and belongingness are absent. Love, approval and belongingness are the next level along with feeling needed and a sense of camaraderie. Products oriented toward social events sometimes appeal to this need and this may include music. Esteem needs drive people to seek validation and status. Self-actualization involves seeking knowledge, self-fulfillment, and spiritual attainment.

Converting browsers to buyers

One of the greatest challenges facing marketers is converting potential customers who are browsing the merchandise into actual purchasers. Retail stores refer to a conversion rate—the percentage of shoppers who actually make a purchase. The actual rate varies and depends upon the level of involvement, with small-ticket items having a higher conversion rate to expensive items such as appliances or jewelry. A mall store will have more walk-in traffic and thus a lower conversion rate than a freestanding "destination" store. Conversion rates can be improved by having the right combination of merchandise, presenting the merchandise correctly and offering a level of service that meets or exceeds expectations. When shoppers don't buy, it's often because they were unable to find what they wanted.

Online, conversion rates are improved by offering purchase opportunities with a minimum amount of consumer effort. This is why many online stores such as Amazon.com offer accounts to customers with their address and credit card information stored in the system. The consumer can purchase with one mouse click.

Glossary

ADI – Area of Dominant Influence (see DMA).

Affect – A fairly general term for feelings, emotions, or moods.

Behavioral segmentation – Based on actual customer behavior toward products.

Cognition – All the mental activities associated with thinking, knowing, and remembering.

Conation – represents intention, behavior, and action.

Consumer behavior – The dynamic interaction of affect and cognition, behavior, and environmental events by which human beings conduct the exchange aspects of their lives.

DMA – Designated Marketing Area. The geographic area surrounding a city in which the broadcasting stations based in that city account for a greater share of the listening or viewing households than do broadcasting stations based in other nearby cities.

Geographic segmentation – Dividing the market into different geographical units such as town, cities, states, regions and countries. Markets also may be segmented depending upon population density, such as urban, suburban or rural.

Involvement – The amount of time and effort a consumer invests in the search, evaluation and decision process of consumer behavior.

Market – A set of all actual and potential buyers of a product. The market includes anyone who has an interest in the product and has the ability and willingness to buy.

Market segmentation – The process of dividing a large market into smaller segments of consumers that are similar in characteristics, behavior, wants or needs. The resulting segments are homogenous with respect to characteristics that are most vital to the marketing efforts.

Personal demographics – Basic measurable characteristics of individual consumers and groups such as age, gender, ethnic background, education, income, occupation and marital status.

Psychographic segmentation – Dividing consumers into groups based upon lifestyles, personality, opinions, motives or interests.

Target market – A set of buyers who share common needs or characteristics that the company decides to serve.

User status – The consumer's relationship with the product and brand. It involves level of loyalty.

Bibliography

Arbitron Radio Today: How America Listens to Radio (2004). http://www.arbitron.com/downloads/radiotoday04.pdf.

Bock, W. (2002). http://www.mondaymemo.net/010702feature.htm.

Evans, J. R. and Berman, B. (1992). *Marketing (fifth edition),* New York: Macmillan Publishing.

Finch, J. E. (1996). *The Essentials of Marketing Principles,* Piscataway: Research and Education Association.

Grossman, L. (August 31, 2003). The Quest for Cool, *Time Magazine.*

Hall, C. and Taylor, F. (2000). *Marketing in the Music Industry (4th edition),* Boston: Pearson Custom Publishing.

Howe, N. and Strauss, W. (2000). *Millennials Rising: The Next Great Generation,* New York: Vintage Books.

Kotler, P. (1980). *Principles of Marketing,* Englewood Cliffs: Prentice-Hall.

Look-Look company website (2004). http://www.look-look.com/dynamic/looklook/jsp/What_We_Do.jsp.

Marketing Teacher (2004). http://www.marketingteacher.com/Lessons/lesson_positioning.htm.

Nickels, W. G, and Wood, M. (1997). *Marketing,* New York: Worth Publishers.

Rentfrow, P. J. and Gosling, S. D. (2003). The Do-Re-Mi's of Everyday life: Examining the structure and personality correlates of music preferences, *Journal of Personality and Social Psychology,* **84**, pp. 1236–1256.

The American Marketing Association (2004). www.marketingpower.com.

Vaughn, R. (Feb/Mar, 1986). How Advertising Works, *Journal of Advertising Research.*

3 The U.S. Industry Numbers

Thomas Hutchison

Sales trends

In the early 1990s, the recording industry was in the midst of the *replacement cycle*, where consumers were still replacing their old vinyl and cassette collections with the first digital format—the compact disc (CD). In 1994, there was a 20% increase in sales, to over $12 billion. This boom was fueled by discount retailers such as Best Buy and Circuit City aggressively entering the music

Table 3.1 RIAA data on annual shipments (Source: RIAA)

| | RIAA Data on annual shipments | | |
Year	CD dollar value*	CD units shipped*	Price per unit
1993	$6511.4	495.4	$13.14
1994	$8464.5	662.1	$12.78
1995	$9377.4	722.9	$12.97
1996	$9934.7	778.9	$12.75
1997	$9915.1	753.1	$13.17
1998	$11416.0	847.0	$13.48
1999	$12816.3	938.9	$13.65
2000	$13214.5	942.5	$14.02
2001	$12909.4	881.9	$14.64
2002	$12044.1	803.3	$14.99
2003	$11232.9	745.9	$15.06
2004	$12154.7	814.1	$14.90
			*in millions

market and discounting CD prices. A calculation of the Recording Industry of America Association (RIAA) figures indicates that the average selling price of a CD dropped from $13.14 in 1993 to $12.78 in 1994. Since then, the retail price has increased to an average of $15 in 2003, but CD sales are still sluggish at best.

By 1995, the CD replacement cycle was complete and the impact of the changing retail landscape was beginning to take its toll on the bottom line. Traditional retail stores were struggling to compete with discount stores, and sales gains in the industry overall were modest (from $12.068 billion in 1994 to $12.320 billion in 1995). Blockbuster closed hundreds of stores causing a massive rush of returned product. There was slight improvement in 1996, with sales at $12.5 billion as the industry sought to re-examine its strategy for selling pre-recorded music (RIAA, 1997), while traditional retailers continued to struggle. The industry experienced a decline in 1997 as the RIAA commented that "the industry was responding to a smaller but healthier retail base" (RIAA, 1998). But it was one faced with bankruptcy filings and consolidation. One anomaly in 1997 occurred with a 54.4% increase in the sale of CD singles, attributed to Elton John's remake of *Candle in the Wind*, a tribute to Princess Diana. Growth returned briefly in 1998 as shipments grew by 5.7%. While shipments of singles dropped, CD units and music video units showed a healthy increase. The RIAA attributed the increase to a steady flow of releases by top artists throughout the year and an increase in the diversity of offerings. The moderate growth continued through 1999 with a 3.2% increase in units, fueled by strong growth in the full-length CD format (12.3% in value), despite a drop-off in music video sales. Credit is given to retailers and suppliers for improved efficiency in inventory management. The RIAA stated that the music industry was successfully competing against the "ever-increasing competition for the consumer's entertainment dollar" (RIAA, 2000).

The new millennium ushered in a period of decline in recorded music sales as the industry struggled against threats on numerous fronts. The market for CD singles plummeted as the Internet took over, and peer-to-peer file-sharing services such as Napster were blamed for much of the downturn. From an industry high of $14.58 billion in 1999, the industry dropped nearly 20% in four years to end at $11.85 billion in 2003.

The decline began in the first half of 2000 and was attributed to the rapid drop in sales of CD singles, brought on by the Internet. (The U.S. economy also experienced a downturn at that time.) Cassettes continued to decline, while full-length CDs grew slightly in dollar value. The decline was more drastic in 2001 as

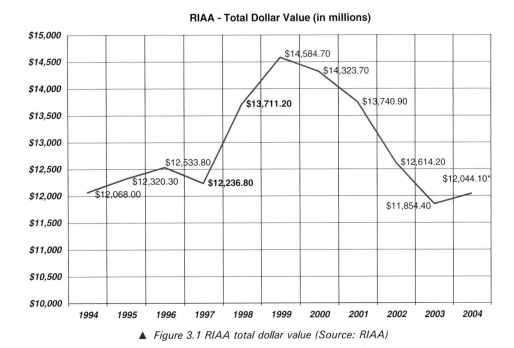

▲ *Figure 3.1 RIAA total dollar value (Source: RIAA)*

total units dropped 10%, with a 6.4% drop in full-length CDs for the first time. This downturn coincided with the rapid adoption of music downloading from illegal peer-to-peer (P2P) services and ownership of CD burners, up drastically in two years.

In 2002, the market dropped another 11.2% in units and 8% in value. Every configuration saw a decrease in sales except DVD. At this point, the RIAA began an aggressive campaign to discourage consumers from using illegal file-sharing services, but other factors also contributed to the downturn. Young consumers were spending their entertainment budget on cell phones, computers, video games, DVD collections, and other forms of entertainment. Consumers began to report (through research studies) that they perceived the cost of CDs to be too high. Upon further analysis, young consumers, who had grown accustomed to cherry-picking songs from albums through P2P services, were opposed to paying retail price for an album just to own the one or two songs they wanted. These consumers were beginning to burn their own personal mix CDs either from their own collection or from P2P services. As a result, they preferred music a-la-carte and were a receptive market for licensed music downloading services. But these services were slow to develop and the offerings were not sufficient to entice droves of consumers into subscription services.

39

Things did not improve for 2003 as the industry fell another 7.2% in units and 6% in value. Again full length CDs, the industry's moneymaker, fell 6.7% in value and 7% in units. The industry began to see some evidence of bottoming out, and even a slight turnaround in 2004 with total sales of $12.154 billion. This turnaround was fueled by a slight increase in CD album sales (1.9% in dollar value), digital singles (139.4 million units), and DVD videos (66% unit increase).

Annual sales trends

Music sales are seasonal, with the greatest majority of sales occurring during the fourth quarter holiday season. For this reason, many of the superstars wait until the fourth quarter to release a new album, so they can take advantage of the holiday shopping season. And conversely, many newer artists will avoid releasing an album in this same time period because competition for retail space is more intense. Valentine's Day is the second largest record purchasing holiday (if Thanksgiving weekend is considered a part of the Christmas holiday season).

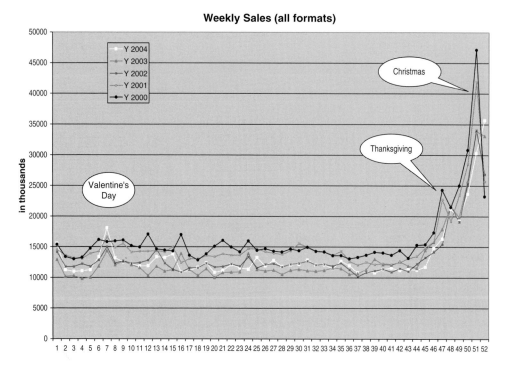

▲ *Figure 3.2 Weekly sales (Source: SoundScan)*

Music genre trends

While the category of rock music has dominated the marketplace for many years, the double-digit lead held by rock music for decades has been reduced from 35% of the market share to near 25% over the past decade. The major shift occurred in 1998 as the steady rise of rap and the return of pop began to siphon off market share.

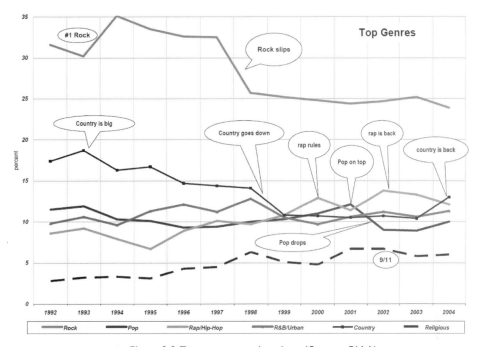

▲ *Figure 3.3 Top genres: market share (Source: RIAA)*

Country music reached a market share peak in 1993 when Garth Brooks was its top seller. Pop music suffered in the mid-1990s but came back strong with the emergence of the "boy bands" and teen pop idols Britney Spears and Jessica Simpson. That trend peaked in 2001, and since then, pop music has suffered a drop in market share. Rap music was on a steady rise throughout the 1990s until emerging as the number two genre in 2000. After a setback in 2001, rap began to rise again in 2002 to hold on to the number two spot. R&B music enjoyed a growth in market share until reaching a peak in 1998. Religious music enjoyed a banner year in 1998 with the *Prince of Egypt* soundtrack, and took another upturn after the attacks of September 11, 2001.

To examine the other genres with smaller market share, a logarithmic scale was created to expand the lower end of the spectrum. The following chart shows the relative market share of these genres.

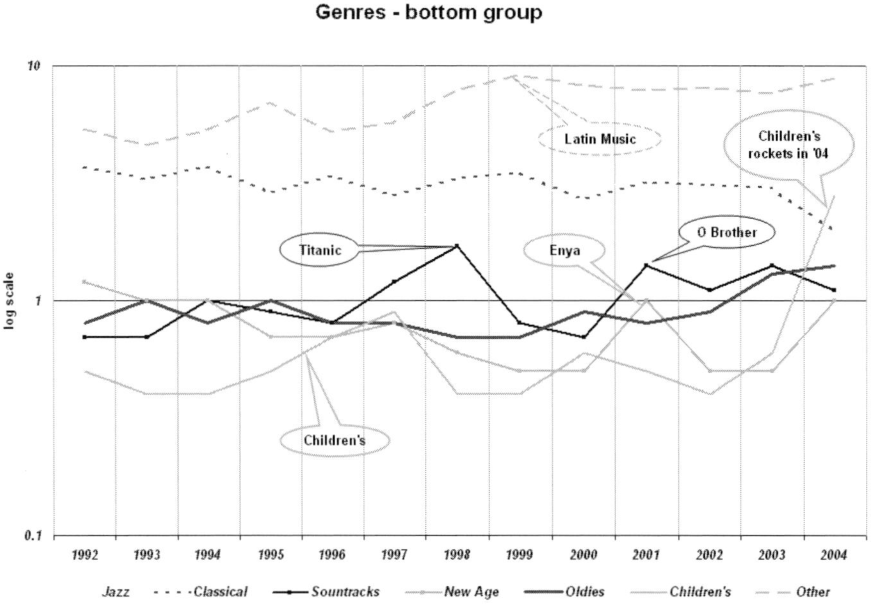

Genres - bottom group

▲ *Figure 3.4 Genre market share: bottom group (Source: RIAA)*

In the smaller genres, a breakout release (one that sells outside the niche market) can elevate the market share for the genre for the duration of the release. For example, a blockbuster album from Enya in 2001 finished the year at number four on the *Billboard 200* year-end chart. Movie soundtracks had strong showings in 1998 and 2001, assisted by the top-selling soundtrack to the *Titanic* in 1998 and *O Brother* in 2001. The growth of Latin music has contributed to the rise in the other category, posting a 12% gain in 1998 and another 8% in 1999 before falling 7% in 2000. The "other" category also includes big band, swing, ethnic, spoken word, holiday, comedy, exercise, electronic, standards and folk. Children's music has been boosted by the introduction of Radio Disney in 1996.

Demographic trends

The RIAA only measures age and gender for their consumer profile studies.

Age

The RIAA reports that record buyers are getting older. The 45+ group has expanded from 15.1 percent in 1994 to 26.6 percent in 2003. Perhaps it is time to further divide this group to gain more insight into differences between what people buy in their 40s to those in their 60s. College age consumers (20–24) began to decline in 1997, a trend that was assisted by peer-to-peer file-sharing services starting in 1999. The decline has continued through 2003. The P2P services have also been blamed for a decline in the teen age group of 15–19 year olds starting in 1999, despite the popularity of acts geared toward teens during this time. A gradual decline in the 25–29 group was halted in 2003, while the 10–14 group has remained relatively steady.

▲ Figure 3.5 Market share by age group (Source: RIAA)

Gender

While males did not exactly dominate the market in the early 90s, they comprised more than half of all record buyers until 1997. The metal, rock and grunge acts of the early 90s sent males to the record stores. This started to change in 1997 as many female artists who appeal to females began to dominate the charts. The Lillith Fair brought female fans to concerts and inspired sales for many of the artists appearing on the tour. The Spice Girls were also popular with younger females as were teen pop acts such as N'Sync, Britney Spears and Backstreet Boys. This changed again in 1999 and 2000 when rap music and rage rock brought male fans back to the record stores despite the fact that Britney Spears finished in the top five both years. Females were once again in the majority in 2001–03 with blockbusters from Enya, Destiny's Child, Alicia Keys, Dixie Chicks, Shania Twain, Norah Jones and Beyonce.

Gender

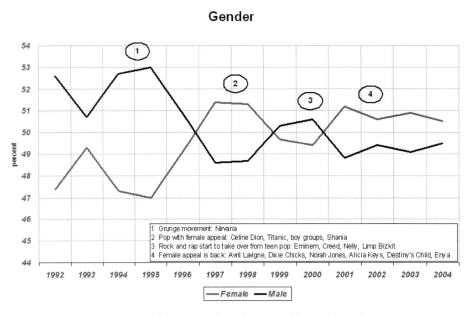

1 Grunge movement: Nirvana
2 Pop with female appeal: Celine Dion, Titanic, boy groups, Shania
3 Rock and rap start to take over from teen pop: Eminem, Creed, Nelly, Limp Bizkit
4 Female appeal is back: Avril Lavigne, Dixie Chicks, Norah Jones, Alicia Keys, Destiny's Child, Enya

▲ *Figure 3.6 Market share by gender (Source: RIAA)*

Format – configuration

The category of configurations, as reported by the RIAA, deals with the different types of recorded music product such as CDs, cassettes, singles in all formats, music video in two formats, the extended-play EP and vinyl LP, Super Audio CD (SACD), and will ultimately include downloads of singles and albums. They are reported as market share in the RIAA consumer profile and as units and dollar value in the year-end sales table. The product life cycle (from Chapter 1) can be used to examine the rise and fall of configurations. At the present time, the CD is the dominant format for albums, yet seems to be in the mature stage of the life cycle. All physical formats for singles are in the decline stage, with the market for downloaded singles entering the growth stage. The new formats of DVD audio and digital downloads are showing growth since their introduction in 2001. The growth in digital downloads has created a recent turnaround in the category for all singles as noted in the chart below.

Configuration

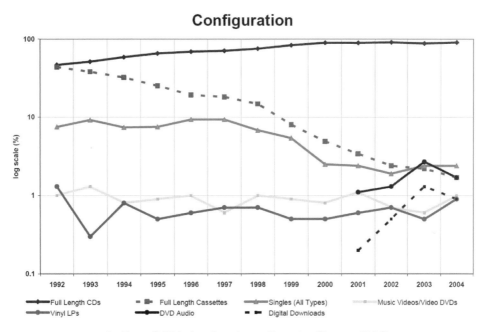

▲ *Figure 3.7 Market share by configuration (Source: RIAA)*

Outlets

The RIAA reports on where records are sold. Trends indicate that traditional retail record stores are no longer the source of the majority of record sales. This loss of market share spells trouble for new artists and indie labels that have traditionally relied on record retail stores to introduce new artists to consumers. Traditional record stores carry a larger variety of titles and are more likely to carry titles that are obscure or geared toward a niche market. With the loss of these stores, many indies and self-promoted artists have turned to the Internet to generate sales and provide product to consumers. Traditional record stores held 60% of the market in 1992, and this has dwindled to just 33% in 2003.

The category of "other stores" includes mass merchants such as Wal-Mart, Target, and K-Mart, as well as "big box" electronic superstores like Best Buy and Circuit City. This segment has grown from less than ¼ of the market in 1992 to over half in 2003. These retailers are less likely to carry the depth of product of the record stores, and the mass merchants concentrate on the top sellers. Record clubs remained popular through the mid-1990s, but have fallen off in recent years as consumers have finished the CD replacement cycle.

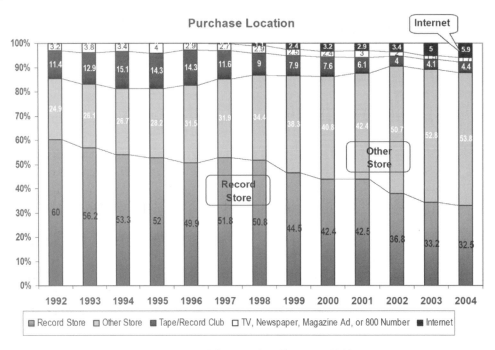

▲ *Figure 3.8 Sales outlets (Source: RIAA)*

Market share of the majors

Label market share is measured for both new releases and for a combination of new releases and recent sales of catalogue titles. Universal has dominated market share since acquiring PolyGram Records in 1998 for $10.4 billion. It is impressive that Universal has managed to grow additional market share since the purchase of PolyGram. The 2004 merger of Sony and BMG has created a second behemoth in the industry and put the joint venture just behind Universal at number two. In addition to Universal, BMG has increased its market share over 1998, with the largest growth coming in 1999 and 2000. BMG was in partnership with Jive Records who, at that time, was responsible for the teen hit sensations N'Sync, Britney Spears and Backstreet Boys. Since then, BMG has struggled to regain market share.

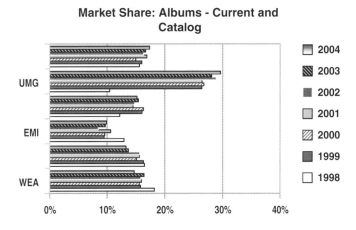

▲ *Figure 3.9 Market share of majors: current and catalog (Source: SoundScan)*

One of the strengths of a record label can be measured by the relative market share for new releases and indicates the vitality of its A&R skills. Labels that are increasing market share of the sales of new releases are those with success in finding and developing new acts and releasing successful new recordings from their established acts. The following table shows that only Universal and the indies have been successfully increasing market share of new releases over the past several years, again with BMG surging for a while and then dropping back. But BMG still holds a larger market share for new releases than it does for the combination of catalog and new releases (see charts below).

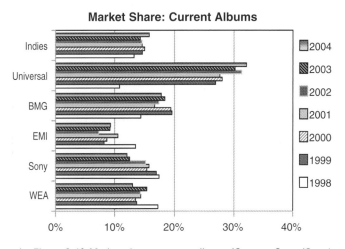

▲ *Figure 3.10 Market share: current albums (Source: SoundScan)*

Comparison of all and current albums

In the following charts, a comparison is made for market share of all albums and market share of new releases for each label. So not only is an increase or decrease in current album market share an indicator, but a comparison of all albums and current albums indicates how much a label relies on catalog sales compared to sales from its new releases.

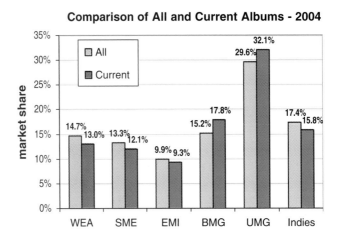

▲ *Figure 3.11 Comparison of all and current albums, 2004 (Source: SoundScan)*

Of the major labels, only BMG and UMG have a larger market share for current releases than for all album releases. When Sony and BMG are combined, the new company shows strength in current titles.

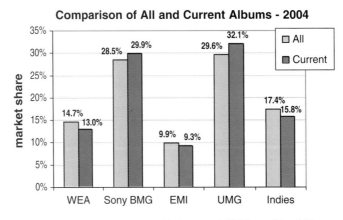

▲ *Figure 3.12 Comparison of market share with Sony and BMG combined (Source: SoundScan)*

Catalog sales

Catalog sales are defined as sales of records that have been in the marketplace for over 18 months. Current catalog titles are those over 18 months old but less than 36 months old. Deep catalog albums are those over 36 months since the release date.

When the compact disc was first introduced in the early 1980s, it fueled the sales of older catalog albums as consumers replaced their old vinyl and cassette collections with CDs. This windfall allowed labels to enjoy huge profits and led to the industry expansion of the 1980s and early '90s. However, catalog sales started to diminish in the mid-1990s as consumers finished replacing their collections. The closure of traditional retail stores also contributed to the decline in catalog sales, with customers having fewer opportunities to be exposed to the older titles.

> "The business was saved by the introduction of the compact disc, or the CD, in the early '80s. That reversed the decline, and created another 15 years of growth and profits for the business as, over the course of that period of time, people converted from buying ... what was then a $6.98 cassette to $13.98 CDs, you know.
>
> "And for the first couple of years that most people got a CD player they would actually replace a lot of their collection. So classical sales went through the roof and old recordings went through the roof, and the business therefore bounced back, and had another 10 to 15 years of double-digit growth,,,"
>
> - Danny Goldberg
> Chairman and CEO, Artemis Records
>
> Published with the permission of FRONTLINE, a documentary series on PBS. This material originally was published on FRONTLINE's web site "The Way the Music Died" www.pbs.org/frontline/shows/music.

▲ Figure 3.13 Impact of CDs on sales (Source: PBS Frontline)

Pricing strategies to sustain catalog sales were introduced as early as 1996. In the late '90s labels sought new ways to promote catalog sales through reissues, compilations, and looking at new formats. Michael Omansky, senior VP of strategic marketing at RCA, stated in 1999, "We [now] put out what I think is substantially better product [on Elvis], with substantially more unreleased material" (Morris, C., 1999). An article in the LA Times in 2001 stated, "companies have watched catalog sales slip from about 50% to 38% over the last decade." (Phillips, C., 2001). In 2002, recent catalog sales fell 14.2% for the year, while deep catalog sales declined 7% (Morris, C., 2003). In 2003, catalog sales fell 7.5% to 232.3 million and deep catalog was down 6.2% to 165.9 million (Hiestand, J., 2004). In 2004, catalog sales made up 35.8% of sales. Major record labels are anticipating that digital downloads and the availability of deep catalog CDs through online retailers will reinvigorate the sale of catalog recordings. Already there is evidence of this. Catalog sales accounted for 46% of digital downloads in 2004, while they accounted for only 35% of sales of physical product.

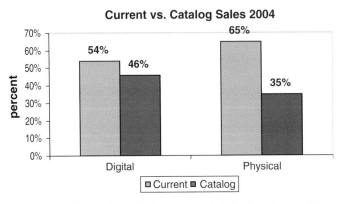

▲ *Figure 3.14 Comparison of physical and digital sales (Source: NARM)*

Industry concentration

More and more, the industry is relying on just a few massive-selling hits to drive the industry, rather than spreading the wealth around with a plethora of profitable releases. In 1999, the top ten country titles accounted for 30% of all country sales, up from 20% in 1995 (Roland, T., 2000). Ed Christman of *Billboard Magazine* reported in 2001 that less than one-half of 1% of releases accounted for more than 50% of all units sold and that "about 3% of the total universe of available albums in the U.S. last year accounted for more than four-fifths of all album sales" (Christman, E., 2001). In 2002, Christman reported that the number of releases had jumped 5.4% from 2001 to 33,433, while sales of those releases dropped by 10% (Christman, E., 2003). There were 38,269 releases in 2003, and 44,476 releases in 2004. Of the 265 million unit sales of albums released in 2004, only 100 titles made up nearly 50% of unit sales. Over 70% of the 265 million album sales in 2004 came from less than 1% of the releases.

According to Christman, new releases for 2002 averaged 7,871 sales per title, down from 9,291 sales per title in 2001 for a 15.3% drop. (In 2000, the average was 8,350 units scanned per title.) More than half of albums released in 2004 sold less than 100 units each, and 81% sold less than 1,000 units each (SoundScan, 2005).

While these findings are not conclusive, they suggest the fact that the sale of recordings is becoming more concentrated—a top-heavy structure—with just a few major titles accounting for a vast majority of sales in the industry.

Conclusions

This chapter is meant to illustrate trends within the U.S. recording industry. While the numbers and percentages will change from year to year, the goal here is to present the tools from which to draw conclusions in the future as new data becomes available. By looking at trends for music genres, demographics of buyers, configurations, outlets where music is sold, label market share, the proportion of catalog to current releases, and the proportion of blockbuster hits to the total number of releases, one can make inferences about the state of the industry. Updated information can be found annually at www.RecordLabelMarketing.com or www.riaa.com.

Glossary

BMG – Bertelsmann Music Group. One of the major labels and is now in partnership with Sony.

Catalog – Older album releases that still have some sales potential. Recent catalog titles are those released for over 18 months but less than 36 months. Deep catalog: those titles over 36 months.

Dollar value – The monetary worth of a stated quantity of shipped product multiplied by the manufacturers suggested retail price of a single unit. The value of shipments is given in U.S. dollars.

EMI – Electrical and Mechanical Industries. One of the major labels.

Logarithmic scale – A scale of measurement that varies logarithmically with the quantity being measured. A constant ratio scale in which equal distances on the scale represent equal ratios of increase. A method of displaying data (in powers of ten) to yield maximum range, while keeping resolution at the low end of the scale.

Market share – A brand's share of the total sales of all products within the product category in which the brand competes. Market share is determined by dividing a brand's sales volume by the total category sales volume.

P2P (Peer-to-peer) – Electronic file swapping systems that allow users to share files, computing capabilities, networks, bandwidth and storage.

Product configuration – Any variety of "delivery system" on which pre-recorded music is stored. Various music storage/delivery mediums include the full-length CD album, CD single, cassette album or single, vinyl album or single, mini-disc, dual-disc, DVD audio, DVD or VHS music video.

Replacement cycle – Consumers replacing obsolete collections of vinyl records and cassettes with a newer compact disc format.

SME – Sony Music Entertainment. One of the major labels and is now in partnership with BMG.

UMG – Universal Music Group. One of the major labels.

Units shipped – The quantity of product delivered by a recording manufacturer to retailers, record clubs, and direct and special markets, minus any returns for credit on unsold product.

WEA – The distribution arm of WMG. (Stands for Warner, Elektra, Atlantic)

WMG – Warner Music Group. One of the major labels.

Bibliography

Christman, E. (April 26, 2003). Average Sale of Albums Dropped in '02 as Labels Released More, Sold Less, *Billboard Magazine.*

Christman, E. (April 28, 2001). SoundScan Numbers Show .35% of Albums Account for More Than Half of All Units Sold, *Billboard Magazine.*

Hiestand, J. (Jan. 02, 2004). Music sales off in '03, but decline slows, *Hollywood Reporter. http://www.hollywoodreporter.com/thr/article_display.jsp?vnu_content_id=2059949.*

Morris, C. (April 26, 2003). Biz Seeks Boost for Ailing Catalog Sales, *Billboard Magazine.*

Morris, C. (December 11, 1999). Labels Search For New Ways to Jump-Start Catalog Sales, *Billboard Magazine.*

Phillips, C. (May 31, 2001). Record Label Chorus: High Risk, Low Margin, *Los Angeles Times.*

RIAA (1997). 1996 Yearend Marketing Report on US Recording Shipments, *RIAA press release.*

RIAA (1998). 1997 Yearend Marketing Report on US Recording Shipments, *RIAA press release.*

RIAA (2000). Recording Industry Releases 1999 Yearend Marketing Report, *RIAA press release.*

Roland T. (January 6, 2000). Country Music's Sales lag Persists, *The Tennessean.*

SoundScan, (2005). aud.soundscan.com.

4 Record Label Operations

Amy Macy

Label operations

Every record label is uniquely structured to perform at its best. Often times, the genre of music along with the "talent pool" of actual label personnel dictate the organization and inner-workings of the company. As talent is signed to a label, the "artist" will come in contact with nearly every department in the process of creating a music product for the market place.

A typical record label has many departments with very specific duties. Depending on the size of the label, some of these departments may be combined, or even outsourced, meaning that the task the department fulfills is hired out to someone not on staff of the label. But the end result is the same—to create a viable music product for the marketplace. In the following structure, there is usually a general manager or senior vice president of marketing who coordinates all the marketing efforts.

Getting started as an artist

As talent is being "found" or developed, the first contact with a record label is usually the artist & repertoire department.

Timeline
of the "artist" as they come in contact with every department

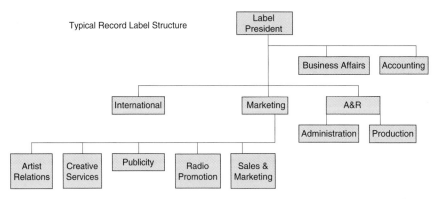

Typical Record Label Structure

▲ *Figure 4.1 Typical record label structure*

The A&R department is always on the hunt for new talent as well as songs for developing and existing artists to record. But before the formal A&R process occurs, the talent has to be signed to the label.

A contract may take months to negotiate. Depending on the agreement and nature of the deal, the artist/manager/lawyer may sign a contract a year or more prior to street date of the first release

Business affairs

The *business affairs department* is where the lawyers of a label reside. Record company lawyers are to negotiate in the label's best interest. Most often, new talent will have a manager and lawyer working on their behalf, with the contract in the middle. Clearly, the label wants to protect itself and hopes to reduce risk by maximizing the contract.

Besides being the point person for artist contracts, label lawyers negotiate and execute many other types of agreements, including:

- License of recordings and samples to third parties
- Negotiate for the right to use specific album art

"The label's business affairs attorney negotiates agreements between the label and other parties, such as recording artists and independent producers (who may have deals with particular artists whereby the producers own the master recordings). The attorney also negotiates distribution agreements between the label and wholesale distributors that sell CDs to retail stores. Licensing agreements are also key, including contracts with online and mobile music services that offer on-demand streaming and downloads. And from day to day, the attorney advises label executives about legal issues related to marketing, advertising and promotion."

Vincent Peppe
Director and Counsel, Internet and
New Media Licensing, SESAC

▲ *Figure 4.2 Business affairs quote*

- Point person when an artist asks for an accounting or audit of royalties
- Renegotiates artist contracts
- Contractual disputes such as delivery issues
- Conflicts with contract such as being a guest artist on a project for another label
- Vendor contracts and relations

Often times, the accounting department falls within business affairs, since the two are related regarding contractual agreements and financial obligations. The accounting department is the economic force driving all the activity within the record company. It takes money to make money, and the accounting department calculates the budgets for each department as it aligns with the forecast of releases. Most record label accounting departments have sophisticated forecasting models that calculate profitability. Each release is analyzed to determine the value of its contribution to the overhead of the company. This analysis is examined in the profit & loss (P&L) statement, which acts as a predictor equation as to the break-even of a release and its future value over time. (P&L is discussed at great length in another chapter.) Additionally, the accounting department acts as an accounts payable/receivable clearinghouse, managing the day-to-day business of the company.

Artist & repertoire

The A&R person's job is "to direct the long and complicated creative process with the artist: choosing a producer, finding the right studio, deciding which songs you'll record, finishing the record with an engineer, sequencing the record, and then mastering it," according to Madelyn Scarpulla (2002).

The ultimate job of the *A&R department* is to acquire masters for the label to market. To do so, labels obtain masters in several ways. As in any business, labels need to manage risk. Repackaging and remastering previous recordings is the safest way to produce a master—by creating compilations of known artists with successful sales histories. How it is marketed eventually determines its success, but the label

Once an agreement is reached, the accounting department will distribute "advances" to the artist, as outlined in the contract.

The A&R process, from securing the talent to finding the musical content, can take months. Ideally, masters should be delivered 4 months prior to street date knows what it has from the start. A successful example of repackaging would be the Beatles #1 reissue—still the number one selling artist on the EMI roster.

Purchasing a master already produced is another way to manage risk. Many artists will work in conjunction with producers to create a recording, with the idea that the master will be marketed and purchased by a label, and that the artists will secure a recording contract. Again, the label knows what it's selling, since it has the final product in hand. An example is Matchbox Twenty, produced by Matt Serletic.

"It's a seven day a week job. I get into the office around 10:30 am, checking email & voice mail, lining up the day's events which include our weekly A&R meeting, marketing meetings as well as artist and manager meetings. In our weekly A&R meeting, we present new music and artists along with making sure that we prioritize all the shows that we've been invited to view. Easily, we receive 3-15 demos a day, which are officially logged to insure against "unsolicited materials." Depending on the act, I will sometimes reach out to publishers and managers of songwriters, looking for material. I follow the acts that I'm responsible for through the process - from the choosing of the songs, the recording sessions, the photo shoot and the development of the marketing plan. I work weekends too - going to shows and traveling with the artists is all a part of the job."

Louie Bandak
Senior Director of Artist and Repertoire
Capitol Records - Los Angeles

▲ *Figure 4.3 Artists and repertoire quote*

The riskiest of masters is that of signing an unknown artist (Hull, G., 2004). Record labels must feel that the talent warrants a contract, with an ear in the creation of the master. Labels take many approaches to this process. But each project has similar determining qualities, such as which songs should be recorded, how many songs will be included, and who will produce the sessions. Along with the artist, the A&R department designates a producer or producers for a project. Trust is placed in the producer, who needs to be compatible with the artist and hold the same vision that the record label has conceived for the act. Various projects are pivotal on who produces. Some producers are very "hands on" with regards to the creative input, by sharing the co-writing role or playing on the master itself. These producers tend to lend credence to the project and assist its marketability. A current example is Alicia Keyes and Ludacris' debut releases, both produced by Kanye West, who became a notable producer on Jay-Z's albums, and who has become an artist in his own right.

Other artists may need less of a heavy hand, but more like subtle guidance from both the A&R department as well as the producer. Songwriting artists have a unique quality, since both the content and the creation are open to direction. And these artists can have strong ideas as to their music and how they want it produced. But A&R representatives can take very active roles in the direction of the project, by aiding the songwriting process and nurturing the recording through the entire timeline. It all depends on what the label is looking for in the final product.

Artist development/relations

Once a contract is signed, an artist usually assigned an AD representative 12 months prior to Street Date

Sometimes called the *product development department* or *product managers*, this department manages the artist through the maze of the record company and its needs from the artist. Madelyn Scarpulla describes them as, "the product manager in effect is your manager within the label" (Scarpulla, M., 2002). Artist Development specialists hold the hand of the artist, helping them clarify their niche within the company. Artist development usually cultivates strong relationships with the artist and artist manager, with other departments in the company looking to artist development as a clearinghouse, and helping to prioritize individual department needs with the artist. In some situations, the artist *development* person is in charge of the imaging and works with creative services early in the project, whereas the artist *relations* person is the post-release contact liaison at the label for tour support, and so forth.

The artist development/product manager has been compared to the hub of a wheel where the spokes are the various marketing and development functions that occur within a label. Everything is funneled through the AD. Artist Development not only manages the artist through the process, such as the delivery of the recording, photo and video shoots, and promotional activities, but also looks for additional marketing opportunities that maximize the unique attributes of the act.

As the A&R process evolves, a photo shoot is needed to represent the content of the recording. 6 months prior to Street Date

Creative services

Depending on the company, the *creative services department* can wear many hats. Artist imaging begins with creative services assisting in the development of style, and how that style is projected into the marketplace. Special care is taken to help the artist physically reflect their artistry. Image consultants are often hired to assist in the process. "Glam Teams" are employed to polish the artist, especially for high-profile events such as photo and video shoots, and personal appearances.

The creative services department often manages photo and video shoots, setting the arrangements and collaborating on design ideas and concepts with the artist. Once complete, images are selected to be the visual theme of the records and the design process begins. In some cases, creative services contain a full design team that is "in-house" and a part of label personnel. Such in-house teams can ensure quality and consistency in imaging of the artist, and that the album cover art is the image used on promotional flyers, sales book copy, and advertisements. When there isn't an in-house design crew, design of album cover art and support materials is farmed out to outside designers. Interestingly, the use of subcontractors can enhance unique design qualities beyond the scope of in-house designers, but there can be a lack of cohesive marketing tools if not managed properly.

Creating a press kit for advance awareness occurs shortly after photo shoot; 5 months prior to Street Date

Publicity

The priority of the *publicity department* is to secure media exposure for the artists that it represents. The publicity department is set into motion once an artist is signed and helps define the message that will represent the artist. The biography of the artist via an interview is created. Other tools such as photos from the current photo shoot, articles and reviews, discography, awards and other credits, are collected into one folder, creating a press kit for each artist. These press kits are tools that are used by the publicity department to aid in securing exposure of their artists and are often sent to both trade and consumer outlets.

Pitching an artist to different media outlets can be a challenge. As an artist tours, the ideal scenario would be that the local paper would review the album and promote the show. Additional activities would be to obtain interviews with the artist in magazines and newspapers. Booking television shows and other media outlets falls to the publicity department as well. Since label publicists work with more than one act at a time, on occasion, artists will hire their own publicists to focus on them and assist in creating higher profile events for the act. These publicists try to work with label publicists to enhance in-house efforts and build on relationships already established.

Amy Willis, Media Coordinator at Sony Music Nashville, says the publicist should be a first-rate salesperson with good time-management skills, and also needs to be creative and flexible.

"Day-to-day tasks [of a publicist] include securing coverage, meeting the needs of media (sending press kits, photos, music, etc.), controlling the budget, managing independent publicists, communication with managers, booking agents, other label departments etc., reading newspapers & magazines, maintenance of contact database, research, and hiring hair & makeup artists for TV & magazine appearances."

Amy Willis
Media Coordinator
Sony Music Nashville

▲ *Figure 4.4 Publicity quote*

Radio promotion

In most record companies, the number one agenda for the promotion department is to secure radio airplay. Survey after survey continues to conclude that music consumers learn about new product and artists via the radio. Typically, radio promotion staffs divide up the country into regions, and each promotion representative is responsible for calling on specified radio stations in that region, based on format. Influencing the music director and radio programmer is key in securing a slot on the rotation list of songs played. These communications often take place on the phone, but routinely, radio promotion staffs visit stations, sometimes with the artist in tow, to help introduce new music, and secure airplay.

With the consolidation of radio stations continuing, developing an influential relationship with individual stations is getting harder and harder. But there still

Depending on the genre, the 1st radio single needs to be released to garner airplay and create demand. Artist visits with radio can enhance airplay.
3 months prior to Street Date

exists a level of autonomy within each station to create its own playlist as it reflects their listenership. To strengthen the probability of radio airplay, promotion staffs conceive and execute radio-specific marketing activities such as contests, on-air interviews with artists, listener appreciation shows, and much more. Independent radio promotion will be addressed in Chapter 8.

Sales and marketing

The *sales and marketing department* sell product into retail and create visibility of the product at the consumer level. If all the other departments have done their job correctly, selling the music to retail should be easy, since there will be a pent-up demand for the release. But to ensure sales success, the sales and marketing department must create awareness to the gatekeepers.

By visiting retail buyers, artists can assist in the set-up and sell-through of their record. Solicitation process begins 2 months prior to Street Date

Within the majors, the distribution company is part of the conglomerate, and is currently a vital conduit within the vertical integration of entertainment companies. Independent labels look to their distributors with the same vitality, but the enhanced financial relationship is not as prevalent. In either scenario, the sales departments of these labels look to their distributor not only as their partner, but also their first line of customers. The distribution company needs to be well-informed as to new releases and the marketing plan that goes with them so that they can represent the product to retailers. To do so, sales departments educate their distributor by often visiting distribution offices, sharing with them detailed marketing plans for upcoming releases.

"I have one of the best jobs on the planet. I work with great people, help develop the careers of many artists, act as the "hub of the wheel" by synchronizing the efforts of sales, promotion, publicity and creative with artists and their managers. I stay in the creative process by personally developing and executing the marketing strategies of several acts including Dave Matthews Band, David Gray, and The Strokes. It's long hours: I'm in the office from 9 am to 8 pm, five days a week, plus shows and dinners after work - not to mention the extensive traveling that goes with the gig. It's not just a lifestyle - my job is my passion."

Hugh Surratt
Senior Vice President of Marketing and Creative Services
RCA Records - New York

▲ *Figure 4.5 Marketing and creative services*

The second line of customers is retail. Sales departments continue the education process by informing retailers as to new releases and marketing plans. In tandem, record label sales representatives, along with the distribution represent-ative, will visit a specific retailer together (and on occasion, may bring an artist by to visit with the retail buyer). In the standard retail visit, the amount of music that is to be pur-chased by the retailer, along with any specific deal and dis-count information, is discussed. Co-op advertising is usually secured at this time as well, with pricing and positioning vehicles along with other in-store marketing efforts concluded.

New media

Although not always freestanding, a burgeoning entity is the New Media Department. Often housed within other depart-ments, New Media is responsible for the "face" of the label, designing and maintaining the label's website and the image it wants to project. But it can be a retail site, and/or a B2B exchange, along with other job responsibilities that are synerg-istic to the entrie label, both interal and external. A thorough examination of the internet is discussed in Chapter 14.

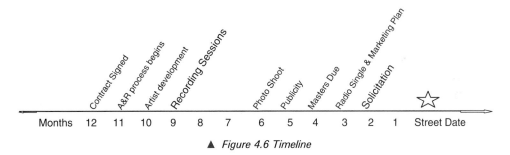

▲ *Figure 4.6 Timeline*

Glossary

A&R department – Charged with finding new talent for the label and helping that talent develop.

Artist development – Sometimes called the *product development department* or *product managers*, this department manages the artist through the MAZE of the record company and its needs from the artist.

Business affairs department – Where the lawyers and accountants reside at a record label.

Creative services department–In charge of imaging, graphics, photos and design.

Promotion department – Secures radio airplay.

Publicity department – Secures media exposure for the artists and writes materials such as press releases, bios, and so on.

Sales and marketing department – Sells product into retail and creates visibility of the product at the consumer level.

Bibliography

Hull, G. (2004). *The Recording Industry*, Routledge, p. 139.
Scarpulla, M. (2002). www.starpolish.com. A&R.
Scarpulla, M. (2002). www.starpolish.com. Product Manager/Marketing Department.

5 The Profit and Loss Statement

Amy Macy

Profitability … or the lack thereof!

Record companies use profit and loss (P&L) statements to both predict the success of a record prior to its release as well as analyze a project as it nears the end of its lifecycle. To understand the "math," let's look at all the components of a P&L, recognizing the financial significance of every line item.

SRLP – suggested retail list price

The suggested retail list price (SRLP) is set by the record label and is based on cost of the recording project, the artist's status, genre of music, competitive landscape, and what the market will bear. Although the royalty models are changing, many labels still pay artist royalties based on the SRLP. The SRLP has a correlating wholesale price, which is usually structured by the distribution company.

Card price

This line item is the wholesale price. The *card price* is the dollar figure that is set by the distribution company that sells the product. For fair trade practices, wholesale card prices are published entities that are the basis of further financial negotiations between seller and buyer. The chapter on retail shows an actual published rate card of wholesale prices.

Discount

Music product regularly receives a discount, which is set by the label and is administered via the distributor. (In some situations, it can be specified by the artist's contract.) In mainstream music, the discount is applied to the wholesale price. In some genres of music such as Christian music, discounts are applied to the retail price, and negotiations are then based on retail prices. But the majority of music is sold on wholesale pricing strategies.

Discounts are based on many variables. If a label has a really hot artist that is a big seller, and demand at the consumer level is high, a discount may not be offered, since most retailers will buy the product at full price. To entice retailers to purchase a new artist, record labels will offer discounts, which will increase the margin and profitability at the store level. Fair trade practices require that labels and their distributors offer the same discount on the same release to all retail purchasers. But the discount can be changed based on the marketing elements that the retailer may offer.

For example, a label has a new artist and is offering a 5% discount. This discount will increase the potential profitability of the retailer. If the retailer agrees to include this new artist on a "new artist" end cap for an additional 5%, then the discount will be increased to 10%, adding to the potential profitability of the retailer.

Gross sales

Gross sales refers to the wholesale price (value) of the product with the reduction of the discount included. It usually reflects the number of records shipped minus returns. Remember that music retailing is basically a consignment business and that stores can return product back to the distributor and label and receive a credit for this unsold product.

Distribution fee

Depending on the relationship between the label and distributor, the distribution fee is based on sales after the discount. Meaning, it is in the distributor's best interest to keep discounts as low as possible to help increase their profitability. This fee is a percentage of sales and varies greatly by the distributor. The conglomerates that own both the labels and distributor often charge between 14%–16%. Independent distributors structure deals with indie labels and artists that range between 18%–30%, depending on the services being offered.

Gross sales after fee

Once again, the gross sales figure is determined on shipped minus returned product, being *NET*. Net units are multiplied by the per unit price, after discounts and distribution fees are deducted.

Return provision

The industry averages about 20% returns, meaning that for every 100 records in the marketplace, 20 will be returned. To protect their business, record labels insure against returns by "reserving" a percentage of sales. Record labels "reserve" 20% of sales by pocketing these funds in an escrow account. Not until the life of the record has run the majority of its cycle will the reserve be adjusted. In some cases, a record may only have a 5% return in its lifecycle. The record company will then adjust the profitability statement to reflect such a low return, and royalties will then be distributed. In other cases, a record may have a 30% return, which adversely affects the company's overall profitability, since they expected they had sold 10% more of a particular release then previously accounted.

Net sales after return reserve

Basically, net sales after return reserve is the computed net sales minus the 20% reserve deduction.

Returns reserve opp/(risk)

Labels incur returns on records shipped into the marketplace. For accounting purposes, the profit and loss statement reflects these returns within the overall equation. The returns reserve line item is usually computed at the end of the lifecycle of the release. A label will calculate what actual returns occurred and plug this adjusted number into the P&L, determining the ultimate profitability of the project. But labels will use P&Ls as predictor equations to determine potential profitability of a future project and will plug return reserve standard into the overall equation, helping to evaluate a project's future.

Total net sales

The total net sales is the adjusted sales number deducting discounts, distribution fees, and returns.

Cost of sales

The cost of sales are costs associated with making the actual product, including the pressing of the disc, printing of paper inserts, marketing stickers on the outside of the product, all-in royalties and mechanical royalties, and inventory obsolescence. In accounting terms, these are the variable costs since the amount varies based on the number of units produced.

Many of these costs are negotiable, such as the mechanical rate. Labels often receive a reduced mechanical rate on artists that are also the songwriter. Older songs as well as reissued material can often receive lower mechanical rates based on the age and inactivity of the copyrights.

Dealing with the inventory of product that does not sell into the marketplace costs money. So, labels build into the cost of goods an amount that will fund the management of an obsolete inventory. It takes manpower and resources to "scrap" a pile of CDs. This includes moving the inventory off the warehouse floor, pulling the inserts and CDs from the jewel cases (which are recycled for new releases), and breaking/melting the actual CDs into pellets, which can then be used to make new CDs.

Gross margin before recording costs

Gross Margin Before Recording Costs deducts the cost of sales from the total net sales.

Recording costs

Know that the recording costs are initially funded by the record label. But built into most artist contracts, recording costs are recoupable, meaning that once the release starts to make money, the record company will pay itself back prior to the artist receiving royalties.

Also included in recording costs are advances. Advances are monies fronted to the artist to assist them with living expenses. It takes time to record a record, which doesn't allow an artist to make money elsewhere. Advances are recoupable.

Gross margin

Gross margin reflects gross sales minus discounts, distribution fees, returns, cost of sales, and recording costs. These are revenues made prior to marketing expenses.

Marketing costs

To launch an artist's career in today's climate, it takes a lot of money. How a company manages its marketing costs can determine the success of an album ... and artist. Beyond the making of the record, marketing costs include the imaging of an artist, advertising at both trade and consumer publications, publicity, radio promotion, and usually the most expensive element of marketing is retail positioning in the stores.

Account advertising

Even in today's business model, one of the most expensive marketing costs— ranking right up there with radio promotion and video—is the positioning of a record in the retail environment, known as *co-op advertising*. In the grocery business, this is called a *slotting allowance*. Ever notice that premium, name-brand items are placed in the most prominent positions within a grocery store? The same occurs in record retail. The pricing and positioning (P&P) of a record, meaning a "sale" or reduced price along with prime real estate placement in-store, can cost record labels hundreds of thousands of dollars.

Other types of account advertising include print advertising via the store's Sunday circulars, end-cap positioning, listening stations, event marketing such as artist in-store visits, point of purchase materials placement guarantees, and more. Chapter 12 discusses the many aspects of the retail environment.

Advertising

Most schools of advertising include a lesson on internal versus external advertising. Trade advertising is considered an internal promotional activity.

Trade advertising creates awareness of a new product to the decision makers of that industry by using strong imaging of the artist and release along with relevant facts such as sales and radio success, tour information, sales data, and upcoming press events.

In the record business, decision makers include music buyers for retail stores, radio stations and their programmers, talent bookers for television shows, reviewers for newspaper and consumer magazines, and talent buyers for venues. Some prominent trade magazines for the music industry include *Billboard*, *R&R*, *Amusement Business*, *Pollstar*, and *Variety*, to name a few.

Consumer advertising is considered external advertising since it is targeting the "end consumer." Consumer advertising also creates awareness, but to end consumers who will purchase music. This advertising can include artist album art with imaging via print, but the actual product can be "heard" via broadcast advertising, alerting consumers as to a new release now available for purchase.

Video production

Negotiated into the recording contract, artists are usually responsible for 50% of the cost of video production. Although the record company will fund the video shoot, the record company will expect to receive recoupable pay of 50% of the overall costs from record sales, and 100% recoupable from video/DVD sales.

Artist promotion

This line item usually covers the costs associated with introducing and promoting the artist to radio. Still a primary source for learning about new music, labels often take artists to visit with radio stations, including on-air interviews, dinners with music programmers, and Listener Appreciation events. The record company incurs this cost and it is usually not recoupable.

Independent promotion and publicity

Depending on the artist's stature and current competitive climate, record companies will hire the services of independent promotion companies and independent publicists. In addition to the label's efforts, these independent agents should enhance the label's strategy by assisting in creating exposure for their artists via additional radio airplay and media. This line item is not recoupable.

Media travel

Record labels will fund the costs associated with travel for an artist who is doing a media event. Media travel is usually an isolated event where an artist is doing a television talk show or an awards event. Again, an expense to the label and not recoupable.

Album art

The imaging of an artist can take time ... and money. Most artists receive some type of grooming, if not just a polishing of what the artist already represents. Professionals trained at artist imaging are hired to create a "look" that is unique and defining. Such professionals as hair and make-up artists, clothing specialists, even movement/dance professionals are often required to give an artist a specific shine. As a part of the team, photographers are hired to shoot cover artwork as well as press images for publicity use. And then a designer is hired to create the overall album art concept, from artist image use, album title treatment, booklet layout, and so forth. All of these efforts are expenses to the label and not usually recoupable.

No charge records

A marketing tool often used by labels is the actual CD. No charge records are those CDs that are used for promotional use such as giveaways on the radio, or in-store play copies for record retailers. The value of these records and their use must be accounted for, and are not recoupable.

Contribution to overhead

From the gross margin, a label would subtract the marketing costs including account advertising, trade and consumer advertising, video production, artist promotion, independent promotion and publicity, media travel, album art, and no charge records to determine the release's contribution to a record company's overhead.

Note the percentages given within the spreadsheet. The total net sales percentage reflects 100% of money generated by the sale of the project. Each line item subheading also reflects a percentage, causing each department to consider how much is being spent as a reflection of the project as a whole. Although the contribution to overhead may be a large number, its percentage of the whole determines how effective and efficient the project was managed.

Contribution to overhead: example of two different projects

Example:

	Project A	Project B
Total Net Sales	$3,500,000	$5,000,000
Contribution to Overhead	$1,350,000	$1,500,000
Percentage:	38.6%	30.0%

Looking at the numbers, Project B made more money for the company, but as for efficiency, Project A was a more effective release since the company did a better job at managing its expenses.

Ways at looking at project efficiency and effectiveness

Record labels look at the "numbers" in many ways to determine how well they are performing. By "spinning" these figures, a company should analyze where spending is less effective, thus causing the overall project to be less profitable.

Return percentage

Gross Ships − Cumulative Returns = Net Shipment
Cumulative Returns/Gross Ships = Return Percentage

Example: 500,000 units − 35,000 returns = 465,000 units Net Shipment

35,000 units / 500,000 units = 7% return

Again, industry standards reflect an approximate 20% return percentage. Generally, a record that returns more than 20% is not performing at the market average. A couple of decisions could have affected this percentage. The record company over-sold the project and caused the returns, or the record company did not promote and have a hit with the project, thus causing returns. In any case, the record company needs to determine what occurred to ensure that it doesn't happen again.

If a project under performs a market dramatically, such as a lifetime return average of 5%, the project could be considered a great success, minimizing return costs as well as manufacturing costs. But the company should also be sensitive to the fact that they may have undersold the project, thus not realizing the full sales potential of the project too. Again, further analysis is vital to the overall success of future projects and the company.

SoundScan and sell off

SoundScan is an information system that tracks sales of music and music video products throughout the United States and Canada. Sales data using UPC bar codes from point-of-sale cash registers is collected weekly from over 19,000 retail, mass merchant and non-traditional (on-line stores, venues, and so on etc.) outlets. Weekly data is compiled and made available every Wednesday. Now branded with the Nielsen name, SoundScan is the sales source for the Billboard music charts, major newspaper, magazine, TV, MTV and VH1 charts. Chapter 6 is dedicated to SoundScan, its analysis and impact on the industry.

To evaluate inventory, record companies can use a simple equation to know how many units remain in the marketplace:

Net Shipments − SoundScan Sales = Remaining Inventory
SoundScan Sales/Net Shipments = Sell Off Percentage

Example: 465,000 Net Ships - 437,000 units SoundScanned = 28,000 units

437,000 SoundScan/465,000 Net Ships = 94% Sell Off

If this project is a steady seller, quietly moving units every week, a record company could use SoundScan weekly sales to determine how many weeks of inventory are left in the marketplace. If this project sold 2,000 each week:

28,000 remaining inventory/2,000 SoundScan each week = 14 weeks of inventory left

Knowing inventory levels and keeping aware of sales is critical to the success of a project. A record company does not want to run out of records, which is called *can't fill*. If it's not on the retailer's shelves, it cannot be purchased.

Percentage of marketing costs

Isolating marketing costs and analyzing their effectiveness in selling records is best reflected in the following equations:

Percentage of Marketing Cost to Gross: Marketing Costs/Gross Sales $\times$ 100 = %
Percentage of Marketing Cost to Net: Marketing Costs/Net Sales $\times$ 100 = %

Example: $800,000 Marketing Costs/$4,238,000 Gross Sales = 18.9%
$800,000 Marketing Costs/$3,402,000 Net Sales = 23.5%

The lower the percentage reflects, the better performing project. Keeping marketing costs in check and knowing when to stop "fueling the fire" is usually a great determiner of seasoned record labels.

Marketing Costs per Gross Unit:
Marketing Costs/Gross Shipments = Cost per unit

Marketing Costs per Net Unit:
Marketing Costs/Net Shipments = Cost per unit

Example: $800,000/500,000 units Gross Shipped = $1.60
$800,000/465,000 units Net Shipped = $1.72

In all of these equations, the "real" picture is best drawn when using Net Shipments, since that is the ultimate number of units in the marketplace.

As a predictor equation

Record companies can use the profit and loss statement as a predictor of success. Often, labels will "run the numbers" to see how profitable, or not, a potential release could be. By using the spreadsheet and plugging in forecasted numbers, including shipments, cost of sales, recording costs (or the acquisition of a master), and marketing costs, the equation will help a label evaluate and determine whether a project is worth releasing.

Small considerations can dramatically affect the contribution to overhead. List price, discounts, number of pages in the CD booklet, royalties—both artist and mechanical, and the various marketing line items can either make a project profitable or not.

Break-even point

When does a record "break even" in covering the costs that it took to make the project, and when does it start to turn a profit? Depending on the equation, record companies look at this value in several ways.

Without marketing costs, a number can be derived simply by dividing the total fixed costs by price, using wholesale dollars.

Break-even point

$$\text{Break-even point (units)} = \frac{\text{Total fixed costs}}{\text{Price} - (\text{total variable costs/units})}$$

Break even with costs listed

$$BE = \frac{\text{Recording Costs} + \text{Advances}}{\text{Price} - \text{Variable costs}}$$

Break-even example

Example:

$$BE = \frac{\{\$250,000 \text{ recording costs} + \$75,000 \text{ advance}}{\$12.04^a - (\$6.05^b + \$1.69^c)}$$

   [a] *(card price of SRLP $18.98)*   [b] *(based on total cost of sales)*   [c] *(Distr. Fee)*

$BE = 75{,}581$ units without marketing costs

Cost of sales

Cost of Sales

Manufacturing		0.77
Royalties		
Artist: % based on list price	9.00%	1.7082
Mechanical: # of tracks based on ($0.0825 × 10)		0.825
		2.53
Other		0.21
TOTAL COST OF SALES		6.05

List price

List Price	18.98	
Card Price	12.04	Distribution Fee
Distribution Fee	1.69	14.0%

Adding marketing costs changes the outcome of the equation dramatically:

A label has to derive predicted/budgeted marketing costs to add to the equation. Most seasoned labels have an idea as to how much each activity may cost to launch a record. Using the following dollars, check-out the break even analysis:

Marketing costs

Marketing Costs

Account Advertising	100,000
Trade Advertising	10,000
Consumer/Other Advertising	10,000
Video Production	75,000
Artist Promotion (Radio)	100,000
Indy Promotion/Publicity	5,000
Media Travel	2,500
Album Art/Imaging	25,000
No Charge Records	10,000
TOTAL MARKETING COSTS	337,500

Break-even example with marketing costs

Example:

$$BE = \frac{\$250,000 \text{ recording costs} + \$75,000 \text{ advance} + \$337,500 \text{ mkt costs}}{\$12.04^a - (\$6.05^b + \$1.69^c)}$$

[a] (card price of SRLP $18.98)   [b] (based on total cost of sales)   [c] (Distr. Fee)

$$BE = \frac{\$662,500}{\$5,99} = 110,601 \text{ units}$$ just to recover the costs of making and marketing this record. This does not take into account overhead costs and salaries of the employees getting this job done.

Clearly, the job of the profit and loss statement is multifaceted. It can be used as a predictor of success (or not), as well as an evaluation tool of existing projects. Not a part of the equation is the overhead that it takes to operate the business, such as building expenses, salaries, supplies, and so on. But lessons learned from this type of analysis should aid record labels as to the better allocation of funds and resources.

PROFIT AND LOSS STATEMENT
Record companies use Profit and Loss Statements to both predict the success of a record prior to its release as well as analyze a project as it nears the end of its lifecycle. To understand the "math," let's look at all the components of a P&L, recognizing the financial significance of every line item.

SRLP – Suggested Retail List Price
This price is set by the record label and is based on artist status, genre of music, competitive landscape, and what the market will bear. The SRLP is the monetary base from which artist royalties are paid, and it sets the wholesale price.

CARD PRICE
This is the wholesale price that is determined by the SRLP. The term "card price" is the dollar figure that is set by the distribution company that sells the product.

STANDARD DISCOUNT
Music product regularly receives a discount, which is set by the label and is administered via the distributor. In mainstream music, the discount is applied to the wholesale price.

GROSS SALES
This term refers to the wholesale price of the product with the reduction of the discount included. Gross sales usually reflect the number of records shipped minus returns. Remember that music retailing is basically a consignment business and that stores can return product back to the distributor and label and receive a credit for this unsold product.

DISTRIBUTION FEE
Depending on the relationship between the label and distributor, this fee is based on sales after the discount. This fee is a percentage of sales and varies greatly on the distributor. The conglomerates that own both the labels and distributor often charge between 14%-16%.

GROSS SALES AFTER FEE
This figure is determined on shipped minus returned product, being NET. Net units are multiplied by the per unit price, after discounts and distribution fees are deducted.

RETURN PROVISION
The industry averages about 20% returns, meaning that for every 100 records in the marketplace, 20 will be returned. To protect their business, record labels insure against returns by "reserving" a percentage of sales–20%–by pocketing these funds in an untouchable account.

NET SALES AFTER RETURN RESERVE
Computed net sales minus the 20% reserve deduction.

RETURNS RESERVE OPP/(RISK)
This item is usually computed at the end of the lifecycle of the release. A label can calculate what actual returns occurred and plug this adjusted number into the P&L, determining the ultimate profitability of the project.

TOTAL NET SALES
Adjusted sales number deducting discounts, distribution fees, and returns.

COST OF SALES
Costs associated with making the actual product, including the pressing of the disc, printing of paper inserts, marketing stickers on the outside of the product, all-in royalties and mechanical royalties, and inventory obsolescence. These are the variable costs–based on the number of units produced.

GROSS MARGIN BEFORE RECORDING COSTS
Subtracting Cost of Sales from the Total Net Sales generates a dollar figure.

RECORDING COSTS
Recording costs are initially funded by the record label. But built into most artist contracts, recording costs are recoupable, meaning that once the release starts to make money, the record company will pay itself back prior to the artist receiving royalties. Also included in recording costs are advances.

GROSS MARGIN
Gross Sales – Discount, Distribution, Returns, Cost of Sales, and Recording Costs. Monies that are made prior to marketing costs.

MARKETING COSTS
Beyond the making of the record, marketing costs include the imaging of an artist, advertising at both trade and consumer publications, publicity, radio promotion, and usually the most expensive element of marketing is retail positioning in the stores.

▲ *Figure 5.1 The profit and loss summary*

Account Advertising

The most expensive marketing cost is the positioning of a record in the retail environment. The pricing and positioning of a record (P&P), meaning a "sale" or reduced price along with prime real estate placement in store, can cost record labels hundreds of thousands of dollars. Other types of account advertising includes print advertising via the store's Sunday circulars, end cap positioning, listening stations, event marketing such as artist in-store visits, point of purchase materials placement guarantees, and more.

Advertising

Trade advertising is considered an internal promotional activity. Trade advertising creates awareness of a new product to the decision-makers of that industry by using strong imaging of the artist and release along with relevant facts such a sales and radio success, tour information, sales data, etc.

Video Production

Negotiated into the recording contract, artists are usually responsible for 50% of the cost of video production.

Artist Promotion

Usually covers the costs associated with introducing and promoting the artist to radio. Labels often take artists to visit with radio stations, including on-air interviews, dinners with music programmers, and Listener Appreciation events. The record company incurs this cost and is usually not recoupable.

Independent Promotion and Publicity

Record companies will hire the services of independent promotion companies and independent publicists. They assist in creating exposure for their artists via additional radio airplay and media. This line item is not recoupable.

Media Travel

Record labels will fund the costs associated with travel for an artist who is doing a media event. This is an expense to the label and not recoupable.

Album Art

Photographers are hired shoot cover artwork as well as press images for publicity use. And then a designer is hired to create the overall album art concept, from artist image use, album title treatment, booklet layout, etc. All of these efforts are expenses to the label and not usually recoupable.

No Charge Records

A marketing tool often used by labels is the actual CD. No Charge Records are those CDs that are used for promotional use such as give-aways on the radio, or instore play copies for record retailers. The value of these records and their use must be accounted for, and are not recoupable.

CONTRIBUTION TO OVERHEAD

From the Gross Margin, a label would subtract the marketing costs including Account Advertising, Trade and Consumer Advertising, Video Production, Artist Promotion, Independent Promotion and Publicity, Media Travel, Album Art, and No Charge Records to determine the release's contribution to record company's overhead.

Note the percentages given within the spreadsheet. The total net sales percentage reflects 100% of money generated by the sale of the project. Each line item subheading also reflects a percentage, causing each department to consider how much is being spent as a reflection of the project as a whole. Although the Contribution to Overhead may be a large number, its percentage of the whole determines how effective and efficient the project was managed.

▲ *Figure 5.1 Continued*

Good vs. Bad Project P&L
Same Artist: Debut Album vs. Sophmore Release
Actual sales and marketing dollars

	Net Per Unit		Good Project Estimate Units and $s in 000's		Bad Project Estimate Units and $s in 000's	
	Product Mix		Project P&L		Project P&L	
Average Based on Mix	100% CD		Total	%	Total	%
	Amt	%	Gross Units		Gross Units	
List Price	18.98		1,305		478	
Card Price	12.04					
Discount	0.72	6.0%				
			Net Units		Net Units	
Adj. Price to Accounts	11.32		1,214		318	
Gross Sales			14,769		5,410	
Returns Provision (after Fee)			-2,954	20.0%	-1,082	20.0%
Sales After Reserve	11.32		11,816		4,328	
Distribution Fee	1.58	14.0%	1,654	14.0%	606	14.0%
Net Sales After Fee	9.73		10,161		3,722	
Sales Reserve (Opp)/Risk			1,964	13.3%	-633	-11.7%
Total Net Sales	9.73		12,126		3,089	
Cost of Sales						
Manufacturing	0.77	7.9%	942	7.8%	255	8.3%
Royalties	2.08	21.4%	2,527	20.8%	668	21.6%
Other	0.21	2.2%	72	0.6%	127	4.1%
TOTAL COST OF SALES	3.06	31.4%	3,541	29.2%	1,050	34.0%
Gross Margin (Before Recording Costs)	6.67	68.6%	8,585	70.8%	2,039	66.0%
Recording Costs						
Recording & Advances			310	2.6%	310	10.0%
Recording Recoupment			-310	-2.6%	-310	-10.0%
NET RECORDING COST			0	0.0%	0	0.0%
Gross Margin			8,585		2,039	
Marketing Costs						
Account Advertising			582	4.8%	465	15.1%
Trade Advertising			34	0.3%	29	0.9%
Consumer/Other Advertising			89	0.7%	134	4.3%
				0.0%		0.0%
Video Production			175	1.4%	166	5.4%
				0.0%		0.0%
Artist Promotion (Radio)			134	1.1%	102	3.3%
Independent Promotion/Publicity			10	0.1%	2	0.1%
Media Travel			13	0.1%	27	0.9%
				0.0%		0.0%
Album Art/Imaging			38	0.3%	55	1.8%
No Charge Records			22	0.2%	32	1.0%
TOTAL MARKETING COSTS			1,097	9.0%	1,012	32.8%
Contribution to Overhead			7,488	61.8%	1,027	33.2%
Gross Shipped to date			1,305		478	
Cumm Return to date			91		160	
Return Percentage			7.0%		33.5%	
SoundScan			1,141		270	
Sell off Percentage			94.0%		84.9%	
% Mktg Cost to Gross			7.4%		18.7%	
% Mktg Cost to Net			9.0%		32.8%	
Mkg Cost per Gross Unit			$0.84		$2.12	
Mktg Cost per Net Unit			$0.90		$3.18	

▲ *Figure 5.2 Example P&L worksheet: actual sales and marketing dollars*

Good vs. Bad Project P&L
Same Artist: Debut Album vs. Sophmore Release
Sales made equal to show Contribution to Overhead Difference

Average Based on Mix	Net Per Unit Product Mix 100% CD		Good Project Estimate Units and $s in 000's Project P&L		Bad Project Estimate Units and $s in 000's Project P&L	
	Amt	%	Total	%	Total	%
			Gross Units		Gross Units	
List Price	18.98		500		500	
Card Price	12.04					
Discount	0.72	6.0%				
			Net Units		Net Units	
Adj. Price to Accounts	11.32		465		331	
Gross Sales			5,659		5,659	
Returns Provision (after Fee)			-1,132	20.0%	-1,132	20.0%
Sales After Reserve	11.32		4,527		4,527	
Distribution Fee	1.58	14.0%	634	14.0%	634	14.0%
Net Sales After Fee	9.73		3,893		3,893	
Sales Reserve (Opp)/Risk			753	13.3%	-662	-11.7%
Total Net Sales	9.73		4,646		3,231	
Cost of Sales						
Manufacturing	0.77	7.9%	361	7.8%	263	8.1%
Royalties	2.08	21.4%	891	19.2%	693	21.4%
Other	0.21	2.2%	28	0.6%	133	4.1%
TOTAL COST OF SALES	3.06	31.4%	1,280	27.6%	1,089	33.7%
Gross Margin (Before Recording Costs)	**6.67**	**68.6%**	**3,366**	**72.4%**	**2,142**	**66.3%**
Recording Costs						
Recording & Advances			310	6.7%	310	9.6%
Recording Recoupment			-310	-6.7%	-310	-9.6%
NET RECORDING COST			0	0.0%	0	0.0%
Gross Margin			**3,366**		**2,142**	
Marketing Costs						
Account Advertising			275	5.9%	275	8.5%
Trade Advertising			30	0.6%	30	0.9%
Consumer/Other Advertising			90	1.9%	90	2.8%
				0.0%		0.0%
Video Production			175	3.8%	175	5.4%
				0.0%		0.0%
Artist Promotion (Radio)			135	2.9%	135	4.2%
Independent Promotion/Publicity			10	0.2%	10	0.3%
Media Travel			20	0.4%	20	0.6%
				0.0%		0.0%
Album Art/Imaging			40	0.9%	40	1.2%
No Charge Records			25	0.5%	25	0.8%
TOTAL MARKETING COSTS			800	17.2%	800	24.8%
Contribution to Overhead			**2,566**	**55.2%**	**1,342**	**41.5%**
Gross Shipped to date			500		500	
Cumm Return to date			35		169	
Return Percentage			7.0%		33.8%	
SoundScan			437		185	
Sell off Percentage			94.0%		55.9%	
% Mktg Cost to Gross			14.1%		14.1%	
% Mktg Cost to Net			17.2%		24.8%	
Mkg Cost per Gross Unit			$1.60		$1.60	
Mktg Cost per Net Unit			$1.72		$2.42	

▲ *Figure 5.3 Sales made equal to show contribution to overhead difference (return % varies)*

Good vs. Bad Project P&L
Same Artist: Debut Album vs. Sophmore Release
Sales Equal / Discount % increased to 10%

Average Based on Mix	Net Per Unit Product Mix 100% CD		Good Project Estimate Units and $s in 000's Project P&L		Bad Project Estimate Units and $s in 000's Project P&L	
	Amt	%	Total	%	Total	%
			Gross Units		Gross Units	
List Price	18.98		500		500	
Card Price	12.04					
Discount	1.20	10.0%				
			Net Units		Net Units	
Adj. Price to Accounts	10.84		465		331	
Gross Sales			5,418		5,418	
Returns Provision (after Fee)			-1,084	20.0%	-1,084	20.0%
Sales After Reserve	10.84		4,334		4,334	
Distribution Fee	1.52	14.0%	607	14.0%	607	14.0%
Net Sales After Fee	9.32		3,728		3,728	
Sales Reserve (Opp)/Risk			721	13.3%	-634	-11.7%
Total Net Sales	9.32		4,448		3,094	
Cost of Sales						
Manufacturing	0.77	8.3%	361	8.1%	263	8.5%
Royalties	2.08	22.3%	870	19.6%	676	21.9%
Other	0.21	2.3%	28	0.6%	133	4.3%
TOTAL COST OF SALES	3.06	32.8%	1,259	28.3%	1,072	34.7%
Gross Margin (Before Recording Costs)	**6.26**	**67.2%**	**3,189**	**71.7%**	**2,022**	**65.3%**
Recording Costs						
Recording & Advances			310	7.0%	310	10.0%
Recording Recoupment			-310	-7.0%	-310	-10.0%
NET RECORDING COST			0	0.0%	0	0.0%
Gross Margin			3,189		2,022	
Marketing Costs						
Account Advertising			275	6.2%	275	8.9%
Trade Advertising			30	0.7%	30	1.0%
Consumer/Other Advertising			90	2.0%	90	2.9%
				0.0%		0.0%
Video Production			175	3.9%	175	5.7%
				0.0%		0.0%
Artist Promotion (Radio)			135	3.0%	135	4.4%
Independent Promotion/Publicity			10	0.2%	10	0.3%
Media Travel			20	0.4%	20	0.6%
				0.0%		0.0%
Album Art/Imaging			40	0.9%	40	1.3%
No Charge Records			25	0.6%	25	0.8%
TOTAL MARKETING COSTS			800	18.0%	800	25.9%
Contribution to Overhead			**2,389**	**53.7%**	**1,222**	**39.5%**
Gross Shipped to date			500		500	
Cumm Return to date			35		169	
Return Percentage			7.0%		33.8%	
SoundScan			437		185	
Sell off Percentage			94.0%		55.9%	
% Mktg Cost to Gross			14.8%		14.8%	
% Mktg Cost to Net			18.0%		25.9%	
Mkg Cost per Gross Unit			$1.60		$1.60	
Mktg Cost per Net Unit			$1.72		$2.42	

▲ *Figure 5.4 Discount increased to 10%, return percentage varies*

6 SoundScan and the Music Business

Amy Macy with Trudy Lartz, Vice President of Sales and Service, Nielsen SoundScan

Technology and the music business

Technological advancements not only launched the music business, but have continually given it a "shot in the arm" just when it needed it most. From stereo—to recordable tape—to compact discs—to digital files, technology continues to evolve the hardware and software of our industry. Not only has technology changed the format of our business, but technology has changed how the actual business gets done. Technological advancements permeate the core of the record industry, creating the scorecards for success.

SoundScan

Prior to the current charting system, an archaic reporting system based on undocumented sales information was garnered by Billboard to produce the sales charts. Billboard had a panel of "reporting" retail stores that stated the best-selling record in their store, based on genre. Often times, this information was not verified through actual sales data, but was based on what individuals "thought" to be the best seller. Record labels employed retail promotion teams to help influence these reports; hence, the sales charts were not always valid depictions of true sales throughout the nation.

One must remember that this reporting structure was prior to the use of bar code scanning systems and the use of point-of-sales data. So capturing accurate sales data was difficult, even for the retailer. The use of bar codes, or

Insight Insight combines Soundscan sales and BDS airplay in one report. View 3 week sales trends, BDS spins, audience, dayparts and video information. Robo A&R Robo A&R is a web-based A&R research engine that scours the BDS and Soundscan databases for unsigned and upstart indie bands breaking out in sales and airplay based on search criteria that you control. BDSradio.com BDSradio is a subscriber-based service for radio programmers. It is a web application that provides fast accurate airplay information to the radio industry. Indie Insight Airplay and sales information tailored for the Independent. All the customized tools you need to track your record, plus the latest industry conditions for 13 weeks. Market Studies Market Studies is a SQL based software system specifically designed for use in the entertainment industry. Its main purpose is to assist marketers in the targeting and allocation of the products across geographies, within media vehicles, and within retailers. Market Studies software contains a full range of demographic data items in addition to a PRIZM NE based segmentation system specifically designed and refined for use with entertainment products.

▲ *Figure 6.1 Other SoundScan products (Source: Nielsen SoundScan)*

Universal Product Codes (UPC), has greatly affected product management. Not only do UPC bar codes assist in inventory status and reorder generation of hot-selling items, the sales information captured allows the retailer to determine the best-selling item by store and by chain, as confirmed through real sales data.

With the introduction of bar codes and efficient computer management of inventory, a new idea was introduced. Mike Shalett, an ex-record label promotion guy, along with Mike Fine, a statistician who had previously worked with major newspapers and magazines with a focus on surveys, conceived a revolutionary concept that would use this new-found technology to derive the top-selling records of the week. And in 1991, SoundScan was born. The two partners held the company privately until 1998 when they sold a majority share of SoundScan, and sister company VideoScan, to VNU, a company based in the Netherlands that also owned Billboard Publications (BPI), Broadcast Data Systems (BDS) and A.C. Nielsen. This merger formed a nice framework that would allow VNU-owned companies to build new product lines by merging and overlaying airplay, demographic, lifestyle, movie and TV viewing data. This would allow decision makers at labels to see "the big picture" and create more effective marketing strategies.

SoundScan is an information system that tracks sales of music and music video products throughout the United States and Canada. Sales data using UPC bar codes from point-of-sale cash registers is collected weekly from over 19,000 retail, mass merchant and nontraditional (online stores, venues, and so on) outlets. Weekly data is compiled and made available every Wednesday. Now branded with the Nielsen name, SoundScan is the sales source for the Billboard music charts, major newspaper, magazine, TV, MTV and VH1 charts.

Although 19,000 retail outlets sounds like a lot of stores, SoundScan does not capture all music retailers. Through analysis, SoundScan knows which retailers it does not capture, and via statistical equations, is able to derive a fairly accurate sales estimate. The record that sells the most earns the #1 position for the week.

The UPC bar code and registration with SoundScan

The UPC bar code contains a unique sequence of numbers that identifies a product. For music, record companies are designated a 5- to 6-digit number that identifies the label, via the Uniform Code Council. The record company then assigns a 4- to 5-digit product code that identifies the release including artist and title of the album. The eleventh digit designated the configuration of the product. The number "2" designates a CD, "1" is vinyl, and "4" is cassette, and so on. The last digit is known as the "check" digit. When scanned, a mathematical equation occurs, which determines if the product has been correctly scanned. The "check" digit is the "answer" to that equation, verifying an accurate scan. In the U.S., the standard UPC bar code contains 12 digits. For digital tracks, downloadable via the Internet, an International Standard Recording Code (ISRC) is required.

To learn more about UPC bar codes and the Uniform Code Council, contact:

Uniform Code Council, Inc.
7887 Washington Village Drive, Suite 300
Dayton, OH 45459
937-435-3870
937-435-7317
www.uccouncil.org

Once a product has a UPC bar code or ISRC code, the release can be registered with SoundScan so that sales or tracks can be captured on a nationwide basis. To register, the following form along with a copy of the release should be sent to SoundScan.

A look at the data

The Charts

By compiling and organizing sales data, SoundScan derives many charts that help analyze the marketplace. Let's look at a sample of these charts—know

Nielsen
SoundScan

TITLE ADDITION SHEET

To add a title to the SOUND SCAN database, each field on the title addition sheet must be completed in order for it to be accepted. Please use a separate form for each additional title.

Title: _____ Release Date:_____

Artist: _____

Label Information as it applies to this product

Parent Label: _____ Distribution Co.

Sub Label: _____ Label Abbr: __ __ __ __

Please enter **all** digits of the **U.P.C.Code**. (Including **Prefix** and **Suffix**)
To enter identification codes on how your product should be listed please check the example below.

PLEASE PRINT IN ONE CONFIGURATION FOR EACH LINE.

U.P.C.Code

EXAMPLE			PRICE	TYPE	Configuration Types
9	9999999999	9	9.99	_A_	
_	- - - - - - - - -	__	____	__	
_	- - - - - - - - -	__	____	__	
_	- - - - - - - - -	__	____	__	
_	- - - - - - - - -	__	____	__	
_	- - - - - - - - -	__	____	__	
_	- - - - - - - - -	__	____	__	

```
ALBUM                  SINGLES
A=LP12" ALBUM          E=CD SINGLE
B=CASS ALBUM           F=LP12" SINGLE
C=CD ALBUM             G=CASS SINGLE
D=DVD AUDIO            I=CD MAXI

VIDEO
M= VHS
L= DVD
```

PLEASE SELECT ONE GENRE WHICH APPLIES
TO THIS PRODUCT:

____150-R&B	____520-SOUNDTRACK	
____400-COUNTRY	____184-WORLD	VIDEO SUBMISSIONS ONLY
____300-JAZZ	____100-ROCK	
____186-LATIN	____620-COMEDY	____900 MUSICAL PERFORMANCE
____102-METAL	____640-GOSPEL	____901 SPORTS
____200-CLASSICAL	____630-CHRISTIAN	____907 MOVIE
____152-RAP	____625-KARAOKE	____905 OTHER (E.G. EXERCISE, DOCUMENTARY)
____180-REGGAE	____360-NEWAGE	
____690-CHILDREN	____156-DANCE/ELECTRONIC	
____470-BLUES	____178-SKA	

Your Name: _____ PH# ( )____-_____ FAX# ( )____-_____
Email:

Please enter your name, phone number and fax number in case we have any questions.

▲ *Figure 6.2 SoundScan title addition form (Source: Nielsen SoundScan)*

that there are hundreds of charts, subdivided by specific headings. The charts are the actual sales charts that drive the Billboard charts.

The Hot 100 Singles chart is a combination of single sales and radio airplay, as calculated by Billboard. On this chart, the ratio average of single sales is less than 1%, due to the lack of physical production of singles by the labels. A new version of the Hot 100 Singles chart is being introduced in 2005, using download

Weeks On	Label	2W Rank	LW Rank	TW Rank	Artist	Title	TW Sales	% CHG	LW Sales	RTD Sales
4	UME	2	2	1	VARIOUS	NOW 16	207,490	-16	247,313	1,236,052
5	GEFN	1	1	2	SIMPSON*ASHLEE	AUTOBIOGRAPHY	164,026	-38	263,445	1,382,731
18	COL	7	9	3	PRINCE	MUSICOLOGY	98,917	35	73,293	1,584,314
1	TVT			4	213	HARD WAY	95,301	999	482	95,785
22	LAF	4	6	5	USHER	CONFESSIONS	78,491	-11	88,204	5,342,850
67	OCTJ	10	12	6	MAROON 5	SONGS ABOUT JANE	72,448	11	65,554	2,084,977
2	DEF		3	7	SHYNE	GODFATHER BURIED	67,402	-57	157,562	225,380
1	ATLG			8	CABRERA*RYAN	TAKE IT ALL AWAY	66,490	999	299	66,796
13	ARI	6	10	9	LAVIGNE*AVRIL	UNDER MY SKIN	63,968	-9	70,227	1,439,963
16	WAR	8	11	10	BIG & RICH	HORSE OF A DIFFERENT COLOR	63,088	-5	66,282	819,500
6	RCNA	3	7	11	BUFFETT*JIMMY	LICENSE TO CHILL	62,405	-19	77,036	745,059
15	EPIC	5	8	12	WILSON*GRETCHEN	HERE FOR THE PARTY	61,655	-20	76,680	1,563,822
26	EPIC	9	13	13	LOS LONELY BOYS	LOS LONELY BOYS	55,645	-11	62,664	1,041,956
2	JIVE		4	14	MOBB DEEP	AMERIKAZ NIGHT MARE	43,851	-60	108,821	153,490
3	DBV	27	16	15	PRINCESS DIARIES 2	SOUNDTRACK	43,472	-9	47,795	124,746
8	INT	11	15	16	BANKS*LLOYD	HUNGER FOR MORE	42,246	-13	48,584	1,025,219
2	WNDU		5	17	ALTER BRIDGE	ONE DAY REMAIN	42,009	-56	95,373	137,606
68	COL	16	17	18	SWITCHFOOT	BEAUTIFUL LETDOWN	40,683	-2	41,642	1,097,116
61	AAM	21	18	19	BLACK EYED PEAS	ELEPHUNK	39,411	-3	40,638	1,785,182
1	ISL			20	SALIVA	SURVIVAL	38,857	999	111	38,968
57	ARNV	22	22	21	PAISLEY*BRAD	MUD ON THE TIRES	35,129	-6	37,209	1,005,125
22	GEFN	39	35	22	GUNS N'ROSES	GREATEST HITS	32,870	15	28,473	1,106,150
2	CAP		14	23	HOUSTON	IT'S ALREADY WRITTEN	32,410	-34	49,374	81,934
37	ISL	30	32	24	HOOBASTANK	THE REASON	31,737	3	30,878	1,735,787
9	UNIV	18	24	25	JOJO	JOJO	31,487	-12	35,936	486,954
5	WAR	15	26	26	VAN HALEN	BEST OF BOTH WORLDS	31,102	-10	34,609	310,314
4	VIT	12	19	27	TAKING BACK SUNDAY	WHERE YOU WANT TO BE	30,756	-24	40,326	291,558
17	INT	19	27	28	D-12	D12 WORLD	30,667	-11	34,422	1,661,016
4	ATLG	13	21	29	LYTTLE*KEVIN	KEVIN LYTTLE	30,649	-19	37,764	201,419
11	RCA	23	30	30	VELVET REVOLVER	CONTRABAND	30,288	-4	31,589	844,625
29	BNA	24	29	31	CHESNEY*KENNY	WHEN THE SUN GOES DOWN	29,769	-6	31,722	2,428,868
20	EPIC	28	28	32	MODEST MOUSE	GOOD NEWS FOR PEOPLE WHO LOVE	29,625	-9	32,585	762,615
10	ISL	54	48	33	KILLERS	HOT FUSS	29,167	19	24,486	197,415
38	J	36	38	34	KEYS*ALICIA	DIARY OF ALICIA KEYS	29,068	3	28,120	3,281,990
35	CAP	35	33	35	YELLOWCARD	OCEAN AVENUE	28,871	0	28,846	1,003,794
9	INT	17	25	36	JADAKISS	KISS OF DEATH	27,665	-21	35,173	739,615
53	COL	26	20	37	SIMPSON*JESSICA	IN THIS SKIN	26,713	-33	39,950	2,385,556
74	WAR	40	42	38	LINKIN PARK	METEORA	26,533	1	26,398	4,418,677
4	UNRF	14	31	39	TERROR SQUAD	TRUE STORY	26,317	-16	31,448	200,866
77	WNDU	34	37	40	EVANESCENCE	FALLEN	26,102	-8	28,331	5,698,812
28	DEF	29	36	41	WEST*KANYE	COLLEGE DROP OUT	25,358	-11	28,383	2,157,563
1	HOL			42	QUEEN	GREATEST HITS-WE WILL ROCK YOU	24,642	999	177	24,819
8	HOL	47	47	43	BREAKING BENJAMIN	WE ARE NOT ALONE	24,566	-3	25,198	211,420
9	COL	25	39	44	SPIDER-MAN 2	SOUNDTRACK	24,139	-12	27,539	473,358
21	COL	50	49	45	LIL' FLIP	U GOTTA FEEL ME	24,029	-1	24,217	845,201
22	EPIC	42	41	46	FRANZ FERDINAND	FRANZ FERDINAND	23,768	-10	26,452	451,182
10	CAP	43	43	47	BEASTIE BOYS	TO THE 5 BOROUGHS	23,726	-10	26,311	873,896
4	MERN	32	40	48	CLARK*TERRI	GREATEST HITS	23,688	-12	26,901	144,710
8	UNRF	56	62	49	AKON	TROUBLE	22,910	10	20,763	150,072
1	COL			50	TRITT*TRAVIS	MY HONKY TONK HISTORY	22,899	999	334	23,235

Nielsen SoundScan, a division of Nielsen Entertainment LLC

| 1 | 2 | 3 | 4 | 5 | 6 | 7 | 8 | 9 | 10 | 11 |

This is a sample of the Top 200 Album Chart. Each column is identified by number:

1 – Number of weeks the album has been on the Top 200 Chart
2 – Record Label
3 – Chart position 2 weeks ago
4 – Chart position last week
5 – Current chart position
6 – Artist Name
7 – Album Title
8 – Number of units sold this week
9 – Percentage of change from this week to last week
10 – Number of units sold last week
11 – Total number of units sold since the release of the album

▲ Figure 6.3 Example of a Top Current Album chart (Source: Nielsen SoundScan)

Rank	Label	Artist	Title	YTD Sales
1	LAF	USHER	CONFESSIONS	5,342,850
2	BNTE	JONES*NORAH	FEELS LIKE HOME	3,337,475
3	BNA	CHESNEY*KENNY	WHEN THE SUN GOES DOWN	2,428,868
4	WNDU	EVANESCENCE	FALLEN	2,334,074
5	LAF	OUTKAST	SPEAKERBOXX-LOVE	2,205,009
6	DEF	WEST*KANYE	COLLEGE DROP OUT	2,157,563
7	WAR	GROBAN*JOSH	CLOSER	1,993,015
8	COL	SIMPSON*JESSICA	IN THIS SKIN	1,893,591
9	CAP	VA-NOW THAT'S WHAT I CALL MUSI	VOL. 15-NOW THAT'S WHAT I CALL	1,766,684
10	INT	D-12	D12 WORLD	1,661,016
11	ATLG	TWISTA	KAMIKAZE	1,634,562
12	COL	PRINCE	MUSICOLOGY	1,584,314
13	J	KEYS*ALICIA	DIARY OF ALICIA KEYS	1,574,984
14	AAM	CROW*SHERYL	VERY BEST OF SHERYL CROW	1,569,090
15	EPIC	WILSON*GRETCHEN	HERE FOR THE PARTY	1,563,822
16	ISL	HOOBASTANK	THE REASON	1,563,756
17	OCTJ	MAROON 5	SONGS ABOUTJANE	1,533,260
18	ARI	LAVIGNE*AVRIL	UNDER MYSKIN	1,439,963
19	GEFN	SIMPSON*ASHLEE	AUTOBIOGRAPHY	1,382,731
20	DRMW	KEITH*TOBY	SHOCK N Y'ALL	1,313,453
21	UME	VARIOUS	NOW 16	1,236,052
22	JIVE	SPEARS*BRITNEY	IN THE ZONE	1,232,355
23	DEF	JAY-Z	BLACK ALBUM	1,190,347
24	ROAD	NICKELBACK	LONG ROAD	1,176,325
25	GEFN	GUNS N'ROSES	GREATEST HITS	1,106,149
26	COL	BEYONCE	DANGEROUSLY IN LOVE	1,080,017
27	AAM	BLACK EYED PEAS	ELEPHUNK	1,065,815
28	INT	BANKS*LLOYD	HUNGER FOR MORE	1,025,219
29	EPIC	INCUBUS	CROW LEFT OF THE MURDER	976,509
30	EPIC	LOS LONELY BOYS	LOS LONELY BOYS	973,445
31	DBV	DUFF*HILARY	METAMORPHOSIS	970,596
32	CAP	CHINGY	JACKPOT	966,443
33	BNTE	JONES*NORAH	COME AWAYWITH ME	946,754
34	WAR	LINKIN PARK	METEORA	940,316
35	INT	G-UNIT	BEG FOR MERCY	939,753
36	INT	NO DOUBT	SINGLES 1992-2003	929,586
37	VRGN	JACKSON*JANET	DAMITA JO	919,016
38	GEFN	BLINK 182	BLINK 182	876,124
39	CAP	BEASTIE BOYS	TO THE 5 BOROUGHS	873,896
40	J	STUDDARD*RUBEN	SOULFUL	864,858
41	DEF	LUDACRIS	CHICKEN & BEER	859,563
42	EEG	JET	GET BORN	858,990
43	COL	LIL' FLIP	U GOTTA FEELME	845,201
44	RCA	VELVET REVOLVER	CONTRABAND	844,625
45	COL	VA-NOW THAT'S WHAT I CALL MUSI	VOL. 14	844,482
46	COL	CONNICK*HARRY	ONLY YOU	844,055
47	UNCM	JUVENILE	JUVE THE GREAT	830,990
48	WAR	BIG & RICH	HORSE OF A DIFFERENT COLOR	819,500
49	ARNV	JACKSON*ALAN	GRT HITS V2	815,640
50	COL	SWITCHFOOT	BEAUTIFUL LETDOWN	789,586

Nielsen SoundScan, a division of Nielsen Entertainment LLC

| 1 | 2 | 3 | 4 | 5 |

This is a sample of the Year-To-Date Top 200 Chart. Eachcolumn is identified bynumber:

1 – Rank in sales for the Year 2004/week ending 8/22/04

2 – Label

3 – Artist

4 – Title of Album

5 – Number of units sold in the year to date for weekending 8/22/04

▲ Figure 6.4 Example of a Year-To-Date Album chart (Source: Nielsen SoundScan)

Weeks On	Label	2W Rank	LW Rank	TW Rank	Artist	Title	TW Sales	% CHG	LW Sales	RTD Sales
10	J	1	1	1	FANTASIA	I BELIEVE/CHAIN OF FOOLS/SUMME	5,921	−33	8,774	377,671
9	RCA	2	2	2	DEGARMO*DIANA	DREAMS/DON'T CRY OUT LOUD	4,683	−23	6,075	165,295
8	CAP	3	3	3	HOUSTON	I LIKE THAT	2,805	−30	3,979	40,679
24	RCA	4	4	4	AIKEN*CLAY	SOLITAIRE/WAY,THE	1,940	−18	2,372	312,693
4	WAR	5	5	5	LINKIN PARK	BREAKING THE HABIT	1,792	−15	2,110	9,225
1	GEFN			6	NEW FOUND GLORY	TRUTH OF MY YOUTH	1,593	999	20	1,613
14	COL	8	7	7	SIMPSON*JESSICA	TAKE MY BREATH AWAY	1,505	−10	1,677	45,114
8	COL	6	8	8	NAS	THIEF'S THEME	1,191	−26	1,617	12,853
16	WAR	7	9	9	MIS-TEEQ	SCANDALOUS (REMIX)	1,106	−25	1,480	27,675
11	UNRF	9	13	10	TERROR SQUAD	LEAN BACK	1,055	−7	1,133	13,249
18	COL	10	10	11	BEYONCE	NAUGHTY GIRL	1,049	−21	1,336	51,306
12	LAF	24	18	12	CIARA	GOODIES	1,032	16	891	7,494
4	UNIV	12	12	13	NELLY	FLAP YOUR WINGS	1,030	−10	1,149	5,431
10	DEF	14	20	13	L.L. COOL J	HEADSPRUNG	1,030	16	888	8,557
17	ATLG	11	11	15	BRANDY	TALK ABOUT OUR LOVE	979	−21	1,232	39,130
16	EPIC	18	16	16	LOS LONELY BOYS	HEAVEN	941	−8	1,019	18,436
4	INT	15	15	17	YOUNG BUCK	LET ME IN	894	−14	1,041	3,752
8	EMIG	17	14	18	SHEARD*KIERRA KIKI	YOU DON'T KNOW	836	−23	1,089	6,429
13	ATLG	13	6	19	CABRERA*RYAN	ON THE WAYDOWN	818	−58	1,934	8,186
36	ARI	22	17	20	DIDO	WHITE FLAG B/W STONE	813	−13	930	46,495
4	COL	19	18	21	LIL' FLIP	SUNSHINE	798	−10	891	3,293
24	CURB	23	23	22	LOCKE*KIMBERLEY	8TH WORLD WONDER	738	−12	839	73,701
5	INT	16	24	23	JADAKISS	WHY?	729	−7	786	4,081
20	UNRF	28	33	24	AKON	LOCKED UP/LOCKED	688	24	553	8,176
17	GEFN	25	25	25	ROOTS	DON'T SAY NUTHIN	679	−12	772	17,459
16	EPIC	32	27	26	MODEST MOUSE	FLOAT ON	639	−8	696	12,570
52	ZMST	34	29	27	KELIS	MILKSHAKE	619	−5	650	57,364
6	SNCT	21	26	28	MORRISSEY	FIRST OF THE GANG TO DIE	615	−18	748	6,382
18	UNIV	20	22	29	NINA SKY	MOVE YABODY	606	−29	850	42,071
14	EPIC	27	30	30	MICHAEL*GEORGE	AMAZING	570	−7	613	14,402
3	GEFN	26	28	31	CURE	END OF THE WORLD	560	−15	662	2,057
15	EPIC	50	36	32	FRANZ FERDINAND	TAKE ME OUT	547	1	540	8,400
12	VRGN	38	31	33	PERFECT CIRCLE	OUTSIDER, THE	534	−9	589	12,211
11	BSMG	35	32	34	PRESLEY*ELVIS	THAT'S ALL RIGHT	526	−9	577	18,815
20	UNIV	51	43	35	LACHEY*NICK	THIS I SWEAR	477	7	445	13,464
9	J	36	37	36	STUDDARD*RUBEN	WHAT IF	474	−12	539	8,115
21	JIVE	45	42	37	MOBB DEEP	GOT IT TWISTED	447	−0	449	13,155
21	MERN	30	34	38	ROBERTS*JULIE	BREAK DOWN HERE	436	−21	552	17,483
4	VRGN	40	44	39	BEENIE MAN	KING OF THE DANCEHALL	433	−1	438	2,287
25	ISL	65	45	40	MILIAN*CHRISTINA	DIP IT LOW	429	−1	432	6,689

Nielsen SoundScan, a division of Nielsen Entertainment LLC

| 1 | 2 | 3 | 4 | 5 | 6 | 7 | 8 | 9 | 10 | 11 |

This is a sample of the Top 200 Singles Chart. Each column is identified by number:
1 – Number of weeks the single has been on the Top 200 Chart
2 – Record Label
3 – Chart position 2 weeks ago
4 – Chart position last week
5 – Current chart position
6 – Artist Name
7 – Single Title
8 – Number of units sold this week
9 – Percentage of change from this week to last week
10 – Number of units sold last week
11 – Total number of units sold since the release of the single

▲ *Figure 6.5 Example of a Top Singles chart (Source: Nielsen SoundScan)*

Weeks On	Label	2W Rank	LW Rank	TW Rank	Artist	Title
11	UNRF	1	1	1	TERROR SQUAD	LEAN BACK
12	COL	3	2	2	LIL' FLIP	SUNSHINE
11	LAF	6	3	3	CIARA	GOODIES
7	UNIV	9	7	4	NELLY	FLAP YOUR WINGS
16	ATLG	4	5	5	LYTTLE*KEVIN	TURN ME ON
19		2	4	6	JUVENILE FEATURING	SLOW MOTION
20	ISL	5	6	7	MILIAN*CHRISTINA	DIP IT LOW
10	GEFN	13	9	8	SIMPSON*ASHLEE	PIECES OF ME
19	UNIV	8	10	9	NINA SKY	MOVE YA BODY
27	J	10	11	10	KEYS*ALICIA	IF I AIN'T GOT YOU
19		7	8	11	USHER	CONFESSIONS PART II
12	CAP	12	12	12	HOUSTON	I LIKE THAT
12		17	15	13	ALICIA KEYS FEATURI	DIARY
7		20	14	14	MAROON5	SHE WILL BE LOVED
25		11	13	15	HOOBASTANK	THE REASON
8	INT	23	16	16	JADAKISS	WHY?
13	UNRF	28	19	17	AKON	LOCKED UP/LOCKED
22	UNIV	15	17	18	JOJO	LEAVE (GET OUT)
9	DEF	29	27	19	L.L. COOL J	HEADSPRUNG
18	DEF	14	18	20	WEST*KANYE	JESUS WALKS
17	EPIC	16	21	21	LOS LONELY BOYS	HEAVEN
30	OCTJ	19	20	22	MAROON 5	THIS LOVE
6		34	31	23	AVRIL LAVIGNE	MY HAPPY ENDING
23		22	22	24	SWITCHFOOT	MEANT TO LIVE
7	WAR	35	29	25	LINKIN PARK	BREAKING THE HABIT
26		18	23	26	USHER	BURN
11	ATLG	25	25	27	TWISTA	SO SEXY
16		26	26	28	LLOYD FEATURING ASH	SOUTHSIDE
35	LAF	24	24	29	USHER	YEAH
5		40	34	30	BLACK EYED PEAS	LET'S GET IT STARTED
7		37	38	31	FINGER ELEVEN	ONE THING
13		30	32	32	TIM MCGRAW	LIVE LIKE YOU WERE DYING
14		27	28	33	MONICA	U SHOULD'VE KNOWN BETTER
7	INT	36	37	34	YOUNG BUCK	LET ME IN
9	ATLG	44	39	35	T.I.	LET'S GET AWAY
7		42	40	36	KEITH URBAN	DAYS GO BY
36	JIVE	31	36	37	PABLO*PETEY	FREEK-A-LEEK
6	ATLG	52	43	38	CABRERA*RYAN	ON THE WAY DOWN
13		33	33	39	KENNY CHESNEY	I GO BACK
9		45	44	40	TERRI CLARK	GIRLS LIE TOO
11		39	41	41	COUNTING CROWS	ACCIDENTALLY IN LOVE
12		41	42	42	BRAD PAISLEY FEATUR	WHISKEY LULLABY
8	WAR	50	46	43	LIL SCRAPPY	NO PROBLEM/BE REAL
17	INT	21	30	44	BANKS*LLOYD	ON FIRE
10	INT	32	35	45	D12	HOW COME
5		55	49	46	SARA EVANS	SUDS IN THE BUCKET
29		38	47	47	MARIO WINANS FEATUR	I DON'T WANNA KNOW
5		53	52	48	GRETCHEN WILSON	HERE FOR THE PARTY
6		54	50	49	ANDY GRIGGS	SHE THINKS SHE NEEDS ME
5		57	51	50	ALAN JACKSON	TOO MUCH OF A GOOD THING

Nielsen SoundScan, a division of Nielsen Entertainment LLC

▲ *Figure 6.6 Example of a Hot 100 Singles chart (Source: Nielsen SoundScan)*

sales as part of the chart equation, with test charts reporting 60% airplay/40% single sales.[i]

The Hot 100 Singles chart is not genre specific, but attempts to create a chart based on a combination of popularity votes, both airplay and actual sales.

---

[i] Silvio Pietroluongo, Billboard Hot 100 Chart Manager, Interview 1/12/05.

Nielsen SoundScan — a Service of Nielsen Entertainment · vnu

HOME | TITLE | SUMMARY | CHARTS | MARKETING | PICKUP | SETS | ARCHIVES | ACCOUNT | HELP

DMA CHART: Top Albums New York, NY   LOAD TO EXCEL · PRINT

Week Ending: 08/22/2004 · Display: % CHG

Weeks On	Label	2W Rank	LW Rank	TW Rank	Artist	Title	TW Sales	% CHG	LW Sales	RTD Sales
5	GEFN	1	1	1	SIMPSON*ASHLEE	AUTOBIOGRAPHY	13,161	-28	18,381	100,483
4	UME	2	4	2	VARIOUS	NOW 16	12,591	-12	14,319	70,910
1	ATLG			3	CABRERA*RYAN	TAKE IT ALL AWAY	10,310	999	59	10,375
2	DEF		2	4	SHYNE	GODFATHER BURIED	6,622	-62	17,305	24,152
8	OCTJ	7	6	5	MAROON 5	SONGS ABOUT JANE	6,175	5	5,891	155,338
2	JIVE		3	6	MOBB DEEP	AMERIKAZ NIGHT MARE	5,745	-62	14,950	21,040
8	LAF	3	5	7	USHER	CONFESSIONS	5,504	-9	6,038	369,394
2	EPIC		17	8	GARDEN STATE	SOUNDTRACK	4,509	47	3,064	7,583
8	ARI	9	7	9	LAVIGNE*AVRIL	UNDER MY SKIN	4,339	-7	4,661	89,631
8	INT	8	8	10	BANKS*LLOYD	HUNGER FOR MORE	3,764	-15	4,406	104,969
3	DBV	22	13	11	PRINCESS DIARIES 2	SOUNDTRACK	3,605	-6	3,821	10,047
4	ATLG	5	10	12	LYTTLE*KEVIN	KEVIN LYTTLE	3,287	-23	4,250	23,705
4	UNRF	4	11	13	TERROR SQUAD	TRUE STORY	3,078	-25	4,089	30,551
6	RCNA	10	15	14	BUFFETT*JIMMY	LICENSE TO CHILL	3,077	-16	3,653	31,959
4	VIT	6	9	15	TAKING BACK SUNDAY	WHERE YOU WANT TO BE	3,043	-29	4,292	28,754
1	TVT			16	213	HARD WAY	2,957	999	37	2,994
8	AAM	19	18	17	BLACK EYED PEAS	ELEPHUNK	2,858	-5	3,011	116,973
8	J	23	25	18	KEYS*ALICIA	DIARY OF ALICIA KEYS	2,720	8	2,526	334,337
8	EPIC	13	19	19	LOS LONELY BOYS	LOS LONELY BOYS	2,625	-8	2,850	49,698
8	INT	11	16	20	JADAKISS	KISS OF DEATH	2,506	-23	3,252	92,659
8	RCA	16	23	21	VELVET REVOLVER	CONTRABAND	2,403	-9	2,641	65,252
6	BADB	12	21	22	NOTORIOUS B.I.G.	READY TODIE	2,280	-15	2,681	443,899
8	EPIC	32	29	23	FRANZ FERDINAND	FRANZ FERDINAND	2,221	1	2,204	39,733
8	UNIV	18	22	24	JOJO	JOJO	2,203	-17	2,648	33,661
1	FTSK			25	FUTURE SOUNDTRACK FOR AMERICA	FUTURE SOUNDTRACK FOR AMERICA	2,170	999	12	2,182
8	ISL	46	44	26	KILLERS	HOT FUSS	2,147	14	1,879	15,104
8	COL	17	33	27	DE-LOVELY	SOUNDTRACK	2,089	-2	2,142	23,152
8	GEFN	43	43	28	GUNS N'ROSES	GREATEST HITS	2,085	10	1,891	77,387
8	CAP	14	28	29	BEASTIE BOYS	TO THE 5 BOROUGHS	2,046	-12	2,318	74,689
8	WAR	26	36	30	LINKIN PARK	METEORA	2,002	-2	2,044	271,261
8	CAP	40	38	31	YELLOWCARD	OCEAN AVENUE	1,992	-1	2,005	72,595
8	DECC	36	42	32	WICKED	ORIGINAL CAST RECORDING	1,973	4	1,901	82,576
6	GEFN	20	26	33	ROOTS	TIPPING POINT	1,965	-17	2,362	25,914
8	DEF	24	30	34	WEST*KANYE	COLLEGE DROP OUT	1,936	-12	2,197	197,322
8	OCTJ	41	48	35	MAROON 5	1.22.03 ACOUSTIC	1,915	11	1,730	15,005
8	EPIC	33	40	36	MODEST MOUSE	GOOD NEWS FOR PEOPLE WHO LOVE	1,831	-5	1,924	49,865
5	WAR	21	34	37	VAN HALEN	BEST OF BOTH WORLDS	1,812	-14	2,100	18,333
8	COL	30	27	38	SIMPSON*JESSICA	IN THIS SKIN	1,798	-23	2,323	149,235
8	ISL	47	52	39	HOOBASTANK	THE REASON	1,792	7	1,673	91,546
8	UNRF	31	39	40	AKON	TROUBLE	1,786	-8	1,933	17,693

Nielsen SoundScan, a division of Nielsen Entertainment LLC

| 1 | 2 | 3 | 4 | 5 | 6 | 7 | 8 | 9 | 10 | 11 |

SoundScan can reduce the sales data to create charts within specific DMAs. This chart reflects sales in the New York DMA, stating sales of the best selling albums for the week ending 8/22/04, including a full week of sales.

1 – Weeks on chart 3 – Rank 2 weeks prior 5 – Current ranking 7 – Album Title 9 – % change 11 – Sales since release in specific DMA

2 – Record Label 4 – Rank last week 6 – Artist 8 – Sales this week 10 – Sales last week

▲ *Figure 6.7 Example DMA chart (Source: Nielsen SoundScan)*

Archives

Archives – SoundScan archives the previous year's sales data, dating back to 1994. This Top Current Album chart is the top current albums for the current week of 12/28/2003.

```
CHART: Top Current Albums
```

Wks On	Lbl	2 Weeks	Last Week	This Week			Curr. Week Sales(est)	Last Week
14	LAF	6	4	1	*	OUTKAST|SPEAKERBOXX-LOVE	374133	334647
4	J	2	1	2		KEYS*ALICIA|DIARY OF ALICIA KE	371238	369861
8	DRMW	4	2	3		KEITH*TOBY|SHOCK N Y'ALL	314022	369480
8	COL	3	3	4		VA-NOW THAT'S W|VOL. 14	302162	354414
8	AAM	9	7	5		CROW*SHERYL|VERY BEST OF SHERY	255529	266237
7	WAR	5	5	6		GROBAN*JOSH|CLOSER	249517	318296
18	DBV	8	6	7		DUFF*HILARY|METAMORPHOSIS	245559	274641
5	INT	12	9	8		NO DOUBT|SINGLES COLLECTION	217333	235420
3	J	1	8	9		STUDDARD*RUBEN|SOULFUL	211404	244820
6	JIVE	10	13	10		SPEARS*BRITNEY|IN THE ZONE	205026	204956
7	INT	24	15	11	*	G-UNIT|BEG FOR MERCY	199086	155139
7	DEF	23	18	12	*	JAY-Z|BLACK ALBUM	188699	149268
8	ARI	14	12	13		MCLACHLAN*SARAH|AFTERGLOW	181196	211980
6	GEFN	19	16	14	*	BLINK 182|BLINK 182	178250	155092
11	RCA	11	11	15		AIKEN*CLAY|MEASURE OF A MAN	172472	212088
43	WNDU	25	19	16	*	EVANESCENCE|FALLEN	154878	148356
27	COL	31	21	17	*	BEYONCE|DANGEROUSLY IN LOVE	153865	136202
10	J	7	10	18		STEWART*ROD|VOL.II-GREAT AMERI	150633	221646
7	INT	29	25	19	*	2PAC|RESURRECTION	143443	121519
12	DEF	35	28	20	*	LUDACRIS|CHICKEN & BEER	137439	120269
20	ARNV	15	14	21		JACKSON*ALAN|GRT HITS V2 (LTD)	136308	199049
24	CAP	34	29	22	*	CHINGY|JACKPOT	134678	117242
14	ROAD	28	20	23		NICKELBACK|LONG ROAD	130559	141901
13	ARI	26	22	24		DIDO|LIFE FOR RENT	129096	135502
40	WAR	32	23	25		LINKIN PARK|METEORA	129038	135478
6	CAP	20	17	26		BEATLES|LET IT BE...NAKED	126153	151397
59	UNIV	33	24	27		3 DOORS DOWN|AWAY FROM THE SUN	118553	125560
5	UNIV	39	40	28	*	NELLY|DA DERRTY VERSIONS-REINV	116617	90554
7	ATLG	38	31	29		KID ROCK|KID ROCK	114300	114113
96	BNTE	41	30	30		JONES*NORAH|COME AWAY WITH ME	111718	114539
10	EEG	36	32	31		EAGLES|VERY BEST OF THE EAGLES	105403	108087
1	UNCM	---	---	32	*	JUVENILE|JUVE THE GREAT	103407	926
6	WAR	40	34	33		RED HOT CHILI P|GREATEST HITS	103110	101479
1	COL	---	---	34	*	B2K|B2K PRESENTS...YOU GOT SER	102504	747
6	EPIC	42	38	35	*	KORN|TAKE A LOOK IN THE MIRROR	99128	90898
70	CAP	57	49	36	*	COLDPLAY|RUSH OF BLOOD TO THE	89182	79422
6	WAR	51	46	37	*	LINKIN PARK|LIVE IN TEXAS	86674	81134
3	CAP	16	44	38		WESTSIDE CONNEC|TERRORIST THRE	86337	84388
6	COL	44	41	39		DIXIE CHICKS|TOP OF THE WORLD	84227	87608
56	LAVA	56	45	40		SIMPLE PLAN|NO PADS, NO HELMET	83835	82657
20	DBV	37	33	41		CHEETAH GIRLS E|SOUNDTRACK	82656	106204
16	COL	55	51	42		MAYER*JOHN|HEAVIER THINGS	78410	77676
5	EEG	59	67	43	*	ELLIOTT*MISSY|THIS IS NOT A TE	77922	62159
58	MERN	43	36	44		TWAIN*SHANIA|UP!	76983	96416
5	GEFN	62	60	45	*	PUDDLE OF MUDD|LIFE ON DISPLAY	76766	66023
5	WAR	74	53	46		LORD OF THE RIN|SCORE	76465	73526
13	RCNA	60	39	47		MCBRIDE*MARTINA|MARTINA	73529	90729
75	DRMW	47	43	48		KEITH*TOBY|UNLEASHED	71789	85483
65	EPIC	63	55	49		GOOD CHARLOTTE|YOUNG & THE HOP	71449	70930
61	TVT	84	79	50	*	LIL JON & EAST|KINGS OF CRUNK	69743	52658

Week Ending 12/28/03

Nielsen SoundScan, a division of Nielsen Entertainment LLC

▲ *Figure 6.8 Example archives chart: Top Current Albums (Source: Nielsen SoundScan)*

Marketing Reports

SoundScan creates various marketing charts that analyze the marketplace by segmenting sales into many categories.

When reading the *YTD – Sales by Format Genre Album* report, add (000) to the end of the units to accurately depict album sales by genre. Note that the CD format dominates album sales. Additionally, certain releases can be counted twice since they may be considered in more than one genre.

What are not calculated in this report are the genre *percentages* as they relate to the total. This information is in a different SoundScan report.

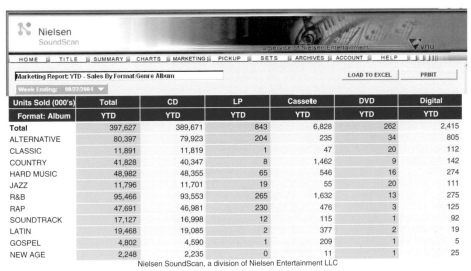

Units Sold (000's)	Total	CD	LP	Cassete	DVD	Digital
Format: Album	YTD	YTD	YTD	YTD	YTD	YTD
Total	397,627	389,671	843	6,828	262	2,415
ALTERNATIVE	80,397	79,923	204	235	34	805
CLASSIC	11,891	11,819	1	47	20	112
COUNTRY	41,828	40,347	8	1,462	9	142
HARD MUSIC	48,982	48,355	65	546	16	274
JAZZ	11,796	11,701	19	55	20	111
R&B	95,466	93,553	265	1,632	13	275
RAP	47,691	46,981	230	476	3	125
SOUNDTRACK	17,127	16,998	12	115	1	92
LATIN	19,468	19,085	2	377	2	19
GOSPEL	4,802	4,590	1	209	1	5
NEW AGE	2,248	2,235	0	11	1	25

Nielsen SoundScan, a division of Nielsen Entertainment LLC

▲ *Figure 6.9 Example marketing report: YTD – Sales by Format/Genre Album (Source: Nielsen SoundScan)*

Label Marketing Share Report

The *Label Share Marketing Report* shows the percentage of business by distribution companies, as well as indies for the week ending 8/22/05. Each distributor, shown as Level 1, has "owned" labels that are part of the conglomerate, and "distributed" labels that have contracted the distributor to place their records into the marketplace, which are noted as Level 2. Label groups have sublabels that are noted as Level 3. Each label is part of the cumulative percentage.

From a competitive standpoint, this data allows distributors and their labels to evaluate their performance as it compares to others. This data is compiled weekly, monthly, and year-to-date.

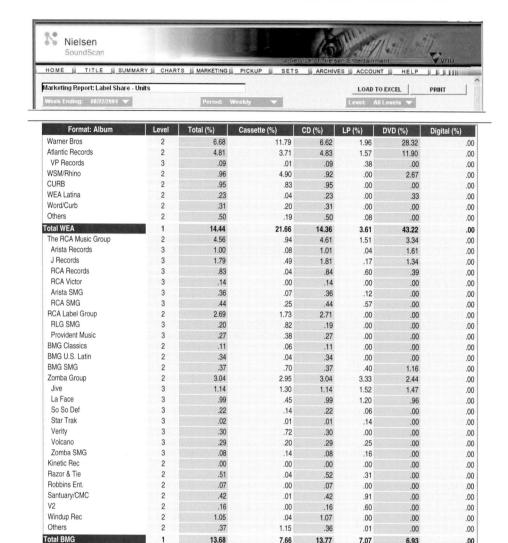

Format: Album	Level	Total (%)	Cassette (%)	CD (%)	LP (%)	DVD (%)	Digital (%)
Warner Bros	2	6.68	11.79	6.62	1.96	28.32	.00
Atlantic Records	2	4.81	3.71	4.83	1.57	11.90	.00
VP Records	3	.09	.01	.09	.38	.00	.00
WSM/Rhino	2	.96	4.90	.92	.00	2.67	.00
CURB	2	.95	.83	.95	.00	.00	.00
WEA Latina	2	.23	.04	.23	.00	.33	.00
Word/Curb	2	.31	.20	.31	.00	.00	.00
Others	2	.50	.19	.50	.08	.00	.00
Total WEA	1	**14.44**	**21.66**	**14.36**	**3.61**	**43.22**	**.00**
The RCA Music Group	2	4.56	.94	4.61	1.51	3.34	.00
Arista Records	3	1.00	.08	1.01	.04	1.61	.00
J Records	3	1.79	.49	1.81	.17	1.34	.00
RCA Records	3	.83	.04	.84	.60	.39	.00
RCA Victor	3	.14	.00	.14	.00	.00	.00
Arista SMG	3	.36	.07	.36	.12	.00	.00
RCA SMG	3	.44	.25	.44	.57	.00	.00
RCA Label Group	2	2.69	1.73	2.71	.00	.00	.00
RLG SMG	3	.20	.82	.19	.00	.00	.00
Provident Music	3	.27	.38	.27	.00	.00	.00
BMG Classics	2	.11	.06	.11	.00	.00	.00
BMG U.S. Latin	2	.34	.04	.34	.00	.00	.00
BMG SMG	2	.37	.70	.37	.40	1.16	.00
Zomba Group	2	3.04	2.95	3.04	3.33	2.44	.00
Jive	3	1.14	1.30	1.14	1.52	1.47	.00
La Face	3	.99	.45	.99	1.20	.96	.00
So So Def	3	.22	.14	.22	.06	.00	.00
Star Trak	3	.02	.01	.01	.14	.00	.00
Verity	3	.30	.72	.30	.00	.00	.00
Volcano	3	.29	.20	.29	.25	.00	.00
Zomba SMG	3	.08	.14	.08	.16	.00	.00
Kinetic Rec	2	.00	.00	.00	.00	.00	.00
Razor & Tie	2	.51	.04	.52	.31	.00	.00
Robbins Ent.	2	.07	.00	.07	.00	.00	.00
Santuary/CMC	2	.42	.01	.42	.91	.00	.00
V2	2	.16	.00	.16	.60	.00	.00
Windup Rec	2	1.05	.04	1.07	.00	.00	.00
Others	2	.37	1.15	.36	.01	.00	.00
Total BMG	1	**13.68**	**7.66**	**13.77**	**7.07**	**6.93**	**.00**
Interscope/A&M/Dream	2	4.36	1.85	4.39	3.31	3.40	.00
Interscope/Dreamw	3	3.23	1.22	3.25	3.31	.00	.00
A&M	3	1.13	.63	1.14	.00	3.40	.00
Geffen	2	5.07	4.86	5.08	3.31	3.16	.00
Island Def Jam Music	2	6.01	5.44	6.02	3.87	2.04	.00
Def Jam/Def Soul	3	1.85	1.90	1.85	2.60	.04	.00
Island	3	3.11	3.36	3.12	.86	2.00	.00
Roadrunner	3	.64	.09	.65	.10	.00	.00
Lost Highway	3	.11	.04	.11	.30	.00	.00
Rounder	3	.29	.06	.30	.00	.00	.00
Distributed Labls	3	.00	.00	.00	.00	.00	.00
Universal Records Gr	2	4.63	4.73	4.64	2.48	2.02	.00
UMG-Nashville	2	1.18	2.88	1.16	.04	1.26	.00
MCA Nashville	3	.57	1.76	.55	.00	.20	.00

Nielsen SoundScan, a division of Nielsen Entertainment LLC

▲ *Figure 6.10 Example marketing report: label share – units (Source: Nielsen SoundScan)*

Mercury Nashville	3	.62	1.11	.61	.04	1.06	.00
The Verve Group	2	.66	.02	.66	.46	3.77	.00
Verve	3	.54	.01	.54	.22	3.67	.00
GRP	3	.12	.01	.12	.24	.10	.00
Classics	2	.84	.12	.85	.00	2.34	.00
Universal Latino	2	.42	.03	.42	.00	.00	.00
Bohemia	3	.00	.00	.00	.00	.00	.00
Protel	3	.02	.01	.02	.00	.00	.00
Platano	3	.00	.00	.00	.00	.00	.00
Latino	3	.00	.00	.00	.00	.00	.00
Karen	3	.01	.00	.01	.00	.00	.00
Rodven	3	.03	.00	.03	.00	.00	.00
RMM	3	.01	.00	.01	.00	.00	.00
Lideres	2	.08	.00	.08	.00	.00	.00
OLE	2	.04	.00	.04	.00	.00	.00
Univision Music Grp	2	2.04	.17	2.07	.00	.00	.00
Univision Records	3	.43	.01	.43	.00	.00	.00
Fonovisa Records	3	.80	.11	.81	.00	.00	.00
Disa	3	.81	.04	.82	.00	.00	.00
Viva Discos	2	.00	.00	.00	.00	.00	.00
Hollywood Records	2	1.48	.05	1.50	.13	.00	.00
Hollywood	3	1.21	.03	1.22	.13	.00	.00
Mammoth	3	.01	.00	.01	.00	.00	.00
Lyric Street	3	.27	.02	.27	.00	.00	.00
Disney/Buena Vista	2	1.50	.42	1.52	.01	2.47	.00
H.O.L.A.	2	.00	.00	.00	.00	.00	.00
Concord Records	2	.16	.00	.16	.00	.00	.00
Peak Records	3	.04	.00	.04	.00	.00	.00
CURB	2	.03	.00	.03	.01	.00	.00
Curb/MCA Nashvill	3	.01	.00	.01	.00	.00	.00
Curb/MCA Los Ange	3	.01	.00	.01	.00	.00	.00
Curb/Universal	3	.00	.00	.00	.00	.00	.00
Curb/Mercury Nash	3	.00	.00	.00	.00	.00	.00
Curb/Lost Highway	3	.01	.00	.01	.01	.00	.00
Thump	2	.06	.03	.06	.08	.00	.00
Universal Music Ente	2	.93	3.61	.91	.01	.00	.00
Hip-O	3	.67	.02	.68	.00	.00	.00
Now	3	.01	.00	.01	.00	.00	.00
Special Products	3	.25	3.59	.22	.01	.00	.00
Varese	2	.21	.00	.21	.00	.00	.00
Trinity Records	2	.00	.00	.00	.00	.00	.00
Mock & Roll	2	.00	.00	.00	.00	.00	.00
Beyond Music	2	.02	.16	.01	.01	.00	.00
Ark 21	2	.03	.00	.03	.00	.00	.00
Bungalo Records	2	.01	.00	.01	.00	.00	.00
Northsound	2	.00	.01	.00	.00	.00	.00
VI Music	2	.14	.00	.14	.00	.00	.00
Others	2	.16	.78	.15	.80	.12	.00
Total UMVD	1	**30.05**	**25.16**	**30.14**	**14.51**	**20.58**	.00
Angel/Bluenote	2	.94	.05	.95	.31	.10	.00
Capitol Nashville	2	.75	.43	.76	.00	.00	.00
Capitol Records	2	3.82	3.27	3.83	3.15	1.85	.00
Christian Music Grou	2	.70	.12	.71	.00	.00	.00
EMI Latin	2	.43	.05	.44	.00	.00	.00
Aries Records	3	.04	.00	.05	.00	.00	.00
Max Mex Records	3	.01	.00	.01	.00	.00	.00
Venevision	3	.02	.00	.02	.00	.00	.00
NOW Compilations	2	.01	.00	.01	.00	.00	.00
Virgin	2	1.43	.59	1.44	1.41	1.41	.00
Right Stuff/Special	2	.42	3.51	.38	.01	.14	.00
Unidentified	2	.25	.06	.25	.00	.37	.00
Others	2	.21	.10	.21	.63	.00	.00
Total EMM	1	**8.96**	**8.19**	**8.98**	**5.50**	**3.87**	.00
Columbia	2	7.03	13.54	6.97	2.39	.00	.00

Nielsen SoundScan, a division of Nielsen Entertainment LLC

▲ *Figure 6.10 Continued*

Epic	2	4.85	8.09	4.82	3.14	.00	.00
Sony Classical	2	.36	.39	.36	.00	.00	.00
Sony Discos	2	1.03	.82	1.04	.00	.00	.00
SMCMG	2	.19	1.54	.17	.00	.00	.00
Others	2	.15	.08	.15	.08	.00	.00
Total Sony	1	**13.61**	**24.47**	**13.50**	**5.61**	**.00**	**.00**
Madacy Records/Excel	2	.91	2.10	.90	.00	.00	.00
Tommy Boy	2	.04	.02	.04	.19	.00	.00
Loud	2	.00	.00	.00	.00	.00	.00
Laserlight/Delta	2	.17	.00	.17	.00	.00	.00
KTEL/qwil	2	.07	.02	.07	.00	.00	.00
Essex	2	.01	.00	.01	.00	.00	.00
American Gramaphone	2	.01	.00	.01	.00	.04	.00
TVT	2	1.38	.19	1.40	1.02	.00	.00
Rounder	2	.01	.00	.01	.00	.00	.00
Fantasy	2	.13	.31	.12	.60	.00	.00
Ryko	2	.08	.00	.08	.15	.00	.00
Warlock/Quality	2	.02	.01	.02	.00	.00	.00
Telarc	2	.08	.01	.08	.00	.26	.00
Caroline Records	2	.04	.00	.04	.72	.00	.00
Gusto	2	.01	.23	.00	.00	.00	.00
Sub pop	2	.20	.00	.20	2.28	.00	.00
Malaco/Muscle Shoals	2	.04	.29	.04	.00	.00	.00
Masters	2	.03	.00	.03	.00	.00	.00
Alligator	2	.05	.00	.05	.00	.00	.00
Nervous Records	2	.00	.00	.00	.16	.00	.00
Michele	2	.00	.05	.00	.00	.00	.00
Matador	2	.08	.00	.08	1.87	.00	.00
DM Records	2	.02	.01	.02	.00	.00	.00
VP Records	2	.06	.01	.05	2.07	.10	.00
Select	2	.01	.00	.01	.00	.00	.00
Metal Blade	2	.09	.00	.09	.00	.00	.00
Vanguard-pop	2	.01	.00	.01	.00	.00	.00
Oh boy	2	.02	.00	.02	.00	.00	.00
R.A.D.	2	.00	.00	.00	.01	.00	.00
Oglio	2	.02	.00	.02	.00	.00	.00
Earache	2	.02	.00	.02	.05	.00	.00
Fat Wreck Chords	2	.10	.00	.10	.51	.00	.00
Lil Joe	2	.01	.00	.01	.05	.00	.00
Kung Fu Records	2	.04	.00	.04	.02	.26	.00
Sugar Hill	2	.09	.01	.09	.01	.00	.00
Turn Up The Music	2	.05	.10	.05	.00	.00	.00
Radikal	2	.01	.00	.01	.00	.22	.00
Spitfire Records	2	.05	.00	.05	.03	.00	.00
Shrapnel	2	.01	.00	.01	.00	.00	.00
Streetbeat	2	.00	.00	.00	.00	.00	.00
Black Market	2	.01	.03	.01	.00	.00	.00
CMH	2	.06	.00	.06	.00	.00	.00
Musicrama	2	.27	.00	.27	.00	.00	.00
Ultra	2	.08	.00	.08	.01	.00	.00
Balboa	2	.34	.02	.34	.00	.00	.00
Classified Records	2	.00	.00	.00	.00	.00	.00
NAXOS	2	.14	.01	.14	.00	.55	.00
Astralwerks	2	.08	.00	.08	1.10	.00	.00
Simitar	2	.00	.02	.00	.00	.00	.00
Collectables Records	2	.10	.00	.10	.00	.00	.00
Wreckshop	2	.01	.02	.01	.00	.00	.00
Cleopatra	2	.09	.00	.09	.00	.00	.00
Primitive/Rawkus	2	.00	.00	.00	.00	.00	.00
Edel America Records	2	.03	.00	.03	.00	.00	.00
Metropolitan Records	2	.00	.00	.00	.00	.00	.00
Original Sound Rec	2	.05	.04	.05	.00	.00	.00

Nielsen SoundScan, a division of Nielsen Entertainment LLC

▲ *Figure 6.10 Continued*

Victory Records	2	.45	.00	.45	.02	.00	.00
Joy/JoeyRecords	2	.01	.14	.01	.00	.00	.00
Nitro Records	2	.04	.00	.04	.06	.00	.00
Peter Pan Industries	2	.00	.00	.00	.00	.00	.00
Mute	2	.03	.00	.03	.62	.00	.00
Strictly Rhythm	2	.00	.00	.00	.00	.00	.00
Groovilicious	2	.01	.00	.01	.00	.00	.00
Beggars Banquet	2	.05	.00	.05	.56	.00	.00
Touch & Go	2	.04	.00	.04	1.21	.00	.00
Six Degrees	2	.06	.00	.06	.00	.00	.00
Blix Street	2	.05	.00	.05	.00	.00	.00
Time Life Music	2	.00	.00	.00	.00	.00	.00
Jellybean Recordings	2	.00	.00	.00	.00	.00	.00
Prime Cuts	2	.01	.08	.01	.00	.00	.00
Cold Front	2	.00	.01	.00	.00	.00	.00
Eagle Records/Vision	2	.02	.00	.02	.00	.00	.00
Koch Entertainment	2	2.11	1.26	2.12	4.68	1.06	.00
Koch Records	3	.75	1.20	.75	1.96	.00	.00
Shanachie	3	.07	.00	.07	.00	.00	.00
Moonshine	3	.03	.00	.03	.00	.00	.00
Artemis	3	.19	.00	.19	.03	.00	.00
Epitaph	3	.30	.01	.31	.73	.63	.00
Compendia	3	.12	.03	.12	.00	.43	.00
Koch Other	3	.66	.01	.66	1.97	.00	.00
Others	2	11.18	7.82	11.15	45.74	22.92	.00
Total Others	1	19.26	12.86	19.25	63.69	25.39	.00

Nielsen SoundScan, a division of Nielsen Entertainment LLC

▲ *Figure 6.10 Continued*

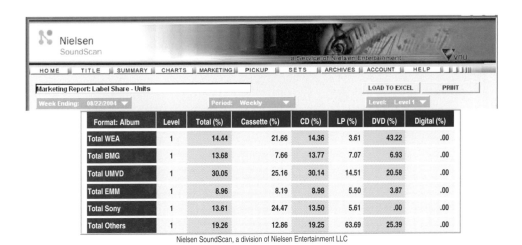

Nielsen SoundScan, a division of Nielsen Entertainment LLC

▲ *Figure 6.11 Example marketing report: market share summary (Source: Nielsen SoundScan)*

Marketing Report: YTD Percent Sales by DMA/Genre

The *YTD Percentage of Sales by DMA/Genre* calculates the percentage of music sold in a specific DMA by genre. DMA stands for *dominant market area* with the regions being derived by Nielsen, the company that calculates television ratings.

Example: New York DMA purchased 6.95% of all records sold. For every 100 records sold, nearly 7 are purchased in the New York DMA.

Additionally, New York DMA purchased 12.03% of all classical records sold, making New York a ripe market for classical music. Again, for every 100 classical records sold, over 12 records are purchased in the New York DMA.

This report helps marketers to look at DMAs as a whole. Labels can compare the percentage of their artists' sales to that of the overall market, seeing if an artist's sales have over or under performed.

Nielsen SoundScan Marketing Report

Marketing Report: YTD - % Sales By DMA/Genre Album

Week Ending: 08/22/2004 | DMA Type: Full

Units Sold (000's) DMA	Total YTD	Alternative YTD	Classical YTD	Country YTD	Hard Music YTD	Jazz YTD	R & B YTD	Rap YTD
Total	100.00	100.00	100.00	100.00	100.00	100.00	100.00	100.00
New York, NY	6.95	6.76	12.03	2.59	6.01	9.97	7.33	6.32
Los Angeles, CA	6.46	6.96	5.91	2.82	6.07	7.27	6.34	5.98
Chicago, IL	3.53	3.51	4.34	2.12	3.31	4.08	4.06	3.87
Philadelphia, PA	2.70	2.74	3.08	1.67	2.57	3.38	3.11	2.78
SF-Okland-San Jose	3.34	3.23	5.07	1.71	2.48	5.74	3.02	2.91
Boston, MA	2.50	2.84	3.45	1.55	2.69	2.80	2.23	2.36
Dallas-Ft. Worth, TX	1.90	1.81	1.91	2.32	1.70	1.66	1.74	1.68
Detroit, MI	1.65	1.58	1.72	1.12	1.58	2.09	2.27	2.09
Washington, DC	2.57	2.22	3.62	2.03	2.04	3.62	3.23	2.81
Houston, TX	1.67	1.48	1.35	1.72	1.51	1.41	1.73	1.56
Cleveland, OH	1.11	1.15	1.09	1.11	1.24	1.27	1.24	1.30
Atlanta, GA	2.00	1.67	1.60	2.11	1.49	2.10	2.63	2.42
Minneapolis-St. Paul, MN	1.46	1.69	1.33	1.57	1.59	1.42	1.22	1.25
Tampa-St.Petersburg, FL	1.18	1.12	1.15	1.34	1.15	1.09	1.10	1.08
Seattle-Tacoma, WA	2.00	2.32	2.55	1.80	1.90	2.86	1.65	1.82
Miami, FL	1.36	.98	1.69	.55	.91	1.53	1.42	1.34
Pittsburgh, PA	.81	.87	.83	1.04	.92	.78	.70	.73
St. Louis, MO	1.05	1.07	1.03	1.19	1.09	1.38	1.06	1.01
Denver, CO	1.61	1.84	1.58	1.49	1.63	1.77	1.26	1.41
Phoenix, AZ	1.51	1.74	1.26	1.37	1.80	1.10	1.35	1.60
Sacramento-Stockton, CA	1.13	1.17	.99	.96	1.20	1.12	1.12	1.13
Baltimore, MD	1.27	1.33	1.10	1.00	1.26	1.67	1.59	1.47
Hartford-New Haven, CT	.87	.86	1.02	.63	.91	.89	.97	.96
San Diego, CA	1.41	1.71	1.39	.75	1.42	1.64	1.17	1.25
Orlndo-Daytona Bch-Mlbrne	1.15	1.24	1.05	1.09	1.21	.98	1.08	1.09
Indianapolis, IN	.96	.93	.83	1.23	.93	.84	1.04	1.14
Portland, OR	1.01	1.08	1.59	1.06	.97	1.48	.64	.68
Milwaukee, WI	.67	.72	.67	.62	.66	.58	.69	.72
Kansas City, KS-MO	.86	.87	.85	1.11	.86	.76	.86	.90
Cincinnati, OH	.66	.70	.51	.83	.72	.60	.69	.73
Charlotte, NC	.76	.69	.49	1.02	.74	.53	.92	.90
Nashville, TN	.78	.71	.46	1.38	.71	.56	.81	.82
Raleigh-Durham, NC	.84	.72	.63	.96	.68	.66	1.03	.95
Columbus, OH	.69	.73	.56	.81	.77	.71	.68	.69
Greenvle-Sprtnbrg-Ashvll	.56	.51	.44	.96	.57	.39	.54	.56
New Orleans, LA	.58	.49	.47	.58	.50	1.14	.71	.63
Grnd Rpds-Klmzo-Bttle Crk	.56	.62	.51	.71	.70	.44	.49	.53
Buffalo, NY	.46	.51	.46	.42	.51	.45	.42	.43
Memphis, TN	.48	.36	.28	.60	.39	.41	.68	.56
Oklahoma City, OK	.51	.52	.42	.85	.56	.34	.44	.47
Salt Lake City, UT	.82	1.01	1.07	1.09	.98	.55	.53	.60
Nrflk-Prtsmth-NwptNws-Hmp	.71	.64	.50	.65	.68	.56	1.04	.94
San Antonio, TX	.71	.68	.51	.90	.77	.48	.64	.69
Providence-New Bedford,RI	.60	.68	.61	.38	.72	.56	.62	.69
Harrsbrg-Yrk-Lncstr-Lbnon	.50	.56	.56	.65	.62	.41	.42	.44
Louisville, KY	.61	.65	.33	.90	.63	.54	.53	.59
Birmingham, AL	.42	.39	.30	.66	.39	.31	.45	.43
Charleston-Huntington, WV	.30	.27	.15	.75	.37	.12	.23	.25
Greensbro-WnstnSalm-HiPnt	.44	.36	.36	.64	.39	.37	.48	.45
WstPlmBch-FrtPerc-VeroBch	.56	.49	.74	.47	.47	.64	.52	.53
Albuquerque, NM	.69	.67	.65	.84	.63	.56	.63	.76
Dayton, OH	.39	.41	.31	.53	.47	.41	.41	.41
Albany-Schnctady-Troy, NY	.45	.50	.58	.47	.57	.45	.38	.40
Wilkes-Barre-Scranton, PA	.44	.53	.38	.52	.63	.23	.32	.36
Mobile-Pensacola, AL-FL	.46	.45	.32	.66	.50	.33	.53	.59
Jacksonville, FL	.56	.51	.35	.68	.53	.50	.73	.73
Little Rock, AR	.37	.32	.20	.67	.39	.20	.38	.37
Tulsa, OK	.33	.32	.27	.60	.35	.23	.29	.31
Flint-Saginaw-BayCity, MI	.32	.31	.23	.44	.39	.24	.33	.34
Richmond, VA	.70	.52	.45	.71	.53	.66	1.21	1.21
Wichita-Hutchison, KS	.32	.32	.30	.52	.40	.20	.24	.27
Fresno-Visalia, CA	.43	.42	.28	.38	.46	.27	.41	.44

Nielsen SoundScan, a division of Nielsen Entertainment LLC

▲ *Figure 6.12 Sample marketing report YTD – % sales by DMA/genre album (Source: Nielsen SoundScan)*

Nielsen SoundScan Marketing Report

Units Sold (000's) DMA Total	Current YTD 100.00	Top Seller YTD 100.00	Soundtrack YTD 100.00	Latin YTD 100.00	Gospel YTD 100.00	New Age YTD 100.00	Catalog YTD 100.00	Deep Catalog YTD 100.00
New York, NY	6.61	6.37	7.12	4.31	5.80	7.13	7.52	7.80
Los Angeles, CA	6.35	5.79	6.48	13.02	3.60	5.49	6.65	6.67
Chicago, IL	3.62	3.62	3.72	3.62	4.08	3.26	3.40	3.41
Philadelphia, PA	2.65	2.72	2.76	1.26	3.92	2.79	2.78	2.83
SF-Okland-San Jose	3.15	2.62	3.26	4.40	1.76	3.55	3.66	3.74
Boston, MA	2.42	2.46	2.63	.80	1.35	2.76	2.63	2.69
Dallas-Ft. Worth, TX	2.01	1.98	2.03	3.52	2.15	1.86	1.72	1.66
Detroit, MI	1.71	1.79	1.73	.51	1.99	1.48	1.56	1.54
Washington, DC	2.62	2.61	2.54	2.32	3.93	2.76	2.50	2.48
Houston, TX	1.69	1.62	1.65	4.02	2.69	1.48	1.63	1.60
Cleveland, OH	1.15	1.26	1.09	.23	1.17	1.27	1.05	1.05
Atlanta, GA	2.16	2.25	1.81	2.49	4.19	1.69	1.72	1.69
Minneapolis-St. Paul, MN	1.51	1.60	1.63	.74	.51	1.49	1.37	1.34
Tampa-St.Petersburg, FL	1.17	1.20	1.10	1.62	1.09	1.52	1.19	1.19
Seattle-Tacoma, WA	1.95	1.85	2.03	1.17	.73	2.32	2.09	2.12
Miami, FL	1.33	1.16	1.21	4.42	1.33	1.94	1.42	1.40
Pittsburgh, PA	.81	.88	.88	.07	.52	.85	.81	.82
St. Louis, MO	1.03	1.09	1.14	.25	.96	1.06	1.10	1.12
Denver, CO	1.57	1.46	1.70	1.65	.43	2.05	1.68	1.69
Phoenix, AZ	1.53	1.52	1.52	2.44	.59	1.73	1.48	1.45
Sacramento-Stockton, CA	1.15	1.17	1.22	1.63	.62	1.25	1.09	1.08
Baltimore, MD	1.22	1.23	1.18	.25	1.96	1.11	1.36	1.38
Hartford-New Haven, CT	.88	.92	.92	.45	.83	1.02	.87	.88
San Diego, CA	1.37	1.25	1.31	2.18	.45	1.56	1.48	1.48
Orlndo-Daytona Bch-Mlbrne	1.18	1.19	1.19	1.26	1.05	1.35	1.11	1.09
Indianapolis, IN	1.04	1.08	.96	.54	1.16	.95	.85	.80
Portland, OR	.96	.90	1.05	.78	.27	1.32	1.09	1.11
Milwaukee, WI	.71	.74	.76	.35	.48	.75	.62	.61
Kansas City, KS-MO	.87	.91	.91	.56	.78	1.06	.83	.82
Cincinnati, OH	.69	.74	.68	.24	.88	.59	.62	.60
Charlotte, NC	.81	.88	.73	.75	1.31	.59	.69	.68
Nashville, TN	.83	.87	.71	.49	.86	.51	.71	.70
Raleigh-Durham, NC	.89	.91	.75	1.11	1.96	.71	.75	.73
Columbus, OH	.72	.76	.67	.42	.68	.60	.64	.62
Greenvile-Sprtnbrg-Ashvll	.58	.60	.54	.50	.96	.49	.54	.54
New Orleans, LA	.56	.55	.51	.23	1.09	.50	.62	.64
Grnd Rpds-Klmzo-Bttle Crk	.57	.61	.59	.33	.24	.60	.55	.55
Buffalo, NY	.45	.47	.47	.06	.84	.42	.46	.46
Memphis, TN	.49	.51	.43	.41	1.23	.26	.46	.46
Oklahoma City, OK	.50	.53	.53	.36	.32	.50	.52	.51
Salt Lake City, UT	.82	.86	1.21	.69	.13	1.07	.81	.79
Nrflk-Prtsmth-NwptNws-Hmp	.70	.75	.68	.36	2.61	.57	.71	.70
San Antonio, TX	.74	.72	.73	1.77	.41	.73	.67	.64
Providence-New Bedford,RI	.59	.64	.64	.13	.35	.61	.63	.64
Harrsbrg-Yrk-Lncstr-Lbnon	.51	.55	.54	.15	.24	.61	.50	.49
Louisville, KY	.63	.60	.52	.26	.36	.39	.59	.59
Birmingham, AL	.45	.47	.39	.32	.69	.25	.37	.37
Charleston-Huntington, WV	.31	.35	.29	.02	.15	.17	.29	.29
Greensbro-WnstnSalm-HiPnt	.46	.48	.42	.48	1.34	.37	.40	.39
WstPlmBch-FrtPerc-VeroBch	.55	.55	.53	.97	.46	.82	.57	.58
Albuquerque, NM	.65	.63	.70	.87	.19	.85	.76	.77
Dayton, OH	.41	.44	.40	.06	.37	.41	.37	.36
Albany-Schnctady-Troy, NY	.43	.45	.45	.07	.17	.52	.49	.50
Wilkes-Barre-Scranton, PA	.43	.48	.46	.06	.10	.38	.44	.44
Mobile-Pensacola, AL-FL	.46	.50	.45	.20	.69	.43	.46	.45
Jacksonville, FL	.58	.63	.51	.30	1.45	.48	.51	.51
Little Rock, AR	.36	.39	.34	.34	.73	.26	.37	.37
Tulsa, OK	.34	.36	.36	.22	.24	.32	.33	.33
Flint-Saginaw-BayCity, MI	.33	.37	.31	.05	.76	.26	.29	.28
Richmond, VA	.73	.70	.56	.30	2.90	.43	.66	.62
Wichita-Hutchison, KS	.32	.34	.34	.37	.17	.42	.31	.31
Fresno-Visalia, CA	.44	.44	.42	1.18	.18	.31	.40	.39
Toledo, OH	.34	.38	.34	.13	.46	.32	.29	.28
Knoxville, TN	.41	.44	.39	.16	.24	.34	.41	.42
Shrvport-Txrcana,AR-LA-TX	.28	.29	.21	.28	.97	.13	.24	.23

Nielsen SoundScan, a division of Nielsen Entertainment LLC

▲ *Figure 6.12 Continued*

National Sales Summary Report

The *National Sales Summary Report* denotes sales by store type, geographic region, and population density. Weekending data is compared to prior week sales, with YTD totals also included. A look from the same week this year to last year helps marketers understand seasonality as well as the impact of special events, such as award shows or national disasters.

Note that chain stores sold the most, with mass merchants not far behind. Chain stores include electronic superstores as well as retail chains.

Although cities have higher population density per square mile, rural areas out-purchase the city geographic marketplace, with the suburbs being the largest subsection of music purchasers.

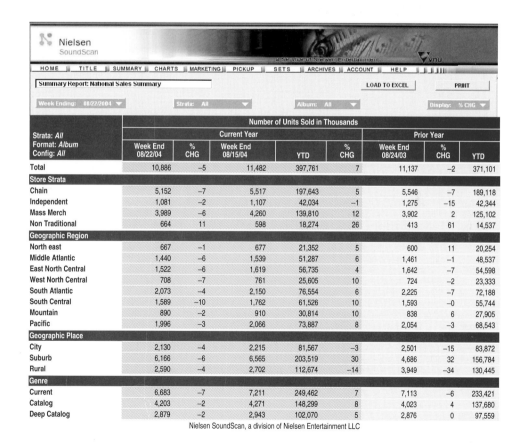

Strata: *All* Format: *Album* Config: *All*	Number of Units Sold in Thousands							
	Current Year					Prior Year		
	Week End 08/22/04	% CHG	Week End 08/15/04	YTD	% CHG	Week End 08/24/03	% CHG	YTD
Total	10,886	−5	11,482	397,761	7	11,137	−2	371,101
Store Strata								
Chain	5,152	−7	5,517	197,643	5	5,546	−7	189,118
Independent	1,081	−2	1,107	42,034	−1	1,275	−15	42,344
Mass Merch	3,989	−6	4,260	139,810	12	3,902	2	125,102
Non Traditional	664	11	598	18,274	26	413	61	14,537
Geographic Region								
North east	667	−1	677	21,352	5	600	11	20,254
Middle Atlantic	1,440	−6	1,539	51,287	6	1,461	−1	48,537
East North Central	1,522	−6	1,619	56,735	4	1,642	−7	54,598
West North Central	708	−7	761	25,605	10	724	−2	23,333
South Atlantic	2,073	−4	2,150	76,554	6	2,225	−7	72,188
South Central	1,589	−10	1,762	61,526	10	1,593	−0	55,744
Mountain	890	−2	910	30,814	10	838	6	27,905
Pacific	1,996	−3	2,066	73,887	8	2,054	−3	68,543
Geographic Place								
City	2,130	−4	2,215	81,567	−3	2,501	−15	83,872
Suburb	6,166	−6	6,565	203,519	30	4,686	32	156,784
Rural	2,590	−4	2,702	112,674	−14	3,949	−34	130,445
Genre								
Current	6,683	−7	7,211	249,462	7	7,113	−6	233,421
Catalog	4,203	−2	4,271	148,299	8	4,023	4	137,680
Deep Catalog	2,879	−2	2,943	102,070	5	2,876	0	97,559

Nielsen SoundScan, a division of Nielsen Entertainment LLC

▲ *Figure 6.13 Example Summary Report: National Sales Summary (Source: Nielsen SoundScan)*

A look at SoundScan title reports

Title lookup

In the current interface, SoundScan has created comprehensive look-up charts for individual artists. After requesting a specific artist, using last name first (for example, Jones*Norah), all releases for this artist will be listed, with release-to-date (RTD) sales included. Each title has a drop-down menu where the researcher can look at specific Title reports.

Nielsen SoundScan, a division of Nielsen Entertainment LLC

▲ *Figure 6.14 Title lookup (Source: Nielsen SoundScan)*

Title report: National sales

Reports are generated that are similar to national marketing reports. Norah Jones' "Come Away with Me" is a huge sales success, with 9 million units sold. The kind of store and geographical regions are reflected in the *National Sales* Title report.

Chain stores, including electronic superstores, outsell the mass merchants, but not by much. Nationally, independents comprise nearly 11% of sales, but in

Jones' case, independents sell only 5.7%, with these leftover sales being picked up by mass merchants. And by doing some fast math, percentage of business can be calculated by dividing independent sales by total sales (see overlay chart, which was calculated separately, but not supplied by SoundScan). Doing the math, a percentage of sales to the total by store strata, geographic region, and geographic place reveals how this record performs against the national averages.

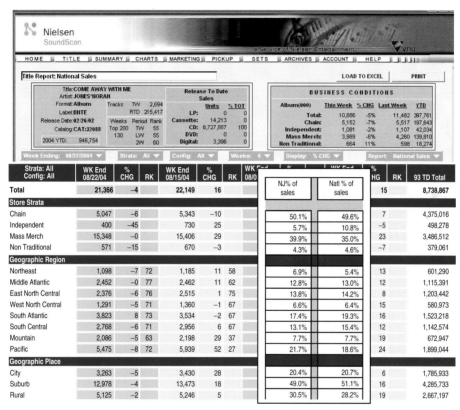

Nielsen SoundScan, a division of Nielsen Entertainment LLC

▲ *Figure 6.15 Example Title Report: National Sales (comparison of NJ to national calculated separately)*
(Source: Nielsen SoundScan)

Title report: DMA

DMA Title Reports shows sales by dominant marketing area. Again, doing some fast math, marketers can analyze the data to show what markets are the best performing, based on population, based on overall sales of the market, based on genre specific sales, and so forth. Marketers, who include label personnel, managers, booking agents, as well as advertisers, look to this type of analysis to create a blueprint—a map of to where to market, tour, create promotions, and so on. The desired result would be increased record sales, but could enhance the artist's profile with increased ticket sales and bigger product endorsements.

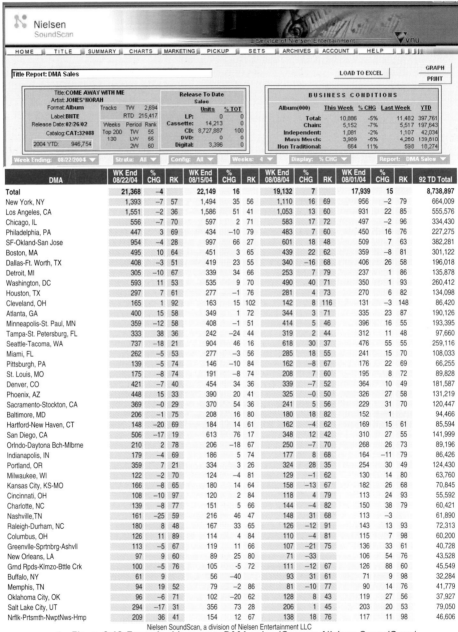

▲ *Figure 6.16 Example title report: DMA sales (Source: Nielsen SoundScan)*

San Antonio, TX	149	9	49	137	−21	72	174	20	45	145	31	66	50,068
Providence-New Bedford, RI	104	−26	78	141	33	69	106	−26	79	143	68	59	56,554
Harrsbrg-Yrk-Lncstr-Lbnon	102	−4	69	106	49	82	71	−10	89	79	32	93	36,574
Louisville, KY	80	−18	100	98	−2	85	100	27	92	79	44		45,035
Birmingham, AL	126	−17	26	152	42	24	107	16	44	92	23	58	34,637
Charlesbro-Huntington, WV	40	−5		42	−18	99	51	6	76	48	78	90	16,352
Greensbro-WnstnSalm-HiPnt	68	−6	76	72	0	89	72	−19	85	89	48	68	34,914
WstPlmBch-FrtPerc-VeroBch	123	−3	44	127	−5	50	133	11	47	120	19	51	55,397
Albuquerque, NM	182	10	35	166	15	39	144	−1	45	145	20	57	53,028
Dayton, OH	58	5	83	55	28		43	−20		54	−24		28,155
Albany-Schnctady-Troy, NY	115	−10	54	128	5	49	122	2	45	120	29	52	43,748
Wilkes-Barre-Scranton, PA	64	28	92	50	−14		58	−30	100	83	77	72	25,083
Mobile-Pensacola, AL-FL	83	−23	75	108	6	57	102	13	66	90	2	76	35,361
Jacksonville, FL	115	31	54	88	−27	86	120	18	67	102	17	76	39,496
Little Rock, AR	68	10	65	62	−18	80	76	52	62	50	−11	100	25,976
Tulsa, OK	68	−8	64	74	−17	66	89	48	43	60	−6	82	28,388
Flint-Saginaw-BayCity, MI	46	−22	95	59	20	81	49	−4	91	51	6	93	21,952
Richmond, VA	112	30	82	86	−25		114	5	86	109	21	89	40,121
Wichita-Hutchison, KS	59	−6	65	63	21	70	52	−21	84	66	74	65	21,526
Fresno-Visalia, CA	104	−9	55	114	18	47	97	14	50	85	35	59	31,743
Toledo, OH	32	14		28	−53		59	44	71	41	−16		22,199
Knoxville, TN	87	−15	53	102	5	54	97	18	56	82	−5	78	32,090
Shrvport-Txrcana, AR-LA-TX	39	−9	89	43	−2	84	44	33	99	33	−15		12,508
Des Moines, IA	67	29	75	52	−32	96	76	43	59	53	6	85	28,954
Green Bay-Appleton, WI	44	−4	85	46	−6	87	49	−21	76	62	32	66	24,767
Syracuse, NY	93	66	47	56	−22	87	72	11	58	65	25	64	28,414
Roanoke-Lynchburg, VA	60	5	69	57	−2	76	58	−13	77	67	24	70	21,884
Lexington, KY	54	−23	87	70	32	69	53	−24	90	70	35	80	23,667
Austin, TX	122	−8	49	133	−1	56	135	6	63	127	27	63	74,103
Rochester, NY	92	6	32	87	26	41	69	−12	50	78	47	45	30,986
Omaha, NE	61	9	78	56	−2	96	57	−26	80	77	43	67	32,962
Portland-PolandSpring, ME	75	−9	70	82	0	58	82	0	58	82	−7	58	37,759
Sprngfld-Decatr-Chmpgn, IL	54	−14	81	63	17	70	54	23	74	44	5	100	23,143
Pdch-CpGrdu-Hrsbg-Mrion, KY-IL	51	13	58	45	45	74	31	−35		48	78	79	14,056
Spokane, WA	115	−18	32	141	64	27	86	−12	49	98	29	45	34,588
Davnprt-RcklsInd-Molin, IL	42	−13	79	48	37	62	35	−17	85	42	14	81	16,676
Tucson, AZ	117	−13	26	134	44	23	93	41	43	66	2	70	32,186
Hntsvlle-Decatr-Flornc, AL	44	−31	70	64	14	46	56	22	60	46	39	72	15,425
CdarRpds-Wtrloo-Dubuqu, IA	38	−30	94	54	20	70	45	−8	76	49	23	71	19,764
Columbia, SC	54	2	72	53	2	79	52	−12	83	59	84	73	20,458
Springfield, MO	65	16	74	56	−27	96	77	15	60	67	−3	68	20,763
Chattanooga, TN	60	28	51	47	12	85	42	−21	95	53	−5	73	15,679
Southbend-Elkhart, IN	44	7	80	41	−51	92	83	36	33	61	42	53	19,405
Jackson, MS	31	−18	91	38	−36	85	59	40	48	42	0	81	15,213
Brstl-Kngsprt-JhnsnCty, TN	24	−27	97	33	−3	83	34	21	75	28	17	94	10,392
Johnstown-Altoona, PA	33	50	71	22	−19		27	−10	87	30	20	87	10,211
Youngstown, OH	23	28		18	−42		31	19	90	26	24		12,058
Madison, WI	58	−13	73	67	14	62	59	18	68	50	39	77	30,815
Las Vegas, NV	147	−8	63	159	10	66	145	4	76	140	10	82	50,750
Brlngtn-Plattsbrgh, VT-NY	102	−8	22	111	2	22	109	−12	23	124	100	23	33,581
Evansville, IN	40	38	66	29	−29	99	41	21	63	34	17	83	12,585
Baton Rouge, LA	43	−19	79	53	33	73	40	14	89	35	−44		16,767
Lincoln-Hastings-Kearney	45	18	68	38	−5	83	40	−33	76	60	54	52	16,157
Ft. Myers-Naples, FL	39	−11	74	44	−8	40	48	9	77	44	5	89	20,551
Waco-Temple-Bryan, TX	53	47	64	36	−38		58	81	69	32	−29		16,733
Springfield, MA	41	−16	74	49	4	69	47	−19	64	58	49	56	22,785
Colorado Sprngs-Pueblo, CO	65	−11	61	73	−10	58	81	1	50	80	14	54	26,477
Hawaii, HI	190	−8	27	207	83	21	113	19	54	95	−6	65	52,673
Other	2,577	−8		2,800	13		2,476	3		2,404	14		914,837

Nielsen SoundScan, a division of Nielsen Entertainment LLC

▲ *Figure 6.16 Continued*

Title report: Store demos

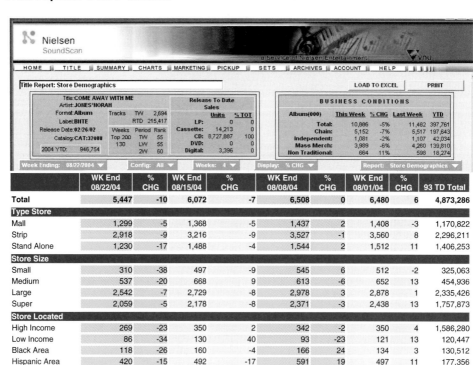

	WK End 08/22/04	% CHG	WK End 08/15/04	% CHG	WK End 08/08/04	% CHG	WK End 08/01/04	% CHG	93 TD Total
Total	**5,447**	**-10**	**6,072**	**-7**	**6,508**	**0**	**6,480**	**6**	**4,873,286**
Type Store									
Mall	1,299	-5	1,368	-5	1,437	2	1,408	-3	1,170,822
Strip	2,918	-9	3,216	-9	3,527	-1	3,560	8	2,296,211
Stand Alone	1,230	-17	1,488	-4	1,544	2	1,512	11	1,406,253
Store Size									
Small	310	-38	497	-9	545	6	512	-2	325,063
Medium	537	-20	668	9	613	-6	652	13	454,936
Large	2,542	-7	2,729	-8	2,978	3	2,878	1	2,335,426
Super	2,059	-5	2,178	-8	2,371	-3	2,438	13	1,757,873
Store Located									
High Income	269	-23	350	2	342	-2	350	4	1,586,280
Low Income	86	-34	130	40	93	-23	121	13	120,447
Black Area	118	-26	160	-4	166	24	134	3	130,512
Hispanic Area	420	-15	492	-17	591	19	497	11	177,356

Nielsen SoundScan, a division of Nielsen Entertainment LLC

▲ *Figure 6.17 Example title report: store demographics (Source: Nielsen SoundScan)*

Again, a similar look to the national *Store Demo* report, an artist's report will show where records are being purchased. Importantly, label marketers could funnel marketing dollars into the most active stores to increase sales—or coop advertising dollars might be supplied to stores that are lagging behind in sales, with the hope of boosting their percentages.

The index report

The *Index report* reveals how well an artist is performing in a specific DMA. To read this report, SoundScan statistically evaluates each DMA by genre and gives the DMA a par score of 100, meaning that the genre sales average index score in that specific market equals 100. If an artist index number is over 100,

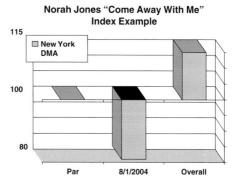

▲ *Figure 6.18 Example of indexing*

Nielsen
SoundScan

a Service of Nielsen Entertainment · vnu

| HOME | TITLE | SUMMARY | CHARTS | MARKETING | PICKUP | SETS | ARCHIVES | ACCOUNT | HELP |

Title Report: DMA Index

LOAD TO EXCEL

GRAPH
PRINT

Title: COME AWAY WITH ME				Release To Date Sales			BUSINESS CONDITIONS				
Artist: JONES'NORAH											
Format: Album	Tracks	TW	2,694	Units	% TOT		Album(000)	This Week	% CHG	Last Week	YTD
Label: BNTE		RTD	215,417	LP: 0	0		Total:	10,886	-5%	11,482	397,761
Release Date: 02/26/02	Weeks	Period	Rank	Cassette: 14,213	0		Chain:	5,152	-7%	5,517	197,643
Catalog: CAT:32088	Top 200	TW	55	CD: 8,727,887	100		Independent:	1,081	-2%	1,107	42,034
	130	LW	55	DVD: 0	0		Mass Merch:	3,989	-6%	4,260	139,810
2004 YTD: 946,754		2W	60	Digital: 3,396	0		Non Traditional:	664	11%	598	16,274

Week Ending: 08/22/2004 ▼   Index: * Current ▼   Weeks: 4 ▼   Report: DMA Index ▼

DMA	WK End 08/22/04	RK	WK End 08/15/04	RK	WK End 08/08/04	RK	WK End 08/08/04	RK	92 TD Total
Total	**21,368**		**22,149**		**19,132**		**17,939**		**8,738,897**
New York, NY	99	57	102	56	88	69	81	79	115
Los Angeles, CA	114	36	113	41	87	60	82	85	100
Chicago, IL	72	70	74	71	84	72	77	96	106
Philadelphia, PA	79	69	74	79	95	60	95	76	98
SF-Okland-San Jose	142	28	143	27	100	48	90	63	139
Boston, MA	96	64	84	65	95	62	83	81	142
Dallas-Ft. Worth, TX	95	51	94	55	88	68	113	58	112
Detroit, MI	83	67	90	66	77	79	77	86	91
Washington, DC	106	53	92	70	98	71	74	93	114
Houston, TX	82	61	74	76	87	73	89	82	91
Cleveland, OH	67	92	64	102	65	116	64	148	86
Atlanta, GA	87	58	73	72	83	71	86	87	101
Minneapolis-St. Paul, MN	111	58	122	51	143	46	146	55	147
Tampa-St. Petersburg, FL	133	36	93	44	143	44	149	48	96
Seattle-Tacoma, WA	177	21	209	16	166	37	136	55	152
Miami, FL	92	53	94	56	112	55	101	70	93
Pittsburgh, PA	80	74	81	84	105	67	121	69	94
St. Louis, MO	80	74	84	74	106	60	106	72	100
Denver, CO	125	40	131	36	113	52	129	49	132
Phoenix, AZ	137	33	115	41	111	50	119	58	98
Sacramento-Stockton, CA	150	29	145	36	110	56	111	70	120
Baltimore, MD	79	75	77	80	77	82	69		89
Hartford-New Haven, CT	79	69	94	61	96	62	107	61	111
San Diego, CA	173	19	202	17	133	42	126	55	119
Orlndo-Daytona Bch-Mlbrne	83	78	79	67	111	70	127	73	86
Indianapolis, IN	81	69	81	74	89	68	88	79	95
Portland, OR	175	21	157	26	176	35	147	49	148
Milwaukee, WI	80	70	79	81	95	62	102	80	103
Kansas City, KS-MO	89	65	93	64	95	67	117	68	93
Cincinnati, OH	73	97	79	84	89	79	91	93	92
Charlotte, NC	80	77	84	66	93	82	103	79	85
Nashville, TN	91	59	117	47	93	68	76		85
Raleigh-Durham, NC	95	48	85	65	74	91	90	93	93
Columbus, OH	82	89	71	84	80	81	89	98	96
Greenvlle-Sprtnbrg-Ashvll	91	67	93	66	96	75	131	61	80
New Orleans, LA	81	60	72	80	66		106	76	89
Grnd Rpds-Klmzo-Bttle Crk	82	76	83	72	102	67	123	60	91
Buffalo, NY	63		56		108	61	88	98	82
Memphis, TN	90	52	73	86	86	77	102	76	98
Oklahoma City, OK	90	71	92	62	134	43	133	56	87

Nielsen SoundScan, a division of Nielsen Entertainment LLC

▲ *Figure 6.19 Example title report: DMA index (Source: Nielsen SoundScan)*

Market								
Salt Lake City, UT	168	31	196	28	131	45	138	53	110
Nrflk-Prtsmth-NwptNws-Hmp	140	41	99	67	103	76	93	98	76
San Antonio, TX	94	49	84	72	123	45	109	66	77
Providence-New Bedford, RI	82	78	108	69	94	79	135	59	110
Harrsbrg-Yrk-Lncstr-Lbnon	94	69	94	82	73	89	86	93	82
Louisville, KY	59	100	70	85	83	92	70		82
Birmingham, AL	131	26	153	24	124	44	114	58	88
Charlesbro-Huntington, WV	60		61	99	86	76	86	90	60
Greensbro-WnstnSalm-HiPnt	69	76	71	89	82	85	108	68	87
WstPlmBch-FrtPerc-VeroBch	105	44	104	50	126	47	122	51	115
Albuquerque, NM	131	35	115	39	116	45	124	57	93
Dayton, OH	66	83	61		55		73		79
Albany-Schnctady-Troy, NY	125	54	134	49	148	45	156	52	116
Wilkes-Barre-Scranton, PA	70	92	52		71	100	108	72	67
Mobile-Pensacola, AL-FL	84	75	106	57	116	66	109	76	88
Jacksonville, FL	93	54	69	86	108	67	98	76	78
Little Rock, AR	88	65	78	80	110	62	77	100	83
Tulsa, OK	94	64	98	66	137	43	98	82	96
Flint-Saginaw-BayCity, MI	65	95	81	81	78	91	86	93	76
Richmond, VA	72	82	53		82	86	83	89	63
Wichita-Hutchison, KS	86	65	89	70	85	84	115	65	77
Fresno-Visalia, CA	111	55	117	47	115	50	108	59	83
Toledo, OH	44		37		91	71	67		75
Knoxville, TN	99	53	112	54	124	56	111	78	90
Shrvport-Txrcana, AR-LA-TX	65	89	69	84	82	99	66		51
Des Moines, IA	98	75	73	96	124	59	92	85	104
Green Bay-Appleton, WI	71	85	72	87	88	76	119	66	98
Syracuse, NY	124	47	72	87	108	58	104	64	93
Roanoke-Lynchburg, VA	88	69	80	76	95	77	117	70	78
Lexington, KY	67	87	83	69	73	90	103	80	71
Austin, TX	79	49	83	56	98	63	98	63	118

Nielsen SoundScan, a division of Nielsen Entertainment LLC

▲ *Figure 6.19 Continued*

the artist is over-performing in that DMA. If the index is under 100, the artist is under-performing in the market.

The index number "week ending 8/1/04" in New York is 81, showing that sales under performed, but the overall index for the record since its release in New York is 115, making it a strong market overall for Jones' "Come Away with Me." By re-sorting any column in descending order, you can obtain a list of top markets weekly, or release-to-date.

Title report: Artist history

The *Artist History* report is a great overview for every artist, listing all releases with pertinent information such as label, release date, and weekly and overall sales. When online, most reports are hyperlinked, so that the viewer can instantly connect to the detail of the data. This report will allow a label to look at the trend of an artist's career.

Nielsen
SoundScan

a Service of Nielsen Entertainment   vnu

| HOME | TITLE | SUMMARY | CHARTS | MARKETING | PICKUP | SETS | ARCHIVES | ACCOUNT | HELP |

Title Report: Artist History   ADD TO SET   AMACY ▼   LOAD TO EXCEL   PRINT

Title: COME AWAY WITH ME
Artist: JONES'NORAH

Format: Album	Tracks	TW	4,141
Label: BNTE		RTD	329,442
Release Date: 02/26/02	Weeks	Period	Rank
Catalog: CAT:32088	Top 200	TW	
2005 YTD: 177,275		LW	
		2W	

Release To Date Sales

	Units	% TOT
LP:	0	0
Cassette:	15,009	0
CD:	9,267,710	100
DVD:	0	0
Digital:	11,755	0

BUSINESS CONDITIONS

Album(000)	This Week	% CHG	Last Week	YTD
Total:	10,378	-16%	12,291	145,210
Chain:	4,786	-8%	5,199	66,510
Independent:	1,010	-0%	1,014	13,195
Mass Merch:	3,801	-29%	5,319	55,716
Non Traditional:	780	3%	761	9,789

Week Ending: 04/03/2005 ▼   Report: Artist History ▼

Title	Label	Release Date	Format	TW Sales	YTD Sales	2004	2003	RTD Sales
COME AWAY WITH ME	MURA	2003-04-01	Album	0	40	1,629	22,914	24,583
COME AWAY WITH ME	BNTE	2002-02-26	Album	8,484	177,275	1,318,457	5,137,468	9,294,482
COME AWAY WITH ME	PID	2003-11-18	Album	0	0	3	1	4
COME AWAY WITH ME	PID	2004-03-30	Album	0	0	0	0	0
COME AWAY WITH ME	PID	2002-04-15	Album	0	0	8	3	13
COME AWAY WITH ME	PID	2002-12-30	Single	0	0	1	0	1
COME AWAY WITH ME	PID	2003-03-04	Album	0	0	1	119	120
COME AWAY WITH ME	PID	2003-03-25	Album	4	39	3,626	7,647	11,312
COME AWAY WITH ME	CNBL	2004-04-20	Album	65	513	934	0	1,447
COME AWAY WITH ME	MSI	2002-12-02	Single	0	0	0	0	0
COME AWAY WITH ME	MSIE	2002-10-08	Single	2	19	279	654	1,241
COME AWAY WITH ME	MSIE	2002-09-02	Single	0	2	31	20	53
COME AWAY WITH ME	MSIE	2003-03-25	Album	0	0	0	5	5
COME AWAY WITH ME(BONUS	EMIA	2003-03-18	Album	0	0	20	161	181
COME AWAY WITH ME (ENHANCED	MSI	2003-03-18	Album	2	108	485	4,276	4,869
COME AWAY WITH ME E.P.	PID	2002-10-04	Single	3	36	372	447	855
COME AWAY WITH ME SINGLE	MGDA	2002-11-26	Album	0	0	0	291	291
DON'T KNOW WHY	PID	2003-09-02	Single	0	1	49	71	121
DON'T KNOW WHY	PID	2002-05-23	Single	7	146	1,122	3,775	5,752
DON'T KNOW WHY	PID	2002-09-16	Single	0	0	3	13	17
DON'T KNOW WHY	PID	2003-02-18	Single	3	31	524	2,200	2,755
DON'T KNOW WHY	MSI	2003-09-08	Single	0	0	96	230	326
DON'T KNOW WHY	MSIE	2000-07-17	Album	0	17	364	687	1,068
FEELIN THE SA	MSIE	2002-10-22	Single	2	45	167	361	740
FEELIN THE SAME WAY	MSIE	2002-07-23	Single	0	0	0	5	12
FEELIN' THE SAME WAY	PID	2002-08-02	Single	0	9	37	113	199
FEELS LIKE HOME	CLRD	2004-05-04	Album	0	2	28	0	30
FEELS LIKE HOME	BNTE	2004-02-03	Album	7,752	187,324	3,842,920	0	4,030,244
FEELS LIKE HOME	PID	2004-02-10	Album	0	0	0	0	0
FEELS LIKE HOME	PID	2004-02-24	Album	0	1	61	0	62
FEELS LIKE HOME	CNBL	2004-03-16	Album	2	148	870	0	1,018
FEELS LIKE HOME	BNTE	2004-09-20	Video	0	7,472	82,368	0	89,840
LIVE IN NEW ORLEANS	BNTE	2003-02-25	Video	388	6,175	55,952	169,478	231,605
MAXIMUM	PID	2003-05-12	Album	7	72	695	1,096	1,863
MAXIMUM NORAH JONES	MURA	2003-05-27	Album	0	41	483	698	1,222
SING-A-LONG	BCIM	2003-03-18	Album	11	229	4,385	9,520	14,134
SLEEPLESS NIGHTS	PID	2005-01-18	Album	0	0	0	0	0
SUNRISE	PID	2004-02-17	Single	3	73	878	0	951
SUNRISE	PID	2004-03-30	Single	0	0	17	0	17
TURN ME ON	BNTE	2003-12-02	Single	74	996	27,542	17,401	45,939
TURN ME ON	PID	2003-08-05	Single	2	31	422	432	885
UNTITLED	PID	2004-02-17	Album	0	0	36	0	36
WHAT AM I TO YOU?	PID	2004-05-25	Single	0	15	70	0	85
WHAT AM I TO YOU? (JAPAN EP)	PID	2004-05-18	Single	0	0	2	0	2
Total			Album	16,327	365,809	5,175,005	5,184,886	13,386,984
			Single	96	1,404	31,612	25,722	59,951
			Video	388	13,647	138,320	169,478	321,445

Nielsen SoundScan, a division of Nielsen Entertainment LLC

▲ *Figure 6.20 Title report: artist history (Source: Nielsen SoundScan)*

Chart History

The *Chart History Report* documents sales from the first week of release to current day. Each week, sales along with ranking of title on the Top 200 chart is included. Additional information such as ranking at retail, mass merchants, as well as the specific genre is also included. This data drives the life cycle sales earlier discussed in this text.

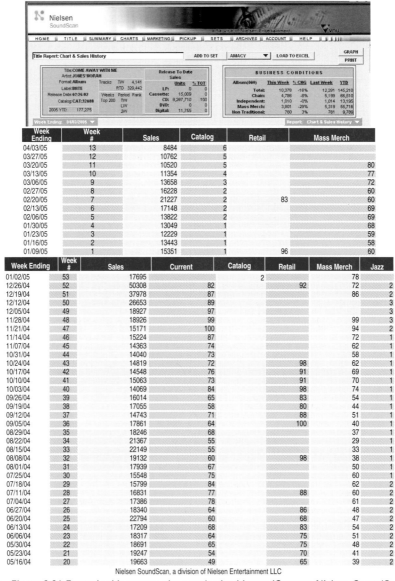

Week Ending	Week #	Sales	Catalog	Retail	Mass Merch
04/03/05	13	8484	6		
03/27/05	12	10762	5		
03/20/05	11	10520	5		80
03/13/05	10	11354	4		77
03/06/05	9	13658	3		72
02/27/05	8	16228	2		60
02/20/05	7	21227	2	83	60
02/13/05	6	17148	2		69
02/06/05	5	13822	2		69
01/30/05	4	13049	1		68
01/23/05	3	12229	1		59
01/16/05	2	13443	1		58
01/09/05	1	15351	1	96	60

Week Ending	Week #	Sales	Current	Catalog	Retail	Mass Merch	Jazz
01/02/05	53	17695		2		78	
12/26/04	52	50308	82		92	72	2
12/19/04	51	37978	87			86	2
12/12/04	50	26653	89				3
12/05/04	49	18927	97				3
11/28/04	48	18926	99			99	3
11/21/04	47	15171	100			94	2
11/14/04	46	15224	87			72	1
11/07/04	45	14363	74			62	1
10/31/04	44	14040	73			58	1
10/24/04	43	14819	72		98	62	1
10/17/04	42	14548	76		91	69	1
10/10/04	41	15063	73		91	70	1
10/03/04	40	14069	84		98	74	1
09/26/04	39	16014	65		83	54	1
09/19/04	38	17055	58		80	44	1
09/12/04	37	14743	71		88	51	1
09/05/04	36	17861	64		100	40	1
08/29/04	35	18246	68			37	1
08/22/04	34	21367	55			29	1
08/15/04	33	22149	55			33	1
08/08/04	32	19132	60		98	38	1
08/01/04	31	17939	67			50	1
07/25/04	30	15548	75			60	1
07/18/04	29	15799	84			62	2
07/11/04	28	16831	77		88	60	2
07/04/04	27	17386	78			61	2
06/27/04	26	18340	64		86	48	2
06/20/04	25	22794	60		68	47	2
06/13/04	24	17209	68		83	54	2
06/06/04	23	18317	64		75	51	2
05/30/04	22	18691	65		75	48	2
05/23/04	21	19247	54		70	41	2
05/16/04	20	19663	49		65	39	2

Nielsen SoundScan, a division of Nielsen Entertainment LLC

▲ *Figure 6.21 Example title report: chart and sales history (Source: Nielsen SoundScan)*

Week Ending	Week #	Sales	Current	Retail	Mass Merch	Jazz
05/09/04	19	30179	36	47	27	2
05/02/04	18	21906	44	67	29	2
04/25/04	17	22856	44	65	27	1
04/18/04	16	23152	43	59	29	1
04/11/04	15	28479	51	64	41	1
04/04/04	14	27494	44	61	31	1
03/28/04	13	29353	39	55	25	1
03/21/04	12	29235	33	45	25	2
03/14/04	11	31769	28	47	22	2
03/07/04	10	35054	32	40	24	2
02/29/04	9	41245	26	34	23	2
02/22/04	8	46026	23	33	22	2
02/15/04	7	79974	18	23	15	2
02/08/04	6	42674	27	36	22	2
02/01/04	5	34224	25	36	20	1
01/25/04	4	32269	24	40	18	1
01/18/04	3	32558	27	42	15	1
01/11/04	2	34122	27	42	17	1
01/04/04	1	43773	34	45	22	1
12/28/03	52	111718	30	32	25	1
12/21/03	51	114539	30	33	28	2
12/14/03	50	71198	41	44	36	2
12/07/03	49	50823	40	49	37	2
11/30/03	48	53352	46	56	44	2
11/23/03	47	49653	35	59	31	2
11/16/03	46	38853	35	49	30	2
11/09/03	45	40098	32	44	25	1
11/02/03	44	29849	34	47	27	1
10/26/03	43	30982	36	48	26	1
10/19/03	42	33294	31	45	26	1
10/12/03	41	33415	29	45	21	1
10/05/03	40	37187	27	41	20	1
09/28/03	39	36027	30	45	21	1
09/21/03	38	37963	23	33	12	1
09/14/03	37	40469	17	26	10	1
09/07/03	36	45886	13	20	10	1
08/31/03	35	49071	14	20	8	1
08/24/03	34	55546	12	20	11	1
08/17/03	33	56894	9	11	9	1
08/10/03	32	59576	7	8	5	1
08/03/03	31	59257	10	11	7	1
07/27/03	30	56378	13	14	11	1
07/20/03	29	61118	9	9	8	1
07/13/03	28	61484	8	7	10	1
07/06/03	27	63137	9	11	11	1
06/29/03	26	67687	9	12	8	1
06/22/03	25	77605	7	8	8	1
06/15/03	24	90317	10	11	7	1
06/08/03	23	74543	11	8	10	1
06/01/03	22	84569	5	6	7	1
05/25/03	21	83099	7	7	8	1
05/18/03	20	87947	6	9	4	1
05/11/03	19	138547	2	3	2	1
05/04/03	18	93610	5	7	6	1
04/27/03	17	98851	6	8	8	1
04/20/03	16	142649	8	7	7	1
04/13/03	15	116824	9	5	8	1
04/06/03	14	132103	5	4	5	1
03/30/03	13	149089	5	5	4	1
03/23/03	12	175897	2	2	1	1
03/16/03	11	243314	2	2	1	1
03/09/03	10	336621	2	2	1	1
03/02/03	9	621030	1	1	1	1
02/23/03	8	144254	3	3	5	1

Nielsen SoundScan, a division of Nielsen Entertainment LLC

▲ *Figure 6.21 Continued*

Week Ending	Week #	Sales	Current	New Artist	Retail	Mass Merch	Jazz
02/16/03	7	140354	3	2	7		1
02/09/03	6	100586	4	2	6		1
02/02/03	5	100632	2	2	4		1
01/26/03	4	112002	1	1	3		1
01/19/03	3	114284	1	1	3		1
01/12/03	2	108296	1	1	6		1
01/05/03	1	118354	2	2	7		1
12/29/02	52	217895	7		2	13	1
12/22/02	51	257271	11		4	19	1
12/15/02	50	170464	15		7	21	1
12/08/02	49	100167	17		10	29	1
12/01/02	48	95484	26		22	36	1
11/24/02	47	68097	20		17	31	1
11/17/02	46	62354	23		19	31	1
11/10/02	45	62698	21		14	26	1
11/03/02	44	58922	18		16	19	1
10/27/02	43	62270	13		10	16	1
10/20/02	42	65729	11		11	14	1
10/13/02	41	63109	12		12	11	1
10/06/02	40	63099	13		14	15	1
09/29/02	39	67334	11		13	11	1
09/22/02	38	69443	8		7	11	1
09/15/02	37	63751	7		7	10	1
09/08/02	36	74835	6		7	10	1
09/01/02	35	72636	11		11	18	1
08/25/02	34	70896	10		8	16	1
08/18/02	33	63903	13		11	17	1
08/11/02	32	55358	15		14	23	1
08/04/02	31	57034	16		13	26	1
07/28/02	30	50374	17		14	32	1
07/21/02	29	46721	15		12	30	1
07/14/02	28	42126	21		15	35	1
07/07/02	27	41083	20		14	43	1
06/30/02	26	39518	22		16	39	1
06/23/02	25	39239	26		18	40	1
06/16/02	24	42984	26		17	48	1
06/09/02	23	43614	22		14	30	1
06/02/02	22	42402	19		10	37	1
05/26/02	21	38042	26		15	74	1
05/19/02	20	31271	35		22	60	1
05/12/02	19	44886	17		11	62	1
05/05/02	18	30760	37		18	94	1
04/28/02	17	25428	46		30		1
04/21/02	16	28595	37		20	97	1
04/14/02	15	24568	42		26		1
04/07/02	14	20224	51		35		1
03/31/02	13	21100	72		44		1
03/24/02	12	22037	58		41		1
03/17/02	11	20048	62		42		1
03/10/02	10	13792	92		61		2
03/03/02	9	9664	139	5	88		2

Nielsen SoundScan, a division of Nielsen Entertainment LLC

▲ *Figure 6.21 Continued*

A deeper look at SoundScan

SoundScan data gives both summaries as well as in-depth analysis as to overall sales of music, a particular genre of music, a particular artist, a particular market, and many more aspects of the business as it pertains to sales of music. By manipulating SoundScan beyond the scope of their pre-determined charts, marketers can better understand the marketplace and its drivers.

Seasonality and record sales

Like most products, there is seasonality to the sales of music. Every year, sales trends show a similar pattern with sales spikes on Valentine's Day, Easter, a lull through summer months, and then a steady rise through the fall going into the holiday selling season. Using the weekly sales charts of total album sales, this overlay of years of weekly sales crystallized seasonal sales trends.

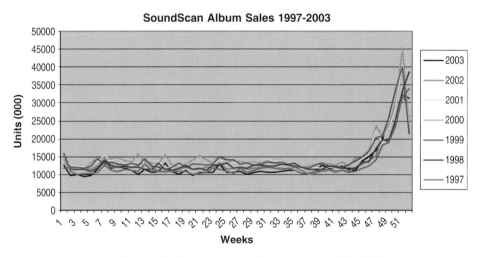

▲ *Figure 6.22 SoundScan Weekly Album Sales 1997–2003*

As students of the business, record company executives analyze when the best time of year would be to release new artist verses that of a platinum act. Strictly based on seasonality alone, most would say to release a new act shortly after the New Year to take advantage of the spring sales spikes and summer touring. And if the act has radio success, the fall selling season would be healthy for the current release already in the marketplace. As for the superstar act, a fourth-quarter release would be perfect timing to capitalize on shoppers, while minimizing long-term advertising dollars. Seasonality is a major factor, but not the only factor; touring and the product life cycle of the artist also play a role.

Business climate and predictor

With some analysis of the numbers, SoundScan can tell a story or paint a picture as to the current business climate. Looking at sales figures of the genre charts before, during, and after 9/11 (2001), industry watchers could predict that the year-end sales figures would not be positive. This chart reflects the percentage of sales change by specific genre, by week. Up until 9/11, the year looked up. But once the tragedies took place, one could watch the trends evolve with the spike in sales in both the country and contemporary Christian genres. The year ended on a down note for the industry as a whole, but country was up, though minimally, and contemporary Christian gained 13.5% in sales for the year.

To determine Percentage of Change, the equation:

This Period (sales) — Last Period (sales) / Last Period (sales) = Percentage of Change

Example:

This Week Sales: 150 $\dfrac{150 - 100 =}{100}$ $\dfrac{50}{100}$ =   50% increase in sales
Last Week Sales: 100

This chart takes weekly sales by genre and applies the previous equation, deriving percentage of change in sales, by week.

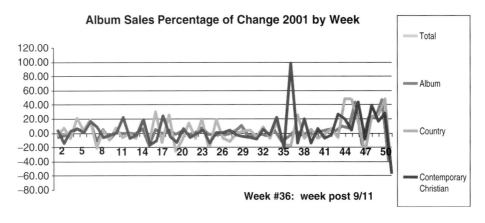

▲ Figure 6.23 Album sales percentage of change in 2001 by week (Source: Nielsen SoundScan)

Life cycles

As discussed earlier in this text, there are four stages within the traditional product life cycle: introduction, growth, maturity, and decline. As a product is adopted into use, and as others learn of its availability, its sales will grow. Eventually, the product will hit maturity, level off in sales, and decline. Either its maker will "reinvent" the product as "new and improved" and evolve it in some manner, or it will no longer exist.

Classic product life cycles

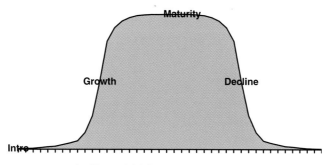

▲ *Figure 6.24 Classic product life cycle*

Product life cycles occur in music too. Although there is an occasional exception to the rule such as a second or third single from a release being the song that drives sales, most releases have a similar sales pattern. Once an artist is established, sales patterns rarely vary, which is why the first few months of a release are so critical to the success of a record. In many cases, the sales success of a release is dependent on how many units are shipped initially into retail. And how many units are shipped is usually dependent on the pent-up demand felt from the marketplace including radio airplay, publicity, touring, press, and other marketing events. Recognize that retailers buying music also look at the track record of the artist and their previous sales as well.

Looking at these examples, most of the releases show a similar pattern in sales. Note that they are different artists, genres, sales plateaus, time of year, and yet the similarity in sales trend is unmistakable. An established artists' sales trend shows an undeniable peak in sales early in the life cycle that tapers off within the first 6 weeks and 3 months, which is why most record labels pack their marketing strategies into this small window of time.

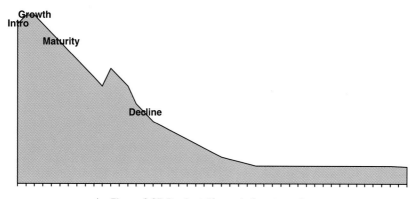

▲ *Figure 6.25 Product life cycle for new releases*

The exceptions to the rule are the artists that are "breaking." Check out the first record of these now superstar acts. Most of their initial releases had a "slow boil" effect, meaning that sales did not catch on at a street date, but later, as consumers became knowledgeable about the artist.

For example, Norah Jones' debut album, "Come Away With Me," is a perfect example of a "slow boil." The single, "Don't Know Why," was released in summer 2002 on the AC, Adult Top 40, Hot 100, and Top 40 mainstream charts, and did not peak until Jan/Feb 2003. The album sales chart shows that over-the-counter activity began to pick up during this time frame, with a peak in sales in early 2003 as consumers sought out this new artist's release.

Note that Jones' sophomore album, "Feels Like Home," reflects the classic established artist's sales trend, with the bulk of sales occurring early in the release. Blue Note, Jones' record label, released this album one week prior to Valentine's Day 2004, and capitalized on seasonality of sales as it relates to their artist.

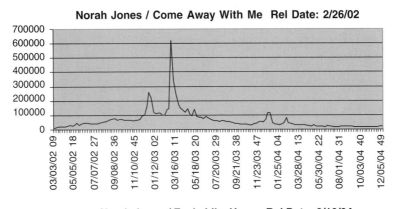

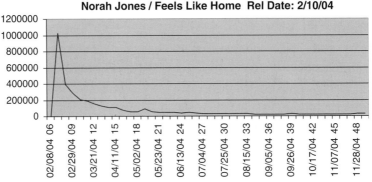

▲ *Figure 6.26 Comparison of sales patterns for Norah Jones' first and second releases (Source: Nielsen SoundScan)*

Released in 2001, "Room for Squares" was John Mayer's "official" first album release. The hit single, "No Such Thing" peaked at radio, depending on the format, in mid-2002, with the bigger hit, "Your Body Is A Wonderland," driving sales even higher. Note the "slow boil" affect with the first release.

"Any Given Thursday" is a live album release that was to bridge sales demand while Mayer recorded the next album. Although a classic "established" artist sales profile occurs, the actual volume was much lower than that of Mayer's initial release. "Heavier Things" continued the sales trend of an established act, though showing a quick die-off in sales, mostly because of the lack of a big single on radio.

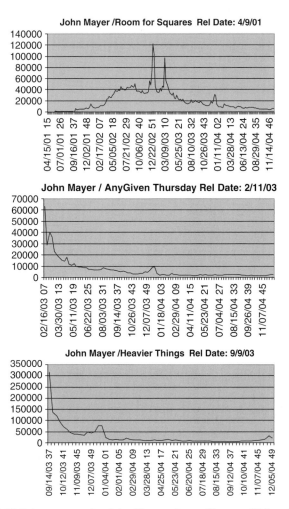

▲ *Figure 6.27 Sales patterns for John Mayer releases (Source: Nielsen SoundScan)*

DMAs and market efficiencies

The designated market area (DMA) is A.C. Nielsen's geographic market design, which defines each television market. DMAs are composed of counties (and possibly also split counties), and are updated annually by the A.C. Nielsen Company based on historical television viewing patterns. Every county or split county in the United States is assigned exclusively to one DMA.

Table 6.1 Comparison of number of DMAs needed to get 33 percent of the market for Jazz and Country genres (Source: Nielsen SoundScan)

	YTD % of Sales by DMA/Genre 2004	Jazz	Country
1	New York	10.1	2.59
2	Los Angeles	7.13	2.81
3	Chicago	4.05	2.2
4	Philadelphia	3.41	1.72
5	SF - Oakland	5.59	1.65
6	Boston	2.82	1.56
	TOTAL	33.1	
7	Dallas	1.6	2.3
8	Detroit	2.1	1.15
9	Washington	3.76	2.03
10	Houston	1.4	1.68
11	Cleveland	1.29	1.16
12	Atlanta	2.11	2.07
13	Minneapolis	1.42	1.62
14	Tampa	1.08	1.33
15	Seattle	2.66	1.77
16	Miami	1.5	0.54
17	Pittsburgh	0.81	1.08
18	St. Louis	1.33	1.17
19	Denver	1.79	1.49
20	Phoenix	1.17	1.38
			33.3

Radio audience estimates for DMAs are published in the Radio Market Reports of all standard radio markets whose metros are located within the DMA *and* whose names are contained in the DMA name. For example, radio audience estimates for the San Francisco-Oakland-San Jose DMA are reported in both the San Francisco and the San Jose radio market reports; however, radio audience estimates for the New York DMA are reported in the New York report, but not in the Nassau-Suffolk report (Katz Media Group Radio Resource Area).

The following data reflects the percentage of music sold in each of the DMAs. SoundScan lists the DMAs in order of population, with New York being the most populated area surveyed, Los Angeles being the second largest populated area surveyed, and so on, with some variances based on radio or television markets. In *Radio & Records* (R&R), populations are based on 12+ age category, being people that live within the surveyed area that are 12 and over in age. When looking at the markets, record labels should consider efficiency of the advertising dollars and marketing efforts.

Each genre of music contains a unique profile; for example, the percentage of music sold in the top six DMAs for jazz equals 33.1%. Think about it— for every 100 jazz records sold, 33 of them are sold in these top six DMA markets. As a marketing department, advertising in these DMAs should have a big bang for the buck, considering the efficiency of targeting to the buyers in these markets. In contrast, to sell 33 records out of 100 in the country genre, the top twenty DMA markets are needed to achieve that percentage. Instantly, to reach buyers of country music, country record labels need to spread their marketing dollars and efforts thinner, or *smarter*, to effectively reach the same percentage of buyers of the genre.

Best-Selling markets vs. strongest markets

Units Sold (000's) DMA	Total YTD	Alternative YTD	Classical YTD	Country YTD	Hard Music YTD	Jazz YTD	R & B YTD	Rap YTD	Current YTD	Top Seller YTD	Soundtrack YTD	Latin YTD	Gospel YTD	New Age YTD	Catalog YTD	Deep Catalog YTD
Total	100.00	100.00	100.00	100.00	100.00	100.00	100.00	100.00	100.00	100.00	100.00	100.00	100.00	100.00	100.00	100.00
New York, NY	8.95	6.76	12.03	2.59	6.01	9.97	7.33	8.32	6.81	6.37	7.12	4.31	5.80	7.13	7.52	7.80
Los Angeles, CA	6.46	6.96	5.91	2.82	6.07	7.27	6.34	5.98	6.35	5.79	6.48	13.02	3.60	5.49	6.65	6.67
Chicago, IL	3.53	3.51	4.34	2.12	3.31	4.08	4.06	3.87	3.62	3.62	3.72	3.62	4.08	3.26	3.40	3.41
Philadelphia, PA	2.70	2.74	3.08	1.67	2.57	3.38	3.11	2.78	2.85	2.72	2.76	1.26	3.92	2.79	2.78	2.83
SF-Okland-San Jose	3.34	3.23	5.07	1.71	2.48	5.74	3.02	2.91	3.15	2.62	3.26	4.40	1.76	3.55	3.66	3.74
Boston, MA	2.50	2.84	3.45	1.55	2.69	2.80	2.23	2.36	2.42	2.46	2.63	.80	1.35	2.76	2.63	2.69
Dallas-Ft. Worth, TX	1.90	1.81	1.91	2.32	1.70	1.66	1.74	1.68	2.01	1.98	2.03	3.52	2.15	1.86	1.72	1.66
Detroit, MI	1.85	1.58	1.72	1.12	1.58	2.09	2.27	2.09	1.71	1.79	1.73	.51	1.99	1.48	1.56	1.54
Washington, DC	2.57	2.22	3.62	2.03	2.04	3.62	3.23	2.81	2.82	2.61	2.54	2.32	3.93	2.76	2.50	2.48
Houston, TX	1.67	1.48	1.35	1.72	1.51	1.41	1.73	1.56	1.69	1.62	1.55	4.02	2.69	1.48	1.63	1.60
Cleveland, OH	1.11	1.15	1.09	1.11	1.24	1.27	1.24	1.30	1.15	1.26	1.09	.23	1.17	1.27	1.05	1.05
Atlanta, GA	2.00	1.67	1.60	2.11	1.49	2.10	2.63	2.42	2.16	2.25	1.81	2.49	4.19	1.69	1.72	1.69
Minneapolis-St. Paul, MN	1.46	1.69	1.33	1.57	1.59	1.42	1.22	1.25	1.51	1.60	1.63	.74	.51	1.49	1.37	1.34
Tampa-St.Petersburg, FL	1.18	1.12	1.15	1.34	1.15	1.09	1.10	1.08	1.17	1.20	1.10	1.62	1.09	1.52	1.19	1.19
Seattle-Tacoma, WA	2.00	2.32	2.55	1.80	1.90	2.66	1.65	1.82	1.95	1.85	2.03	1.17	.73	2.32	2.09	2.12
Miami, FL	1.36	.98	1.69	.55	.91	1.53	1.42	1.34	1.33	1.16	1.21	4.42	1.33	1.94	1.42	1.40
Pittsburgh, PA	.81	.87	.83	1.04	.92	.78	.70	.73	.81	.88	.88	.07	.52	.85	.81	.82
St. Louis, MO	1.05	1.07	1.03	1.19	1.09	1.38	1.06	1.01	1.03	1.09	1.14	.25	.96	1.08	1.10	1.12

Nielsen SoundScan, a division of Nielsen Entertainment LLC

▲ *Figure 6.28 Marketing report: YTD - % Sales by DMA/Genre album (Source: Nielsen SoundScan)*

It sounds confusing—why wouldn't the best selling markets NOT be the *strongest* markets when looking at genre sales and/or title reports of a specific release? Let's look at how record companies can manipulate SoundScan data to be *smarter* marketers.

This data is a new sort of the same numbers. By ranking the DMAs by percentage of sales, marketing experts can now view the best-selling markets in order. (For this book's example, only the Top 20 markets are being analyzed.) But are these the *best* markets for jazz sales? To determine the strongest markets,

Table 6.2 Re-sort YTD % sales by DMA (Source: Nielsen SoundScan)

YTD % of Sales by DMA/Genre 2004

		Jazz
1	New York	10.1
2	Los Angeles	7.13
5	SF – Oakland	5.59
3	Chicago	4.05
9	Washington	3.76
4	Philadelphia	3.41
	TOP 5 MARKETS	**34.04**
6	Boston	2.82
15	Seattle	2.66
12	Atlanta	2.11
8	Detroit	2.1
19	Denver	1.79
7	Dallas	1.6
16	Miami	1.5
13	Minneapolis	1.42
10	Houston	1.4
18	St. Louis	1.33
11	Cleveland	1.29
20	Phoenix	1.17
14	Tampa	1.08
17	Pittsburgh	0.81

using population data in the equation helps to determine where to place marketing efforts.

Here is another look at the same data. By adding 12+ population data and doing a simple ratio, the best markets emerge. The 12+ data comes from Radio and Records (R&R) and DMA ratings information. The equation is sales percentage/12+ population.

Where New York and Los Angeles were ranked number one and number two, respectively, the strongest market for jazz *based on percentage of sales to the population* of the DMA would be Washington D.C., with San Francisco closely behind. *Although Washington D.C. does not sell as much jazz as New York, the propensity of the population to buy jazz in the D.C. marketplace is almost 1/3 greater, making it a better or stronger market for jazz music.*

Table 6.3 Ranking of jazz sales markets based on sales per population (Source: Nielsen SoundScan)

		YTD % of Sales by DMA/Genre 2004		Ratio sales%/pop
		Jazz	12+ Pop	
9	Washington	3.76	4041300	9.30394E-07
5	SF – Oakland	5.59	6012000	9.29807E-07
15	Seattle	2.66	3150300	8.44364E-07
19	Denver	1.79	2150300	8.32442E-07
4	Philadelphia	3.41	4291700	7.94557E-07
6	Boston	2.82	3888800	7.25159E-07
11	Cleveland	1.29	1800600	7.16428E-07
2	Los Angeles	7.13	10609200	6.72058E-07
1	New York	10.1	15340000	6.58409E-07
18	St. Louis	1.33	2210800	6.01592E-07
12	Atlanta	2.11	3750700	5.62562E-07
13	Minneapolis	1.42	2550200	5.56819E-07
8	Detroit	2.1	3859700	5.44084E-07
3	Chicago	4.05	7612100	5.32048E-07
14	Tampa	1.08	2194600	4.92117E-07
16	Miami	1.5	3489800	4.29824E-07
20	Phoenix	1.17	2801300	4.17663E-07
17	Pittsburgh	0.81	2019400	4.01109E-07
7	Dallas	1.6	4576700	3.49597E-07
10	Houston	1.4	4165000	3.36134E-07

Example: Green Day's American Idiot album

This same type of information can be used with Title Report data. By looking at the sales of an artist's specific record, labels can determine where to place marketing and dollars.

This example is Green Day's *American Idiot* album that was released on 9/21/04. Again, ranked by DMA, then ranked by sales, then ranked by sales/population ratio, look at the variance in market strength.

Table 6.4 Green Day's "American Idiot" album ranked by DMA – cumulative sales as of 12/20/2004 (Source: Nielsen SoundScan)

Total	1147215	12+ Pop	sales / pop
New York, NY	102038	15340000	0.67%
Los Angeles, CA	114564	10609200	1.08%
Chicago, IL	38996	7612100	0.51%
Philadelphia, PA	41358	4291700	0.96%
SF-Oakland-San Jose, CA	41916	6012000	0.70%
Boston, MA	35042	3888800	0.90%
Dallas-Ft. Worth, TX	16342	4576700	0.36%
Detroit, MI	19576	3859700	0.51%
Washington, DC	28046	4041300	0.69%
Houston, TX	11888	4165000	0.29%
Cleveland, OH	15206	1800600	0.84%
Atlanta, GA	16381	3750700	0.44%
Minneapolis-St. Paul, MN	24135	2550200	0.95%
Tampa-St.Petersburg, FL	13961	2194600	0.64%
Seattle-Tacoma, WA	25120	3150300	0.80%
Miami, FL	9768	3489800	0.28%
Pittsburgh, PA	11446	2019400	0.57%
St. Louis, MO	9646	2210800	0.44%
Denver, CO	20808	2150300	0.97%
Phoenix, AZ	21248	2801300	0.76%
Top 20 Market Average	617485	90514500	0.68%

Check out the table, *ranked by sales per population*. Based on sales per population, Denver emerges as the number two strongest market for sales of Green Day's *American Idiot* album. By concentrating on markets that have a stronger probability of sales, labels can better manage their marketing dollars through succinct activities that may include radio promotions, in-store events, touring, and so on. The goal is to maximize the market and sell records.

Table 6.5 Green Day's "American Idiot" album ranked by sales per DMA (Source: Nielsen SoundScan)

Total	1147215	12+ Pop	sales / pop
Los Angeles, CA	114564	10609200	1.08%
New York, NY	102038	15340000	0.67%
SF-Oakland-San Jose, CA	41916	6012000	0.70%
Philadelphia, PA	41358	4291700	0.96%
Chicago, IL	38996	7612100	0.51%
Boston, MA	35042	3888800	0.90%
Washington, DC	28046	4041300	0.69%
Seattle-Tacoma, WA	25120	3150300	0.80%
Minneapolis-St. Paul, MN	24135	2550200	0.95%
Phoenix, AZ	21248	2801300	0.76%
Denver, CO	20808	2150300	0.97%
Detroit, MI	19576	3859700	0.51%
Atlanta, GA	16381	3750700	0.44%
Dallas-Ft. Worth, TX	16342	4576700	0.36%
Cleveland, OH	15206	1800600	0.84%
Tampa-St.Petersburg, FL	13961	2194600	0.64%
Houston, TX	11888	4165000	0.29%
Pittsburgh, PA	11446	2019400	0.57%
Miami, FL	9768	3489800	0.28%
St. Louis, MO	9646	2210800	0.44%
Top 20 Market Average	617485	90514500	0.68%

Table 6.6 Green Day's "American Idiot" ranked by sales per population (Source: Nielsen SoundScan)

Total	1147215	12+ Pop	sales / pop
Los Angeles, CA	114564	10609200	1.08%
Denver, CO	20808	2150300	0.97%
Philadelphia, PA	41358	4291700	0.96%
Minneapolis-St. Paul, MN	24135	2550200	0.95%
Boston, MA	35042	3888800	0.90%
Cleveland, OH	15206	1800600	0.84%
Seattle-Tacoma, WA	25120	3150300	0.80%
Phoenix, AZ	21248	2801300	0.76%
SF-Oakland-San Jose, CA	41916	6012000	0.70%
Washington, DC	28046	4041300	0.69%
New York, NY	102038	15340000	0.67%
Tampa-St. Petersburg, FL	13961	2194600	0.64%
Pittsburgh, PA	11446	2019400	0.57%
Chicago, IL	38996	7612100	0.51%
Detroit, MI	19576	3859700	0.51%
Atlanta, GA	16381	3750700	0.44%
St. Louis, MO	9646	2210800	0.44%
Dallas-Ft. Worth, TX	16342	4576700	0.36%
Houston, TX	11888	4165000	0.29%
Miami, FL	9768	3489800	0.28%
Top 20 Market Average	617485	90514500	0.68%

The valuable information contained within SoundScan can help record labels, artists, managers, booking agents, promoters, and distribution companies create a blueprint that is invaluable in the development and ongoing career building activities. As explored in the previous pages, reams of data can be garnered and parsed. However, SoundScan data alone is not the sole answer. It takes creativity on the parts of all contributors to develop a plan that will point an artist in the direction of success.

Glossary

Bar code – See Universal Product Code (UPC).

Breaking – Introducing a new artist into the marketplace.

Chain stores – Multiple stores in a variety of geographic areas that are all owned by one company.

Designated marketing area (DMA) – The geographic area surrounding a city in which the broadcasting stations based in that city account for a greater share of the listening or viewing households than do broadcasting stations based in other nearby cities.

Electronic superstores – Stores such as Circuit City and Best Buy that sell a lot of electronic gear, but also carry recorded music product.

Index number – A number assigned to a value that represents where that number lies in relation to the average, usually expressed with 100 representing the average. For example, a product that has an index of 118 sells 18% more than the average.

International Standard Recording Code (ISRC) – A new, advanced version of bar codes that includes a country of origin code.

Mass merchants – Very large retail chains that sell a variety of goods and depend on volume sales. Wal-Mart and Target are examples.

Point-of-sale – Where the sale is entered into registers. Origination of information for tracking sales, etc.

Product life cycle – The course that a product's sales and profits take over what is referred to as the lifetime of the product.

Slow boil affect – A description for an album whose sales start out slow, but increase astronomically over time.

Uniform Code Council – The regulating body that assigns product codes.

Universal Product Code (UPC) – A standardized bar code used to identify products by laser scanners.

Year-to-date (YTD) – A measure of sales from the beginning of the calendar year until the specified date of the report.

7 How Radio Works

Paul Allen

Radio

Any study of the marketing of recordings must include an understanding of traditional radio and the role it plays in reaching target markets for the music. How important is radio? Veterans of the recording industry estimate that as many as 70% of consumer decisions to purchase CDs can be traced directly back to exposure to the music via radio. As much as consumers complain about the large number of commercials and repeated playing of the same records, radio still is the most important vehicle the recording industry has to showcase its product to the public.

Edison Media Research tells us that Americans spend an average of two hours each day—weekdays and weekends alike—in a vehicle on the road, either as a driver or a passenger. And despite cell phones, tapes, CDs, MP3 players, GPS units, video playback units and other distractions, we still spend a third of our time listening to radio while we are in a vehicle (Edison Research, 2004).

Given the role radio plays in promoting recordings to consumers, it's important to have an understanding of radio and the people who make programming decisions at those stations.

The businesses

One of the best adjectives to describe the relationship shared by the recording industry and radio is "symbiotic." Though it's a term most often used in science, it means the two industries share a mutual dependence on each other for a mutual benefit. Radio depends on the recording industry to provide elements of its entertainment programming for its listeners, and the recording industry depends on radio to expose its product to consumers. No two other industries share a relationship as unique as this. However, the nature of the businesses of a record company and a radio station are very different. For a record company, it's easy to define the business: *to sell recordings*. Money moves from consumers to a record company when recordings are sold.

The business of radio is building an audience that it leases to advertisers. Radio uses music to attract listeners in order to attract advertising revenue. The larger the audience the station attracts, the more it can charge for its advertising. Notice that radio is not in the business of building recording careers, nor is it interested in selling recordings. The number of units that a recording is selling might be of interest to a radio programmer, but that information does not necessarily affect programming decisions.

The radio broadcasting industry

The traditional over-the-air radio broadcast industry in the United States has been consolidating since the Telecommunications Act of 1996 was signed into law. Prior to the new law, radio broadcasting companies were limited in ownership to fewer than 20 stations. The law now allows companies very broad latitude on the number of radio stations they may own, but it typically limits the control of a radio audience to less than 30% for most markets.

This new age of radio ownership has created some of the largest media companies ever. Among them are:

Table 7.1 Radio station groups		
Company	2004 Annual Revenues	Number of Stations
Clear Channel Communications	$3,754,000,000	1,270 in 190 markets
Viacom International	2,096,000,000	185 in 41 markets
Cox Radio, Inc.	438,200,000	80 in 18 markets
Entercom	423,056,000	104 in 20 markets
ABC Radio Incorporated	401,700,000	71 in 29 markets
Citadel Communications	411,501,000	203 in 47 markets

Source: Hoovers Company In-Depth Records, Lexis-Nexis Academic, 2005.

Earnings for the industry have been growing, but slowly. In 2003, revenues for the radio broadcasting industry were $19,705,000,000, or up 1% from the prior year. With an anticipated 6% growth in 2004, the industry was expected to earn about $21 billion when the report was completed (Grillo, J.B., 2004).

Comparing radio with the recording industry, the RIAA reports *shipments* of recordings at retail price for 2003 were $11,854,000,000. The actual value of *sales* after discounts from MSRP was less. Generally, for comparative purposes, the radio industry in our country is nearly twice the size of the recording industry. The FCC licenses stations based upon the available frequencies, and because there has been no growth in frequencies allotted by the commission, the number of stations has remained fairly static. In 2004 in the United States, the FCC reported there were 4,770 AM stations, 6,217 FM commercial stations, and 2,512 FM educational stations for a total of 13,499 (FCC Report, 2004).

The radio station staffing

▲ *Figure 7.1 Typical radio station*

In order to see how decisions are made about music choices at a radio station, it is important to understand the relationships within the station. The *general manager*, or someone with a similar title, is responsible for the business success of the station. Reporting to the general manager is a *manager of administration* who has responsibilities such as accounting, commercial scheduling, and keeping up with regulatory matters. The *sales manager* has a staff of people who sell available commercial time to advertisers. *Promotions* are contests and other sales-oriented activities that are often the collaborative work of the sales team as well as the programming department.

From a record marketing standpoint, the key positions at a radio station are the *program director*, and to a lesser degree, the *music director*. The program director (PD) is genuinely the gatekeeper. Without the "okay" of the programmer, there is no chance that a recording will get on the air at most large radio stations. The PD is directly responsible to the general manager for creating programming that will satisfy the target market and build the existing audience base. The programmer decides what music is played, which announcers are hired, which network services to use, how commercials are produced, and every other aspect of the image the station has within the community it serves.

Critics of radio often say program directors have too much power because they can decide whether a recording is ever exposed to listeners. Large radio chains have group programmers who play an even larger role as a gatekeeper, recommending which music is appropriate for similarly programmed stations owned by the company across the country.

Others say that radio is serving as a filter for the massive amount of recorded music that is created every year. Theoretically, radio finds the most appropriate music for its audience and filters the music by choosing the best selections for the target audience.

Radio audiences

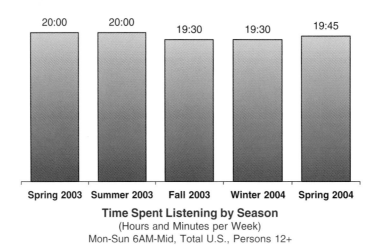

▲ *Figure 7.2 Time spent listening by season (Source: Arbitron Radio Today 2005)*

According to Arbitron in 2004, more than 94% of the U.S. population over 12 years of age listens to radio each week (Arbitron Report, 2005). Americans

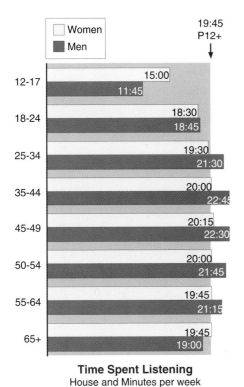

Time Spent Listening
House and Minutes per week
Source: Arbitron Radio Today 2005 Edition

▲ *Figure 7.3 Time spent listening by age group*
(Source: Arbitron Radio Today 2005)

spend about 20 hours per week listening to their favorite stations. This has remained relatively steady despite the growing number of consumer entertainment options. Time spent listening is greatest for males 25–64 and lowest for teens, at less than 12 hours per week for males and 15 hours per week for females. Females spend less time than males across all age categories except 65+.

Radio listening peaks in the morning hours, known as *morning drive time*. Radio listening is divided into *dayparts* of morning drive, midday, afternoon drive, evening, and overnight. From 6:00 a.m. until 9:00 a.m., listening is greatest as commuters wake-up to clock radios, and they continue to listen as they drive to work. Listening picks up again around noon, declines slightly after lunch but remains relatively strong through the *afternoon drive time,* and then tapers off drastically throughout the evening and into the overnight period.

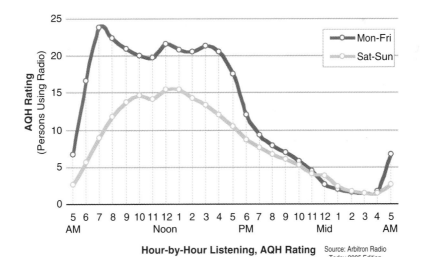

Hour-by-Hour Listening, AQH Rating Source: Arbitron Radio Today 2005 Edition

▲ *Figure 7.4 Hour-by-hour listening (Source: Arbitron Radio Today 2005)*

Radio on the go

Radio listening is very popular away from home. Between 8:00 a.m. and 6:00 p.m. on weekdays, a majority of radio listening occurs outside the home, as much as 72% of all listening. After 6:00 p.m., the majority of listening shifts to in-home. Much of the transient listening occurs either in vehicles or at work. During morning drive time, nearly 40% of listening occurs in the home, with another 36% occurring in vehicles and 23.4% at work. During midday, almost 42% of listening is at work.

	Home	Car	Work	Other
Mon-Sun 6AM-Mid	39.5%	33.4%	24.5%	2.5%
Mon-Fri 6AM-10AM	39.9%	35.2%	23.6%	1.3%
Mon-Fri 10AM-3PM	27.9%	28.2%	41.7%	2.2%
Mon-Fri 3PM-7PM	30.9%	43.0%	23.7%	2.4%
Mon-Fri 7PM-Mid	59.6%	26.1%	10.8%	3.6%
Weekend 10AM-7PM	49.0%	36.4%	10.3%	4.3%

Distribution of AQH Radio LIsteners by Listening Location
Persons 12+

Source: Arbitron Radio Today 2005 Edition

▲ *Figure 7.5 Distribution of AQH radio listeners (Source: Arbitron Radio Today 2005)*

During afternoon drive time, 23% of listening occurs at work, with 43.7% occurring in vehicles and less than 31% at home. After 7:00 p.m., almost 60% of listening occurs at home. Across all age groups, men are more likely than women to listen outside the home.

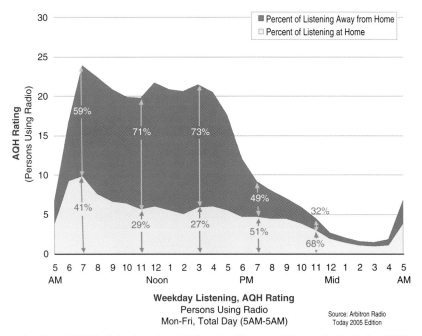

Weekday Listening, AQH Rating
Persons Using Radio
Mon-Fri, Total Day (5AM-5AM)

Source: Arbitron Radio Today 2005 Edition

▲ *Figure 7.6 Weekday listening, AQH rating (Source: Arbitron Radio Today 2005)*

Radio formats

Station owners choose radio formats by finding an underserved audience that is attractive to advertisers. When the format is chosen and developed, a programmer and staff are hired, and the audience develops. The chart shows how the national radio audience shares developed by format in 2004. The horizontal axis reflects the percentage of radio listeners who chose the formats represented in the chart.

The Arbitron audience measurement service reports the national percentages of radio format shares in the chart. The top radio format is news/talk. This format has grown in recent years and now represents about 16% of all listeners.

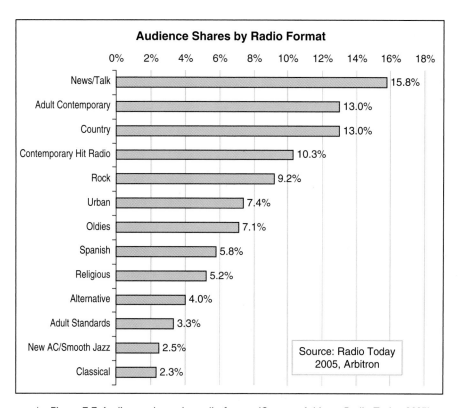

▲ *Figure 7.7 Audience shares by radio format (Source: Arbitron Radio Today 2005)*

The format shares shown in the chart have remained relatively the same over the years 1998–2004 with a few exceptions. Urban listenership has grown by 26%, and listeners to Spanish language formats have grown by 25%. Religious-formatted radio stations have seen a growth of 34% since September 11, 2001.

Arbitron is the only major company that measures the size and demographics of radio audiences. While the audience share chart shows the national audience share, Arbitron measures the same information, radio market by radio market. The share and audience makeup of each individual commercial radio station is measured and reported to subscribing stations and advertising agencies. The size of the station's radio audience is directly related to the amount of money the station can charge for its advertising. The more listeners (or the larger its audience share), the more the station charges companies to access their audience through advertising. Arbitron charges its clients tens of thousands of dollars for its audience measurement services. Since college and other noncommercial stations do not use traditional advertising, their audience shares are not reported.

With this in mind, a programmer is very careful in choosing music for airplay since the objective is to build its target audience. The program director is not inclined to experiment with an unproven recording that will turn an audience off. More about this will be discussed in a later chapter.

Targets of radio formats

In order to be a commercial product, recorded music must find a target that is able and willing to buy it. Finding that target is the first step in the marketing process followed by the development of a strategy to reach the target.

Table 7.2 Formats and target demographic

Format Name	Target Demographic	Artists in the Format
Adult Contemporary	Females 25–54	Rob Thomas, John Mayer, Kelly Clarkson
Active Rock	Men 18–34	Velvet Revolver, NIN, Green Day
Alternative	Persons 18–34	Foo Fighters, Beck, Incubus
CHR*/Pop	Females 18–34	Mariah Carey, Gwen Stephanie, Black Eyed Peas, 3 Doors Down
CHR/Rhythmic	Persons 12–24	50 Cent, Ludacris, Missy Elliott, Game
Country	Persons 25–54	Kenny Chesney, Brooks & Dunn, Reba McEntire, Lonestar
Hot Adult Contemporary	Females 18–24	3 Doors Down, Coldplay, U2, Kelly Clarkson
Urban	Persons 18–34	Destiny's Child, Fantasia, Kanye West

*Contemporary Hit Radio.
Format Name and Target Demographic from MediaBase/RateTheMusic.com; Artists in the Format from www.radioandrecords.com/Formats for the week ending July 8, 2005.

The target market of a particular radio format is the logical consumer target for commercial recordings. As of this writing, there are over 25 specific radio formats. In Table 7.2 on the preceding page are some broad definitions of music formats and their targets.

Radio station group owners often refine these gender and age targets. For example, some country radio stations owned by Clear Channel Communications specifically target females 35-44 for their country programming, while some owners target 12-24 males with the rap and hip-hop sounds of CHR/Rhythmic. The ability to obtain airplay can be a major factor in determining which records get made.

One of the key components of most of these radio market targets is the 18-34 year-old female. Women in this age group heavily influence or actually make the purchase decisions for households, and advertisers highly value this demographic as a target for their messages. Marketers of recorded music should continually consider this reality of radio as they plan to reach their own target market.

The format clock

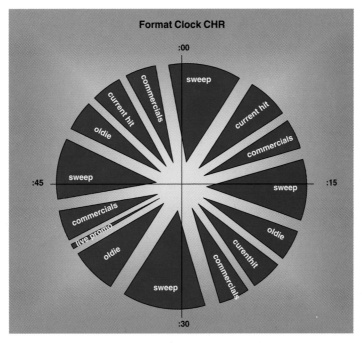

▲ *Figure 7.8 Format clock*

Programmers sometimes use a clock wheel to offer a visualization of how time is allotted to the various broadcast elements. It is a wheel indicating sequence or order of programming ingredients aired during one hour (Tarver, C., 2005). The clock face is divided into pie pieces, and each small section of time in an hour is prescribed a very specific item to be played on the air—from a song in a specific genre to a commercial, or news and weather.

What is important to programmers

Convincing radio to play new music is "selling" in every sense of the word. And in order to sell someone anything, you must know what is important to them and what their needs are. High on that list of important things to radio is Arbitron's measurement of radio audiences because it directly impacts the earnings of the station for its owners. Understanding concepts like this and their importance to programmers will help marketers of recorded music better relate to the needs of radio and its programming gatekeepers.

Ratings research

The term *P-1 listeners* represents one of the prized numbers of radio programming. As Mike McVay with McVay Media puts it, "These first preference listeners ... are referred to in radio station boardrooms, focus groups, and inside the headquarters of Arbitron doing diary reviews." McVay says the term *P-1* describes "the most loyal of radio listeners," and every radio programmer courts this primary core of their radio audiences (www.mcvaymedia.com). The terms *P-2* and *P-3* refer to listeners with a lesser degree of connection and loyalty to a particular radio station.

Cume is a programming term that comes from the word cumulative, and it refers to the total of all different listeners who tune into a particular radio station, measured by Arbitron in quarter-hour segments. In other words, it is the total number of unduplicated persons included in the audience of a station over a specified time period.

AQH, or average quarter-hour, refers to the number of people listening to a radio station for at least five minutes during a fifteen minute period.

TSL means "time spent listening" by radio station listeners at particular times of the day. TSL is calculated by the following formula.

$$\frac{\textbf{Quarter-hours in a time period} \times \textbf{AQH Persons}}{\textbf{Cume Audience}} = \text{TSL}$$

A radio station's *share* refers to the percentage of persons tuned to a station out of all the people using the medium at this time.

$$\frac{\textbf{AQH Persons tuned to a specific station}}{\textbf{AQH Persons in market currently listening to radio}} \times \textbf{100} = \textbf{Share}$$

Rating refers to the percentage of persons tuned to a station out of the total market population.

$$\frac{\textbf{AQH Persons tuned to a specific station}}{\textbf{Persons in market}} \times \textbf{100} = \textbf{Rating}$$

A full review of terms used in audience measurement is available at www.arbitron.com.

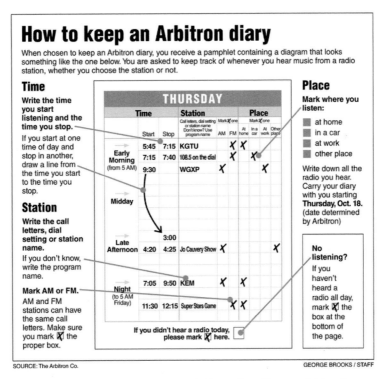

▲ *Figure 7.9 How to keep an Arbitron diary (Source: Arbitron)*

Arbitron measures ratings in 280+ markets in the U.S. Ratings are measured using the diary method. Arbitron selects households at random, and asks members age 12 and above to carry a diary for one week and record their radio listening. Potential diary keepers are first contacted by telephone, and then diaries are sent to the household. Completed diaries are returned to Arbitron and the data are entered into computers and analyzed on the following characteristics:

- Geographic survey area (metro or total survey area)
- Demographic group
- Daypart
- Each station's AQH: the estimated number of persons listening
- Each station's rating: the percent of listeners in the area of study during the daypart
- Each station's share: the percent of one station's total daypart estimated listening audience
- Cume: the total unduplicated audience during the daypart for an average week

Each Arbitron Radio Market Report covers a 12-week period for the specified market and contains numerous pages like the example in Figure 7.10 on the following page. At the top of each page, the target demographic is listed. Beneath that, the dayparts are laid out in columns. Then for each daypart, the AQH, the Cume, the AQH rating and AQH share are listed for each radio station in the area (listed in the left-hand column).

Ratings as defined by the audience share of a station determine advertising rates, and the example of an Arbitron report of the Atlanta radio market makes the point that ratings mean money. The chart in Appendix A (p. 141) shows the call letters, the station format, the owner, and the percentage of listeners in the Atlanta market who choose each station. The table of ratings is shown beginning with the fall of 2003 as an indication of the trends of the ratings for each station in the Atlanta market. The column on the far right is the station rating for fall 2004. The station showing 9.7% of the market's radio audience charges more for advertising than the station with only 0.4% of the audience because it has more listeners, and can therefore charge more for its advertising. Arbitron rating points are the targets of the audience-building efforts of a radio programmer. For example, each one-tenth of a rating point, or 0.1%, is worth $1 million in advertising rates to radio stations in the Los Angeles radio market.

Listener Estimates/Metro

Target Listener Trends

	Persons 12+														
	Monday-Sunday 6AM-MID				Monday-Friday 6AM-10AM				Monday-Friday 10AM-3PM				Monday-Friday 3PM-7PM		
	AQH (00)	Cume (00)	AQH Rtg	AQH Shr	AQH (00)	Cume (00)	AQH Rtg	AQH Shr	AQH (00)	Cume (00)	AQH Rtg	AQH Shr	AQH (00)	Cume (00)	AQH Rtg
WAAA-AM															
SP '01	118	1731	1.9	9.9	167	1118	2.6	10.8	177	923	2.8	10.8	165	1172	2.6
WI '01	123	1980	1.9	10.1	155	908	2.4	9.6	186	1064	2.9	10.9	152	1115	2.4
FA '00	101	2120	1.6	9.0	110	1110	1.7	7.2	130	1207	2.0	8.8	130	1312	2.0
SU '00	115	2238	1.8	9.3	144	1233	2.2	9.3	144	1202	2.2	8.4	148	1264	2.3
4-Book	*114*	*2017*	*1.8*	*9.6*	*144*	*1092*	*2.2*	*9.2*	*159*	*1099*	*2.5*	*9.7*	*149*	*1216*	*2.3*
SP '00	126	2259	2.0	10.5					193	1207	3.0	11.7	153	1492	2.4
WBBB-AM															
SP '01	118	1731	1.9	9.9	167	1118	2.6	10.8	177	923	2.8	10.8	165	1172	2.6
WI '01	123	1980	1.9	10.1	155	908	2.4	9.6	186	1064	2.9	10.9	152	1115	2.4
FA '00	101	2120	1.6	9.0	110	1110	1.7	7.2	130	1207	2.0	8.8	130	1312	2.0
SU '00	115	2238	1.8	9.3	144	1233	2.2	9.3	144	1202	2.2	8.4	148	1264	2.3
4-Book	*114*	*2017*	*1.8*	*9.6*	*144*	*1092*	*2.2*	*9.2*	*159*	*1099*	*2.5*	*9.7*	*149*	*1216*	*2.3*
SP '00	126	2259	2.0	10.5	172	1272	2.7	11.1	193	1207	3.0	11.7	153	1492	2.4
WCCC-AM															
SP '01	118													72	2.6
WI '01	123													15	2.4
FA '00	101													12	2.0
SU '00	115													64	2.3
4-Book	*114*													*16*	*2.3*
SP '00	126													92	2.4
WDDD-AM															
SP '01	118													72	2.6
WI '01	123													15	2.4
FA '00	101	2120	1.6	9.0	110	1110	1.7	7.2	130	1207	2.0	8.8	130	1312	2.0
SU '00	115	2238	1.8	9.3	144	1233	2.2	9.3	144	1202	2.2	8.4	148	1264	2.3
4-Book	*114*	*2017*	*1.8*	*9.6*	*144*	*1092*	*2.2*	*9.2*	*159*	*1099*	*2.5*	*9.7*			
SP '00	126	2259	2.0	10.5	172	1272	2.7	11.1	193	1207	3.0	11.7			

Source: Arbitron

Callout boxes marked (1) (2) (3) (4):

1. During an average quarter hour between 6:00 AM and 10:00 AM for spring 2001, 16,700 people listened to WBBB-AM for a minimum of five minutes.
2. During the 6:00 AM to 10:00 AM daypart, 111,800 different persons listened to WBBB-AM for a minimum of five minutes in a quarter hour.
3. During the average quarter hour in this daypart, 2.6 percent of all persons in this market were listening to this station.
4. During the average quarter hour in this daypart, 10.8 percent of all persons who were listening to any radio station in this market were listening to this station.

▲ *Figure 7.10 Example Arbitron book (Source: Arbitron)*

Programming research

Knowing how radio researches its audience can be helpful to marketers to understand how programmers define benchmarks for their decisions to add or remove music from their playlists, or increase or decrease the frequency songs are played. A pioneer in the area of radio audience research is Ed Salamon, who has a career in radio programming and is also an adjunct faculty member at MTSU. Mr. Salamon provided the following information about audience research.

In the early 1970s, Salamon was at KDKA radio in Pittsburgh and became one of the first radio programmers to research the music preferences of his audience. He began by adding programming questions to a local call out consumer survey, Marketing Information Bank (MIB), that Westinghouse Broadcasting did in each of their radio markets. By 1973, he had found other programmers including Bob Pittman, (at WDRQ, Detroit), Todd Wallace and John Sebastian (both at KRIZ

Phoenix) who had likewise been independently applying research models to radio programming. Their sharing of information resulted in the system of music research that is still in use today.

When asked to explain music research as it is being done by radio today, Salamon provided the following:

> With the demise of the single in favor of the album as the primary consumer unit, music tests became the only way for radio programmers to judge the popularity of an individual song because of the lack of sales information for a specific title.

> In call out research, which is normally used to evaluate current music, respondents are played a short (ten to twenty second) portion of a song, called a "hook," via telephone so that they can recognize it and respond. Approximately thirty titles at a time can be tested in this manner. Originally respondents were read the title and artist, but it was found that it is more accurate to play a portion of the actual recording. Importantly, new and unfamiliar music cannot be tested. A respondent normally needs to have heard a song about three times before it becomes familiar. Natural responses, such as "I don't know it" or "it's my favorite" are assigned numbers according to a semantic differential scale so that they can be tabulated.

Auditorium testing is generally used for noncurrent music, or library titles. It uses similar methodology as call out, but hundreds of titles can be tested in person versus by phone.

> Later, some researchers began asking whether respondents would like to hear a song "more, less, or about the same" instead of open-ended responses. Recently, some music research has been done via the Internet. While many researchers believe that the self-selection of participants invalidates the results, others have adopted it because of lower costs and argue that the ability to play as much of a song as the respondent wants to hear allows new music to be tested for the first time.

He also points out that radio also uses focus group research not to test music, but to determine broad areas of format, personality, and image for the station, and to learn about its competitors. The radio industry has seen a shift away from focus groups recently, according to Edison Media Research (Webster, T., 2004). Instead, radio researchers have begun to include qualitative questions into quantitative projects. Tom Webster, vice president at Edison, warns that while focus groups do not produce answers, they ensure that the right questions are asked on quantitative surveys (see the chapter on research for more information on focus groups).

Radio also uses panels of listeners for programming research. The panel method is a research technique in which the same people are studied at different points in time. Members of the panel are selected to reflect a representative sample of a station or format's listenership, and are periodically surveyed on their opinions of music and programming. The panel members are contacted either by phone or email and asked to respond to song hooks played either over the phone or through the Internet.

Getting airplay

It is the job of the record promoter to get airplay on commercial radio stations. This has become more difficult with the consolidation of radio because there are fewer music programmers, and competition for getting added to the playlist is fierce. The process of record promotion is outlined in the following chapter on charts, airplay and promotion.

The new radio and new technology

Opportunities for marketers of recorded music are improving with the addition of new technology and new services to users of music.

Radio stations have been given approval to convert their signals from older analogue technology to digital signals. What this means to music marketers is that songs played on the radio will deliver near-CD quality audio, and have the ability to display the artist's name and the song title on the radio receiver. Radio announcers infrequently provide artist or song information to listeners, and this new technology will help consumers of recorded music to identify artists and songs. Digital radio is expected to grow since the major broadcast companies have begun converting over 2,000 stations to digital broadcast signals, ultimately making them available to listeners in every major and large radio market. But because of the time and expense necessary for consumers to retrofit their radio receivers, it is unclear how quickly that diffusion into the marketplace will happen.

Major operators such as Clear Channel are acknowledging competitive pressures by continuously seeking new and untried delivery systems for their radio programming content. How traditional radio reacts to new media competitors will define its future.

Satellite radio

Satellite radio is a new opportunity for music makers to get their products exposed to consumers. Two companies, *XM Satellite Radio* and *Sirius Satellite Radio*, are FCC-licensed services that allow subscribers with receivers to receive scores of custom-formatted radio channels. Sirius offers 100 commercial-free music and information channels (Yahoo company profile, 2005), which include the promise of Howard Stern's radio show and every NFL football game. XM offers 150 channels (XM Satellite Radio, 2005), many of which are commercial-free, and their service includes every baseball game played in America. The opportunities for marketers of recordings with satellite services is that they have longer playlists within each genre of music, there are more opportunities for new music to be played, and every song played displays the artist and song title to the listener on the faceplate of the receiver. As a viable service, satellite radio has been slow to be adopted by consumers in part because of the cost of buying special receiving equipment for a vehicle or a home audio system, and because of the monthly fees associated with the services. But after just three and a half years in operation, the combined subscription numbers totaled over five million (Manley, L., 2005).

Internet radio

Internet "radio" is considerably different from radio in its traditional form. A number of browsers and Internet service providers include bundled software, which gives the user access to genre-specific music through their PC. While this is promoted as "radio," it typically does not include the presence of an announcer or noncommercial information. There are, however, commercial terrestrial radio stations that have chosen to stream their off-the-air signal on the station's home page and can often be found at the radio station call letters dot com. One advantage of Internet radio is that the song information is displayed on the computer screen as the song is played, often with a link to an online retail store to facilitate the impulse purchase.

Podcasting

Podcasting is a form of recorded Internet radio that allows users to download a pre-recorded radio program to their iPods, portable music player, or computer and listen at their convenience. A podcast is like an audio magazine subscription; a subscriber receives regular audio programs delivered via the Internet, and can listen to them at their leisure. Several celebrity DJs (such as Adam Curry, former

VJ on MTV) are offering podcasting shows on their websites. The web-based encyclopedia Wikipedia describes podcasting as:

> *A podcast is like an audio magazine subscription: a subscriber receives regular audio programs delivered via the Internet, and can listen to them at their leisure. Podcasts differ from traditional Internet audio in two important ways. In the past, listeners have had to either tune in to web radio on a schedule, or they have had to search for and download individual files from webpages. Podcasts are much easier to get. They can be listened to at anytime because a copy is on the listener's computer or portable music player (hence, the "pod" in "podcasting"), and they are automatically delivered to subscribers, so no active downloading is required (Wikipedia, 2005).*

Podcasting and related technology offer consumers convenient ways to collect and organize new music for MP3 players without having to comb through personal collections and rip and burn individual songs.

Portable people meter

One of the newest technologies in development is the *Arbitron Portable People Meter* system, with Houston and Philadelphia being its first test markets. According to Arbitron:

www.arbitron.com

▲ *Figure 7.11 Arbitron's Portable People Meter System (Source: Arbitron, arbitron.com)*

> "The Portable People Meter (PPM) is a unique audience measurement system that tracks what consumers listen to on the radio, and what consumers watch on broadcast television, cable and satellite TV. The Portable People Meter is a pager-sized device that consumers wear throughout the day. It works by detecting identification codes that can be embedded in the audio portion of any transmission," (www.aribtron.com, 2005).

The meter is the size of a pager, and is worn at all times during the day. At night, the People Meter is placed into a base unit and information is uploaded to a central database. This technology will provide marketers of recorded music the opportunity to learn whether listeners tend to change stations when an artist or a song is played, or whether they tend to remain with a song when it is played. The People Meter could prove to be a key component in the development of pre-release and post-release marketing strategies for the recording industry. The People Meter is undergoing its field-testing in Philadelphia and Houston, but it will be several years before it is available on a nationwide basis (www.arbitron.com, 2005).

Glossary

Add date – This is the day the label is asking that the record be added to the station's playlist.

Average quarter-hour (AQH) – The number of people listening to a radio station during a fifteen minute period as measured by Arbitron is called the *AQH*.

Cume – The total of all different listeners who tune into a particular radio station is its cume.

Format – The kind of programming used by a radio station to entertain its audience is the format.

Heavy rotation – These recordings are among the most popular songs played on a radio station.

Light rotation – These are recordings that are played fewer times on a radio station than songs in heavy rotation.

Playlist – The list of songs currently being played by a radio station makes up a playlist.

P-1 – The primary core of listeners to a specific radio station are P-1s.

Program director (PD) – This is an employee of a radio station or a group of radio stations who has authority over everything that goes over the air.

Share – A share is the radio audience of a specific station measured as a percentage of the total available audience in the market.

TSL – This means "time spent listening" by radio station listeners at particular times of the day.

References

Arbitron, and www.arbitron.com
www.dictionary.com

Federal Communications Commission, http://www.fcc.gov/fcc-bin/audio/DOC-253919A1.doc

McVay, Mike, McVay Media, www.mcvaymedia.com

MidiaBase/Rate the Music (Clear Channel Entertainment)

Radio Advertising Bureau, and rab.org

Radio and Records Magazine and rronline.com

Recording Industry Association of America

Radio Today: How America Listens to Radio (2005). Arbitron Report.

And special thanks to Tom Baldrica, BNA Records; Ed Salamon, Country Radio Broadcasters, Inc; R. J. Curtis, KZLA; Bill Mayne, 903 Music; Erica Farber, *R&R Magazine*; Bob Michaels, Arbitron; Lee Logan; Joe Redmond, Warner Bros. Records; Larry Pareigis, Sony BMG.

Appendix A							
Owner	Station	Format	F03	W04	Sp04	Su04	F04
Cox	WSB-AM	News/Talk	10.3	8.7	9.4	11.6	9.7
Infinity	WVEE-FM	Urban	9.0	7.3	7.9	7.4	7.7
ABC	WKHX-FM	Country	5.3	5.1	6.5	5.7	5.2
Cox	WALR-FM	Urban AC	4.1	4.5	4.0	4.0	4.8
Clear Channel	WLTM-FM	AC	4.8	4.0	2.6	2.8	3.7
Radio One	WPZE-FM	Gospel	5.2	5.8	4.6	5.4	5.0
Cox	WSB-FM	AC	4.1	4.0	4.5	3.1	3.5
Radio One	WHTA-FM	Urban	3.8	5.1	4.3	4.9	4.3
Jefferson-Pilot	WSTR-FM	CHR/Pop	3.9	3.8	4.5	3.5	4.0
Clear Channel	WWVA-FM	Spanish Cont.	1.5	1.0	0.5	0.8	4.6
Salem	WFSH-FM	Christian AC	2.8	3.1	3.3	2.6	3.0
ABC	WYAY-FM	Country	3.4	3.2	3.2	2.8	2.7
Infinity	WZGC-FM	Triple A	2.2	2.8	2.3	2.6	2.9
Clear Channel	WKLS-FM	Rock	2.8	2.8	2.6	2.9	2.3
Radio One	WJZZ-FM	Smooth Jazz	2.9	3.2	2.9	3.0	2.9
Clear Channel	WGST-AM	News/Talk	2.3	2.7	2.7	2.8	2.7
Cox	WBTS-FM	CHR/Rhythmic	2.6	2.8	3.2	3.1	2.4
Susquehanna	WNNX-FM	Alternative	3.4	2.8	3.3	2.9	2.6
Susquehanna	WWWQ-FM	CHR/Pop	2.4	2.2	2.5	2.3	1.6
Clear Channel	WLCL-FM	Oldies	2.0	2.9	2.5	2.2	2.4
Radio One	WAMJ-FM	Urban Oldies	1.7	1.5	1.6	1.3	1.8
Cox	WFOX-FM	Urban	2.0	1.7	2.4	1.5	1.5
Jefferson-Pilot	WQXI-AM	Sports	1.2	1.2	0.8	1.1	1.1
Dickey	WCNN-AM	Sports	0.8	0.8	0.8	1.0	1.1
Davis	WLKQ-FM	Reg. Mex.			0.5	1.2	1.0
Clear Channel	WBZY-FM	Alternative	0.6	0.7	1.1	0.8	0.5
GA-Mex	WAZX-AM	Reg. Mex.	0.8	0.4	0.7	0.4	0.4
La Favorita Inc.	WAOS/WXEM	Reg. Mex.	1.0	0.9	1.1	0.6	0.4

This chart shows the owner of the Atlanta radio station, its call letters, its format, and the percentage of the Atlanta radio audience the station had during the average quarter hour of the radio broadcast day. Ratings are reported in three month periods. For example, F04 is the overall audience "share" for the fall of 2004.[i]

---

[i] Arbitron, used by special permission.

Bibliography

Arbitron Report (2005). How America Listens to Radio , *Radio Today 2005 Report*. *http://www.arbitron.com/home/content.stm.*

Edison Research (2004). www.Edisonresearch.com.

Grillo , J. B. (Feb 16, 2004). Radio Looks for March Upturn, *Broadcasting and Cable*, pp. 32. FCC Report (2004). www.fcc.gov.

Manley, L. (April 5, 2005). As Satellite Radio Takes Off, It Is Altering the Airwaves, *New York Times*, pp. 1.

Tarver, C. (2005). www.udel.edu/nero/Radio/glossary.html.

Webster, T. (2004). Edison Media Research: What's Wrong with Focus Groups? *http://www.edisonresearch.com/home/archives/000426print.html.*

Wikipedia (2005). http://en.wikipedia.org/wiki/Podcasting.

www.arbitron.com (2005). http://www.arbitron.com/portable_people_meters/home. htm. Prior U.S. Market Trials.

www.aribtron.com (2005). http://www.arbitron.com/portable_people_meters/home. htm. The Portable People Meter System.

XM Satellite Radio (2005). http://www.xmsr.com/learn/programming.jsp.

Yahoo company profile (2005). http://finance.yahoo.com /q/pr?s=SIRI.

8 Charts, Airplay and Promotion

Paul Allen

The connection between recorded music and radio is essential to the success of large record labels that compete in the national and international mass markets. This chapter will show this key connection by providing an understanding of how the charts of trade magazines are developed, followed by a view of how those who are responsible for radio promotion at a label can influence those charts.

The trade magazines

The two major trade magazines for the music industry are *Billboard* and *Radio & Records*. Both provide comprehensive weekly views of the recording industry, the music business, and commercial radio.

Billboard Magazine was first introduced in 1894 as a publication providing information about the carnival industry, but began to focus more on music and less on carnivals. In 1936, it published its first "hit parade," which was a term used at the time to rank popular songs and then became a term used by radio to denote its most popular music (Ammer, C., 1997). In 1940, *Billboard* compiled and published its first Music Popularity Chart, and by 1958 the venerable Hot 100 chart became a staple of the magazine (wikipedia.org, 2005).

Gavin and *Cashbox* magazines were key industry trade publications for many years until they were retired for economic reasons. Both of these trades relied upon "reported" airplay by radio stations that gave the publications their airplay charts for the coming week. In October 1973, *Radio & Records* published its first issue[i] and has been *Billboard*'s major competition since.

The importance of the charts

Perhaps the most important piece of real estate a record label can own is a high chart position in trade magazines for its albums and singles. A favorable airplay chart position from record industry trade magazines like *Billboard* and *R&R* has the same effect as a "word of mouth" endorsement of recordings because it is a reflection of the opinions of key radio program directors.

Charts in trade magazines are defined in numerous ways, and it is important to be sure the distinction is made between sales charts and airplay charts. If a label has a number one album, that placement is based on its position on a sales chart. If a label has a number one single on the radio, that position is based upon the number of times a single is played on the radio during a specific week, and how big the cities are in which the song is played.

In the next part of this chapter, we will look at how charts are created and why they are so important to the marketing of recordings.

Creating the airplay charts

Nearly every major music genre has an airplay chart in the two major trade magazines. These charts are a reflection of national airplay of singles on radio stations as detected by two somewhat different systems (except for some specialty charts).

Broadcast Data Systems (BDS)

Broadcast Data Systems (BDS) is the technology used by *Billboard* and *Canadian Music Network* magazines to detect each spin of a recording on radio in cities in which they have installed a computer to monitor airplay. As the spins are detected, the computers upload the number of detections to a main database that is then used to create the weekly airplay charts. Geoffrey Hull cites *Billboard* describing the system as:

> A proprietary, passive, pattern-recognition technology that monitors broadcast waves and recognizes songs and/or commercials aired by radio and TV

---

[i] Lon Helton of *Radio & Record,* personal interview.

stations. Records and commercials must first be played into the system's computer, which in turn creates a digital fingerprint of that material. The fingerprint is downloaded to BDS monitors in each market. Those monitors can recognize that fingerprint or "pattern," when the song or commercial is broadcast on one of the monitored stations," (Hull, G., 2004).

As the computerized airplay monitor "listens" to a song being played on the radio, it compares its digital fingerprint to that on file, and then logs it as a detected play of the song.

BDS has monitors for airplay in 128 radio markets, and claims to listen to over 1,100 stations and detect over 1 million songs each week. These detections are used to compile 25 airplay charts for *Billboard* and its *Airplay Monitor*. Additionally, the service compiles detections of airplay on six cable and satellite channels that feature music videos.

Label marketers must be sure they register their recorded music with BDS or there will be no detections of airplay. BDS provides information on its website on how to register a song and get a digital fingerprint created for the airplay monitoring system. Without the airplay statistics, the radio promotion department will be without the bragging rights they need to continue to promote the single, and the marketing department will be without a key tracking tool (BDS, 2005).

BDS generally describes its market monitoring systems as follows:

> Monitors are strategically positioned in a market to ensure that we get the best reception possible for each station in that market. Each monitor has 10–15 slots. Every slot represents a different signal and is used for either radio or television monitoring. Each station has its own library of song or advertising patterns stored in its memory. These song patterns or "fingerprints" are constantly increasing and are well over 10,000 (and counting) at the monitor site: more than 1,500 titles for country, 3,500 for modern rock and R&B, and over 4,000 for Top 40 (hipnotikent, 2004).

BDS gives the label's marketing department considerable information about which radio stations are spinning a single and how frequently. Combining this information with SoundScan data on sales of singles, label marketers have continuing feedback on the performance of their recorded music projects. And most importantly, this feedback gives marketers the information needed to modify marketing plans in order to draw as much commercial value out of the marketplace as possible.

MediaBase 24/7

MediaBase is a service owned by Clear Channel Entertainment and is under exclusive contract with *Radio & Records (R&R)* to provide data that it then uses to create its airplay charts. It monitors the airplay of recordings on over 1,000 radio stations, including 125 in Canada (SoundSource, 2005).

While the information is very similar to that provided by BDS and is used by record labels in the same ways, there are some differences between the services. Where BDS uses computers to detect airplay, MediaBase employs people to actually listen to radio stations and log the songs played. Employees of the company who are paid to detect airplay are experts in their genres of music, they often work from their homes, and they are provided the necessary hardware and software by the company. Employees who work in airplay detection are often responsible for logging songs for the 24-hour broadcast days of eight radio stations.

Another difference in the services lies in the particular stations whose airplay is monitored. MediaBase monitors an estimated 80% of the same stations as BDS. Some record labels see the need to subscribe to both services to be sure they are getting accurate feedback on the performance of single releases at radio (radio-media, 2005).

The charts

Weekly charts representing airplay, sales, and a combination of both appear in *Billboard* and its related *Airplay Monitor*.

Billboard indicates recognition for accomplishments each week in the charts with the use of special awards. The most commonly known recognition is a bullet, which is awarded for "significant improvement in sales or airplay from the previous week." Other designations are as follows:

Table 8.1 Billboard chart designations

Designation	Description
Greatest gainer	Chart's largest unit increase
Pacesetter	Biggest percentage growth
Hotshot debut	The new album or song with the highest first-week position
Heat seeker impact	Shows removal from heat seeker status that week
●	Gold (500,000 units)
▲	Platinum (one million units)
◆	Diamond (ten million units)

A heat seeker is a special designation by *Billboard* for developing artists and is described on their website as:

> "The Top Heatseekers chart lists the bestselling titles by new and develop-ing artists, defined as those who have never appeared in the Top 100 of The Billboard 200 chart. When an album reaches this level, the album and the artist's subsequent albums are immediately ineligible to appear on the Top Heatseekers chart," (Billboard online, 2005).

"Spins" and "plays" are terms given to denote the number of times a song is played on a radio station in a set time period. *R&R* and *Billboard* magazines report spins for their reporting stations. They also report a "most added" category indicating the new songs that were added to the most stations that week. Songs that are added in light rotation (meaning they are played fewer times) to a station's playlist are those that are either on their way up the music charts of trade magazines or on their way off of the charts. Songs that are in heavy rotation are those most popular with a station's audience. In between, there is medium rotation. The actual number of spins necessary to be placed into one of these categories varies from format-to-format and station-to-station and is subject to change. *Recurrents* are songs that used to be in high rotation at a station, but are now on the way down, reduced to limited spins. *Burnout* (or burn) is the term given to songs that have reached a particular threshold for audience burnout as determined by radio programming research. A station may decide to remove a song from the playlist when the "burnout factor" reaches a certain percentage of the audience.

The Hot 100 has been a part of *Billboard* magazine since 1956, and it has spawned numerous other genre-specific charts. The Hot 100 is a chart that is developed each week using a formula that combines the physical sales of singles, sales of digital downloaded singles, as well as the number of spins a song receives on radio, regardless of the genre of music or the radio format in which the song is programmed.

Other charts published weekly by the magazine include rankings by genre, created by using data gathered for airplay through its BDS reporting system. Airplay charts are then used by radio programmers to give them a basis to compare their audience offerings with those audiences of similar cities. Also, each week *Billboard* publishes its Top 200 chart, which is a compilation of the bestselling albums ranked by the number of unit sales for the previous week. Both of these charts are especially important to the marketing effort by a label on behalf of its active new music projects. Tracking the impact the music is having at radio and at retail gives label marketers information that is helpful to control the success of singles and albums.

The *Billboard* charts are continually modified in order to keep them as an accurate reflection of the needs of the businesses they serve. In early 2005, responding to an increase in the number of audio and video content discs being marketed together, *Billboard* announced:

> *Effective with first-chart week of Nielsen SoundScan's 2005 tracking year, Billboard is amending its dual-charting policy. Going forward, there will be far fewer titles that qualify for both album and music video charts.*
>
> The policy: Most combo or DualDisc titles will appear on either album or music video charts, not both. Titles that offer far more audio content than video content will be considered albums. Those that offer far more video content than audio content will be considered music video titles. Only in cases where the content of both the audio and video programming would both qualify as full-length product will a title appear on both charts *(Nielsen SoundScan, 2005)*.

Another change *Billboard* made to some of its charts is to add a weight to the detection of the airplay of a single. This means that the time of day (audience size), and the size of the city in which a single is played will determine how important that spin is on an airplay chart. For example, a spin of a single by a radio station in Chicago during morning rush hour will be weighted more on the airplay chart than a late night spin on a small town, non-network radio station.

Among the newest charts is *Billboard*'s *Pop 100*. This chart lists only the top songs being played at top 40 radio stations and includes both airplay and sales to determine chart positions. Songs are ranked based on the number of gross impressions. This is compiled by cross-referencing the exact times the songs are played against Arbitron's data on listenership at that time.

R&R's charts are similar to *Billboard*'s in many ways. Both publications include airplay charts based on detected spins by monitored radio stations, and each reports sales data on recorded music. In addition to the charts, the two magazines provide their subscribers with considerable news and information on the music and radio industries each week.

CMJ charts are created through the CMJ Network, which refers to itself as connecting "music lovers with the best in new music through print and interactive media, as well as live events" (cmj.com, 2005). Charts for CMJ are created by reports of airplay from their "panel of college, commercial and noncommercial radio stations," and tend to represent music that is not a part of the commercial mainstream (cmj.com, 2005). Its charts are reported in publications by the company, as well on its website and chart titles. Those charts include:

- CMJ Top 20
- Radio 200

- Hip-Hop Top 20
- Triple A Top 20
- Loud Rock Top 20
- Jazz Top 20
- New World Top 20
- RPM Top 20
- Retail
- In-Store Play

What does this mean to label marketers?

The best way to demonstrate the impact of the airplay charts on label marketing is to look at the number of times people hear a single played on the radio during a week. Below is a listing showing the titles of several airplay charts in *Radio & Records* and the number of "impressions" that a number one record in that genre made during the week of February 4, 2005. "Impressions" means the number of times the number one song in the particular genre was heard by someone on a station monitored by MediaBase, and reported on the weekly airplay chart.

Table 8.2 Radio format and impressions	
Radio Format Chart	**Number of Impressions**
Contemporary Hit Radio/Rhythmic	82,013,300
Contemporary Hit Radio/Pop Top 50	71,877,700
Urban Top 50	54,798,300
Country	42,914,300
Urban Adult Contemporary	12,986,600

Considering the weekly numbers reflected in this chart, it is obvious the immense impact commercial radio has on the exposure of recorded music to targeted potential customers of the label.

Radio promotion

Lobbying and lobbyists have been around as long as any one person has been responsible for a decision or a vote, and people have always wanted to influence that decision or vote in their favor. Everyday, lawmakers at the national and state levels meet with representatives of special interest groups who ask them to vote on matters that are in the best interests of their groups or their clients.

The same thing happens between a radio programmer and a record promoter. The people lobbied are usually the program directors of radio stations that report their charts to the major trade magazines. Program directors, also referred to as PDs, have the ultimate responsibility for everything that a radio station broadcasts—banter by personalities, advertising, information such as news and traffic reports, and all music played by the station. Simply said, the radio programmer can decide whether a record ever gets on the air at their station.

Decisions by radio programmers are the keys to the life of a record and have become the basis for savvy, smart, and creative record promotion. Programming decisions about music determine:

- Whether a new recording is added to the playlist of a radio station that reports its chart and airplay to the major trade magazines.
- Whether the recording receives at least light rotation on the playlist.
- Whether it eventually receives heavy rotation on the station's playlist.

Record promoters are lobbyists in the purest form. They are either on the staff of a record company or they are part of a company specifically hired by the label to promote new music.

Promotion

Traditional marketing texts teach the four P's, one of which is promotion. Those same texts tell us that the promotional mix consists of public relations, advertising, sales promotion, direct marketing, and personal selling.

The definition of record promotion comes closest to being personal selling than any other aspect of the traditional promotional mix, except perhaps the label sales department. Personal selling is one-on-one, and the work of a record promoter is exactly that. The record promotion person at a record label has the responsibility of securing radio airplay for the company's artists. Quite simply, they lobby the station decision makers to add recordings by the label's artists to the station's playlist. With some label executives claiming that radio is responsible for 70% of sales of recorded music, label record promotion can be critical to the success of an album project. It is this close connection between airplay and record sales that creates the need for record promotion to radio as a key element of the marketing plan.

The effectiveness of a record promotion person hinges on the strength of the relationships he or she has with radio programmers. These relationships are built much like anyone in business that has a client base requiring regular service. The promoter makes frequent calls in person and by telephone to the programmer, arranges lunch or dinner meetings, provides the station access to the label's artists, and helps the station promote and market itself with contests and giveaways. The promoter, who has developed a good relationship can then make the telephone call and ask the programmer to treat his current record project favorably.

If the promoter has no relationship with the radio programmer from a reporting station, it is highly unlikely that telephone calls will ever be returned. Programmers today have too many things to do and little time to listen to promotional pitches from people and companies they don't know.

Promoting

First, it is important to understand that promotion by a label is focused on radio stations that program current, new music. Stations airing older music are using record company catalogues as the basis for their entertainment programming. Those albums and their related singles have been around for years, sometimes decades, and there is limited interest on the part of a label to promote them to radio. However, most energy and money is invested by record companies in promoting the newest singles and album projects.

Record promotion to radio by labels typically needs to answer four questions:

1. How does it help us sell records?
2. What does it do for our effort towards developing the artist?
3. What does it do to further develop our agenda and help us market the artist?
4. Does it make sense for the radio station?

If radio stations ask the label promoters for something of value, whatever it is, the first

---

The Contest Promotion

One label discussed its relationship with Clear Channel Communications, the broadcast giant which owns over 1,200 radio stations by describing a national contest. A major artist was preparing to release a new album and go out on tour. The label offered Clear Channel stations a contest which would fly two couples to attend the final dress rehearsal before the concert series began. The winners enjoyed the privilege of being an audience of four rather than four in a crowd of 20,000. The cost was minimal to Clear Channel but provided some interest and excitement around the artist by using the collective power of their stations and web sites.

▲ *Figure 8.1 The contest promotion*

question to the record company becomes, "How does this help me sell copies of the artist's record?"

One label says that everything they do with radio stations to promote an artist "has to really pass the smell test." They require proof that promotional prizes and free concerts by artists are acknowledged on the air as being given by the label. They expect to receive affidavits from the station showing when the announcements were aired and at how much money those announcements were valued. When the payola law was changed in 1960, the burden of disclosure was also placed on the record labels. The major 2005 payola settlements by NY Attorney General Eliot Spitzer mandate this extra care.

As you create a promotional tour for an artist with a new CD, emphasis should be on the most popular radio stations in targeted cities, especially those "reporting" stations that report their airplay charts to major trade magazines. To determine which stations are rated the highest in audience shares, you will find Arbitron's reported findings at www.radioandrecords.com, as well as through other media outlets. Specific information about the station programmer's name, address, and telephone number is available through Bacon's MediaSource.

After a single has been released to radio, nearly everyone at the label tracks its progress. For the promotion staff, the weekly airplay charts and SoundScan data help guide their work. Monitoring tools track the number of downloads of singles on iTunes, Rhapsody, Napster, and similar services to determine if the music is connecting with consumer's pocketbooks. Some labels also have staff members who monitor the "buzz" on their new music in chat rooms. All of this information helps the label determine whether they have a hit on their hands, whether they should step up the promotional effort, or whether it's time to cut their losses and pull the project from the market.

The typical business week for record promoters begins either Tuesday or Wednesday. For country promoters, the airplay charts at *R&R* and *Billboard* close Monday night; for pop music, the airplay charts close on Tuesday night. The chart closings will show the successes and failures of singles from the previous week and give the promotion team the information they need to allocate their time for the next seven days. With the plan in place, the promotion staff then begins its weekly cycle of contacting radio programmers to build airplay for the label's products.

One label executive says his regional label promoters are always on the job. The only time he allows a break is when his promotion team members are on vacation. Jobs as regional promoters with some experience can go to work at a starting salary in the range of $70,000 to $80,000, plus expenses. Vice presidents of label promotion for larger labels can easily earn well over $200,000, plus bonuses and other incentives.[ii]

History

Record promotion and its regulation by the Federal government began not long after the advent of commercial radio broadcasting. In 1934, Congress created and passed the Communications Act, which restricted radio licensees—the stations themselves—from taking money in exchange for airing certain content unless the broadcast was commercial in nature. However, this early act contained nothing that prohibited disc jockeys (DJs) from taking payments in exchange for airplay. During the big band era of the 1940s, and the rock 'n' roll days of the '50s, DJs were routinely taking money from record promoters in exchange for the promise to play a record on the air. Disc jockeys during this time often made their own decisions about which records would be included on their programs, and promoters would approach them directly to influence their record choices.

Lawmakers railed against the rampant bribes being given to DJs to play records. In 1960, Congress amended the 1934 act to include a provision that was intended to eliminate illegal bribes to play music, so called *payola*. Under the revised law, disc jockeys and radio stations were permitted to receive money and gifts to play certain songs, but the amendment placed a requirement that these inducements be disclosed to the public on the air. If this disclosure was not made, it exposed the DJ and management to possible fines and imprisonment (Freed, A., 2005). The change in the law also created the requirement that record labels must report their cash payments and major gifts to the station for airplay. This 1960 amendment continues to guide the radio and records industries today.

Despite the stronger laws against payola, Federal investigators were called upon to investigate scandals within the record promotion business in the 1970s and 1980s. There were no major convictions despite the appearance that money, drugs, and prostitution were being used as leverage by promoters to get radio airplay for recordings (Katunich, L. J., 2002).

[ii] Personal interviews, Fall 2004.

Getting a recording on the radio

The consolidation of radio has concentrated some of the music programming decisions into the hands of a few programmers who provide consulting and guidance from the corporate level to programmers at their local stations. In many cases, local programmers have the ability to add songs to their playlists based upon the preferences of the local audiences. Here is how songs are typically added to station's playlists for those stations that program new music:

1. A record label promoter or an independent promoter hired by the label calls the station music director (MD), or program director (PD) announcing an upcoming release. Radio music directors have "call times." These are designated times of the week that they will take calls from record promoters. The call times vary by station and are subject to change. For example, an MD may take have call times of Tuesdays and Thursdays, 2:00–4:00 p.m.
2. Leading up to the add date, meaning the day the label is asking that the record be added to the station's playlist, the promoter will call again touting the positives of the recording and ask that the recording be added.
3. The music director or program director will consider the selling points by the promoter, review the trade magazines for performance of the recording in other cities, consider current research on the local audience and its preferences, look at any guidance provided by their corporate programmers/consultants, and then decide whether to add the song.
4. The PD will look for reaction or response to adding the song. The "buzz factor" for a song will be apparent in the call-out research and call-in requests, as well as through local and national sales figures.

An important component of promoting a recording to radio is the effectiveness of the record company promotion department or the independent promoters hired to get radio airplay. This would appear to be a simple process, but the competition for space on playlists is fierce. Thousands of recordings are sent to radio stations every year, and the rejection rate is high because of the limited number of songs a station can program for its audience. Some of the recordings are rejected from being included on playlists because they are inferior in production quality, some are inappropriate for the station's format, and many lose their label support if they fail to quickly to become commercial favorites with the radio audience.

However, the marketer's litmus test for the viability of a recording is to honestly compare it with other songs on the charts of trade magazines. If it is not at least as good as those listed on the current trade magazine charts, then it doesn't have any chance at all, even if it has a competitive marketing budget.

Independent promotion

The scandal of the 1980s was about the practice of labels hiring independent promoters, or "indies" to attain airplay at radio. Fred Dannen, in his book, *Hit Men,* found that CBS Records was paying $8–10 million per year to indies to secure airplay for their acts. By the mid-80s he says that amount was $60–80 million for all labels combined. We will discuss independent promotion in the next section.

Label record promotion and independent promoters

Most large labels have a promotion department whose sole purpose is to achieve the highest airplay chart position possible. While most consumers assume a number one song is the biggest seller at retail, the number one song on most *Billboard* and *R&R* charts actually is the song that has the most airplay on radio. The connection between airplay and sales is well-documented, so a high chart position is critical to the success of a recording and becomes the heart of the work of a record promotion department.

Labels often have a senior vice president of promotion who usually reports directly to the label head. The senior vice president of promotion at a pop music label typically has several vice presidents of various music types based on radio formats. These vice presidents then have regional promotion people who are viewed by the label as field representatives of the promotion department. They are liaisons to key radio stations in their region. The vice president and the so-called *regionals* are the front line for the label attaining airplay. It is their responsibility to create and nurture relationships with programmers for the purpose of convincing them to add the company's recordings to their playlists. At a country label, the senior vice president of promotion typically has a national director of promotion and several regionals (see Appendix, p. 159).

> **Radio & Records:** Nominees for Independent Promotion Firm for 2004:
>
> All That Jazz
> The Jesus Garber Company
> Jeff McClusky and Associates*
> McGathy Promotion
> National Music Marketing
>
> *Winner for the past six years

▲ *Figure 8.2 R&R indie nominees*

Labels sometimes hire independent promoters (indies) to augment their own promotional efforts. Since promoting songs for airplay relies on well-developed relationships, indies may have developed stronger relationships with some key stations than the label has, and the record company is willing to pay indies (half of which is often recoupable from the artist) for the value of those relationships.

Independent promoters have typically made their money this way: they sign an agreement with a radio station to be the stations' exclusive consultant on new music for a year, and then pay the radio station for that right. The indie promoter does not require the station to play specific songs, according the typical agreement, but the station does promise to give its playlist for the following week to the indie promoter before anyone else. Then, the promoter sends an invoice back to the record company for $3,000 to 4,000 for each single that is charted on stations that they represent. Because of this "exclusive" arrangement with the radio station, record label promotion people have no dealings with the radio stations represented by indies. One label promotion vice president says the $4,000 can easily turn into $30,000 per single if the indie must provide contest prizes to help promote the single at radio and to pay for other promotional expenses. Many arrangements with indie promoters include provisions for bonuses based on their success at charting records with individual stations (Phillips, C., 2002).

Record Promotion at Industry Conventions: A New Revenue Stream for Radio

Clear Channel became the pioneer of the practice of charging record labels to have an exclusive audience with its key programmers. For example, during the annual Country Radio Seminar in Nashville, Clear Channel has required its key 50 country programmers to be present for five 90-minute "showcases" the week of that convention. Labels that reserved one showcase typically presented their newest acts to the programmers and were billed $35,000 by Clear Channel. For the week, the company received $175,000.

▲ *Figure 8.3 Record promotion at industry conventions*

Independent promoters provide the record company a layer of insulation between themselves and radio. The record companies do not deal directly with certain radio stations in matters of adds and spins; rather, they deal only with independent promoters who promote to these stations. As one executive says, "The use of independent promoters creates a clearinghouse by removing the label one step away from... making any compromises that some might make to get a song on the air." Another says, "This way the money doesn't go directly from the label to the radio station."[iii]

[iii]Personal interviews, Fall 2004.

Critics of the independent promotion system claim it had tended to shut-off access to radio airplay by independent labels and artists. Alfred Liggins is CEO of Radio One, a company that owns nearly 70 radio stations targeting African-American audiences. He acknowledges that their exclusive relationships with independent promoters means that labels without an indie promoter are less likely to get a record played on his stations (ABC Television, 20/20, 2002).

The continuing need for independent record promotion to radio stations is changing; however, Lew Dickey, CEO of Cumulus Broadcasting, said in 2003 that his company was "centralizing" their relationships with independent promoters. He did not want indies to "work program directors... and this is really a pet peeve of mine ... for things that are untoward, and put young men and women in compromising positions at that age and at that level of experience [who] may not know any better. I think it's wrong." Cumulus has since ended its relationships with independent promoters. However, this comment suggests that even new millennium record promotion will always hold the potential to stray into troubling areas for both record labels and radio. The criticism leveled at the practice of independent promotion has caused a large promotion and publicity company to rename its radio promotion services to "radio marketing."

Record promotion has been a big-dollar investment, which made it a key marketing element necessary to stimulate consumers' interest in buying new music. Costs for a label to market and promote a single easily reach $1 million, not including the production of the recording or any advances to the artist or producer. Even in the world of country music where annual sales are often less than 10% of all recorded music, labels invest as much as $300,000 just to get the single of a new artist into the top 20 of an airplay chart.

And, according to information developed by the *Los Angeles Times*, independent promotion has cost record labels an estimated $150 million annually (Phillips, C., 2001).

Clear Channel Communications is among those radio companies that no longer use independent promoters as consultants to its stations. The company instead creates total marketing packages for labels and their artists that can include everything from concert promotion to record promotion. Joining Clear Channel in ending relationships with independent promoters are companies such as Cumulus Media, Enetercom, and others. In 2005, New York Attorney General Eliot Spitzer negotiated multi-million dollar settlements with record labels for alleged violations of payola laws by independent and label promoters, and the FCC began its own investigation of payola violations of radio licensees. While the current trend is away from independent promotion, history has shown that this might just be temporary.

Satellite radio promotion

Satellite radio offers the possibility of an expanding horizon for promoting recordings to radio. The services provided by XM and Sirius have a relatively low monthly subscription fee (less than $15), and offer scores of commercial-free music channels. Though the services require that a vehicle must be equipped with a special receiver, many newer cars and trucks have satellite receivers integrated into their standard radios. Satellite radio delivers an audience in the millions, with the size being equivalent to a major market radio station (though the audience is national in scale). Satellite radio offers more sub-genres and niche formats, which gives labels great opportunity for airplay. Record company promoters actively work with satellite music channel programmers seeking adds to their playlists.

Looking toward the possibility of coupling satellite radio with OnStar® Corporation technology, new opportunities may become available to record labels and their promoters. OnStar is a subscription emergency service offered in some vehicles that wirelessly transmits alerts to customer-service agents. Merging satellite and OnStar-type of technologies, consumers will have the opportunity to: hear a song, see who the artist is on the satellite radio faceplate, press OnStar in their vehicle, order the record delivered to their home, and have the CD added to their monthly OnStar bill. Promoters say this is especially attractive for two reasons. First, it satisfies the impulse to buy the CD at a time when the consumer is involved with the music, and second, it tends to provide opportunities for newer artists. Label promoters say that people who adopt new technology are more likely to be the early adopters of new artists.

Appendix

In Figure 8.5 on the next page is an example of an organizational chart for a typical pop label record promotion department. Note that the VP's of the various formats are specific to the music. However, the individual regional promoters typically promote all current music types marketed by the label.

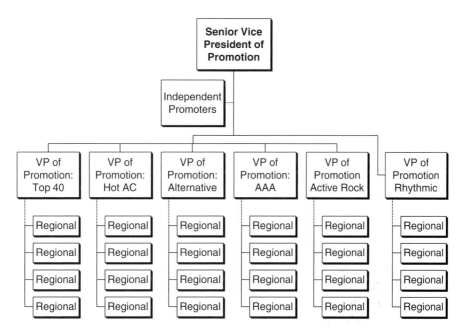

▲ *Figure 8.4 Promotion department organizational chart for pop music*

Organizational chart for a typical country label record promotion department:

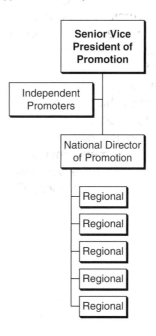

▲ *Figure 8.5 Country label organization chart*

Glossary

Burnout or burn – The tendency of a song to become less popular after repeated playings.

Indie – A shorthand term meaning an independent record promoter who works for radio stations and record labels under contract.

Payola – The illegal practice of giving and receiving money in exchange for the promise to play certain recordings on the radio without disclosing the arrangement on the air.

Playlist – The weekly listing of songs that are currently being played by a radio station.

Recurrents – Songs that used to be in high rotation at a station, but are now on the way down, reduced to limited spins.

Rotation – Mix or order of music played on a radio station.

Spin – This is a reference to the airing of a recording on a radio station one time. "Spins" refers to the multiple airing of a recording.

Trades – This is a reference to the major music business trade magazines.

Bibliography

ABC Television, 20/20 (November 2002). http://www.abcnewsstore.com/store/index.cfm?fuseaction=customer.product&product_code=T020524%2002).

Ammer, C. (1997). *The American Heritage® Dictionary of Idioms,* Boston: Houghton Mifflin.

BDS (2005). http://bdsonline.com/submit.html.

Billboard online (2005). http://www.billboard.com/bb/charts/heatseekers.jsp.

cmj.com (2005). http://www.cmj.com/company/.

cmj.com (2005). www.cmj.com/index.

Freed, A. (2005). http://www.historychannel.com/speeches/archive/speech_106.html.

hipnotikent (2004). http://hipnotikent.com/industryTips/bds.htm.

Hull, G. (2004). *The Recording Industry,* London: Routledge, pp. 201-202.

Katunich, L. J. (April 29, 2002). Time to Quit Paying the Payola Piper, *Loyola of Los Angeles Entertainment Law Review.*

Nielsen SoundScan (2005). aud.soundscan.com.

Phillips, C. (May 24, 2002). Congress Members Urge Investigation of Radio Payola, *LA Times.*
Phillips, C. (May 29, 2001). Logs Link Payments with Radio Airplay, *LA Times.*
radio-media (2005). http://www.radio-media.com/song-album/articles/airplay26.html.
Soundsource (2005). http://soundsource.ca/broadcastservices.asp?subsection=24.
wikipedia.org (2005). http://en.wikipedia.org/wiki/Billboard__magazine#History.

9 Publicity of Recorded Music

Paul Allen

A recording contract, simply stated, says that the artist will create recordings and the label will market the recordings. An important element of marketing planning and execution by a label is promotion, and publicity is typically a part of that overall promotional effort by the label. Labels typically handle publicity for the artist's recording career and for news and press releases about the label itself. Sometimes an artist will hire a personal publicist to handle other areas of their life and career.

This chapter is designed to give an overview of the publicity department at a record label, including its responsibilities and how publicity contributes to the success of a recorded music project.

Label publicity

The objective of label publicity is to place nonpaid promotional messages into the media on behalf of the artist's recorded music project. That can range from a small bullet point in *Rolling Stone* to an appearance on the "Late Show with David Letterman." Appearances in the media contribute to the success of the label's promotional plan to put the artist in print and on the air to support the marketing of the label's music.

The theory is that the more positive impressions consumers receive about a recording, the more likely they are to seek additional information about the recording, and to purchase it. Advertising planners use the term *reach and frequency* as they compile a strategy and its related budget. This means they plan an affordable ad campaign that can "reach" sufficient numbers of their target market with the "frequency" necessary for them to remember the message and act by purchasing. Publicity becomes a nice complement to that strategy without the direct costs of paid advertising.

Label publicity on behalf of a recording artist, like any positive publicity, has a certain credibility that paid advertising does not. After all, a journalist thought the artist was interesting enough to write an article about him or her or a review of the music, and a publisher thought it was interesting enough to make editorial space in a magazine or newspaper to present the story or run the review.

Advertising is the handcrafted, paid message of the record company's marketing department designed to sell recordings. It is someone's crafted message with the intent of getting into consumer's wallets. However, an effective publicity campaign can create an interest by journalists in writing articles about artists and their recordings in the not-so-commercial setting of a feature article. An article in a newspaper or magazine can suggest to the consumer that there is something more to the label's artist than just selling commercial music. Published articles and TV magazine-style stories (for example, "60 Minutes") tend to add credibility to the artist as an "artist" in a way that paid advertising cannot.

There are key differences between publicity from the label and advertising placed by the label. Label publicists generally create and promote messages to the media that are informative in nature and do not have a hard "sell" to them. On the other hand, advertising is designed to influence and persuade the consumer to purchase CDs.

Publicity in the music business—a historical perspective

The earliest music promoters were in the publicity business at the beginning of the last century, primarily helping to sell sheet music that was heard on recording playback devices or at public performances. The title of the job during these times was that of "music pusher." Those who worked in the publicity profession in the early 1900s relied primarily on newspapers and magazines to promote the sale of music.

In 1922, the federal government authorized the licensing of several hundred commercial radio stations, and those in the music business found their companies struggling as a result. People stopped buying as much music because radio was now providing it, and newspapers and magazines were no longer the only way the public got its news. Radio became the entertainer and the informer. But publicists found themselves with a new medium and a new way to promote, and quickly adapted to it, much in the way they did in 1948 with the advent of television as a news and entertainment form.

Today, the label publicist works with print and television for features and live appearances, while radio promotion handles airplay on AM, FM, and satellite radio, and video promotion works with stations and channels that use videos as part of their programming (Lathrop, T., and Pettigrew, J, Jr., 1999).

The label publicity department

The work of a label publicity department (sometimes called the *media relations department*) is very much like a sales department. Rather than selling "things," they are selling ideas and an image. In particular, they are "pitching," or selling story ideas to the media, attempting to stimulate an interest by the correspondent in writing or producing a story about the artist's recording.

Staffing the publicity department

Large labels often employ a director or manager of publicity. The full responsibility of media relations and publicity rests with the director. They are accountable to the media outlets they serve, and they are also accountable to the various departments at the label such as A&R, marketing, and sales. Sometimes that accountability stretches to the artist and the artist management team.

Structurally, some managers of publicity report directly to the president/chief operating officer of the label, and they are often called *director* or *vice president of publicity*, or some similar title. Some labels align the publicity department's accountability directly to the marketing department, while others make publicity a component of the creative services department. (Creative services typically handles design and graphics work for albums and point-of-purchase [POP] material, as well as imaging for the recorded music project.) Those who are hired as independent publicity companies for label projects typically report to the vice president of marketing.

Staff publicists work for the director and handle the day-to-day planning, coordination, and execution of the work of the department. Labels that have a large roster can stretch the time and energy of staff publicists, and sometimes the limitations of staff time require the employment of independent publicists. It is not uncommon to find that staff people are handling ongoing publicity efforts for several active artists at the same time.

Large label publicity departments sometimes hire independent or freelance publicists because these publicists have relationships with key media gate-keepers. They are hired because they have important contacts that the label does not have, and they can be effective in reaching these media outlets on behalf of the artists of the label. In some ways, this is similar to the use of independent radio promoters by record companies to reach programmers that are key to securing airplay for a recorded music project.

Independent and smaller record labels sometimes handle publicity in-house, but they often hire a publicity firm on a project-by-project basis, or pay a retainer fee in order to have access to their services as they are needed. The cost of paying an employee to handle publicity at a small label becomes a financial burden at times when business is slow. During down cycles in the record business, even the major labels have laid-off entire publicity departments and hired independent firms to handle the work for them.

Tools of the label publicity department

The database

The publicity department creates and distributes communications on a regular basis, so the maintenance of a quality, up-to-date contact list is critical to the success of that communication effort. Some publicity departments maintain their own contact lists; some use their own lists, plus lists through a subscription service; others rely entirely on subscription database providers.

An example of a subscription, or "pay" service, for media database management is Bacon's MediaSource. Bacon's updates its online database daily with full contact information on media outlets and subjects on which they report. The value of maintaining a quality database by the publicity department is that the information enables them to accurately target the appropriate media outlet, writer or producer. Services like Bacon's can literally keep a publicity department on target. Most labels and their independent publicist partners maintain several lists within their databases to assure they are not sending news releases to people who are not interested in the subject matter. These lists are used to send out press releases, press kits, promotional copies of the CD for reviews, and complimentary press passes to live performances.

The most effective way to reach media outlets is through email, because "it's inexpensive, efficient, and a great way to get information out very quickly" (Stark, Phyllis, personal interview). Though some outlets still prefer faxes, the challenge with them is to program the machine for delivery at a time that is convenient so that home-based journalists are not awakened in the middle of the night with the alert on a fax machine. A few media outlets still prefer regular mail, but the immediacy of the information is lost. An effective publicist learns the preferred form of communication for each media contact. Bacon's MediaSource provides some of that information but it is always preferable to check with the journalist.

Internet distribution of press information from a label requires the latest software that will be friendly to spam filters at companies that are serviced with news releases. A spam filter is an electronic filter used by many large companies to prevent the sending of unsolicited email to company employees. The most reliable way to assure news releases and other mass-distributed information are received by a media contact is to ask about any spam filters, and ways to bypass them.

The press release

Press releases are a standard tool in public relations; one that works better than letters or phone calls (Spellman, P., 2000). The press release is used to publicize news and events, and is a pared-down news story. Following are five examples of when a press release should be used:

1. To announce the release of an album
2. To announce a concert or tour
3. To publicize an event involving the artist or label
4. To announce the nomination or winning of an award or contest
5. To offer other newsworthy items that would be appealing to the media

The press release should be written with the important information at the beginning. Today's busy journalists don't have time to dig through a press release to determine what it is about. They want to quickly scan the document to determine whether this is something that will appeal to their target audience.

The anatomy of a press release

The press release needs to have a slug line (headline) that is short, attention-grabbing, and precise. The purpose or topic should be presented in the slug line. The release should be dated with contact information including phone and fax numbers, address, and email. The body of text should be double-spaced.

The lead paragraph should answer the five W's and the H (who, what, where, when, why, and how). Begin with the most important information; no unnecessary information should be included in the lead paragraph (Knab, C., 2003a). In the body, information should be written in the inverse pyramid form: in descending order of importance.

The bio

The artist bio provides a window into the artist's persona, and sets the artist apart from others. Before writing the bio, an examination should be done on the artist's background, accomplishments, goals and interests to find interesting and unique features that will set the artist apart from others. (Knab, C., 2003b) Keep in mind the target readership of the bio. Some busy journalists may use portions of the bio in a news story. The bio should be succinct and interesting to read (Hyatt, Ariel., 2004). Create an introduction that clearly defines the artist and the genre or style of music.

The press kit

Another of the primary tools of the department is the press kit they develop on behalf of the artist and the recorded music project. Typical components of the kit include:

- A press release announcing the release of the CD
- The artist's bio, which is often created by someone hired by the label for this specific purpose
- The CD that contains the single or the album
- High-resolution color jpegs of the artist on disc
- A "cut-by-cut," which is a paper attached to the press release listing of the tracks included on the CD with the artist's personal comments about each track
- A discography
- Sometimes it also includes an electronic press kit (EPK), which is a VHS or digital video recording of the artist's music video and also shows them discussing the music and the project. This helps journalists get a sense of what the artist looks like and how they relate other than through their music
- Clippings (tear sheets): articles printed or video stories created by other media

Journalists say that the most useful parts of the press kit, sometimes called a *pitch kit*, are the CD, the cut-by-cut, and the photos.

A cover letter is included with the press kit and should be personal to the journalist. The purpose of the letter must be very clear in the first paragraph of the letter, and it should be specific, not general. From the journalist's perspective, answer the questions: "What do you want?" and "Why is this important to me?" Otherwise, the label publicist risks having the journalist set the kit aside to, perhaps later, figure out what it is all about. And later may never happen.

Photos and video

A good photograph can generate a lot of publicity. It can be the most striking and effective part of a press kit. A good quality photo has a much better chance of being run in print media and is worth the extra effort and expense. An experienced professional photographer can bring out the true personality of the artist in a photograph. Sometimes, location is used to help portray the artist's identity, but studio shots are easier to control (Knab, C., 2001c). Publicity shots are not the same as a publicity photo. A publicity shot is one taken backstage with other celebrities, or at events. The publicity photo is the official photographic representation of the artist. Publicity photos should be periodically updated to keep current with styles and image. But, once a photo is released to the public, it is fair game for making a reappearance at any time in the artist's career, even if they have moved on and revamped their image.

The music video also becomes an important part of the continuing effort of the publicity department to promote the album as consecutive singles are released. Though many labels have video promotion departments, the artist's music video is also a valuable tool used by publicists in securing live and taped appearances in television programs.

Working with the artist's image

Any aspect of the entertainment business relies on the created perception of the artist or event. After all, it is show business. It is important to the label to know the perception of the recording artist in the minds of music buyers. One of the most important contributions a label publicity department can make towards defining the public's perception of the artist is carefully helping the artist develop their image.

There are several key items that contribute to the image of the artist. Among those are:

- The name chosen by the artist
- Physical appearance of the artist
- Their recording style and sound
- Choices of material and songwriting style
- Their style of dress
- The physical appearance of others who share the stage
- The kind of interviews done on radio and TV
- Appearance and behavior when not on stage (Frascogna, X., and Hetherington, L., 2004).

Some labels will hire hair stylists and clothing and costume consultants, some will pay for dental work, and some are rumored to pay for cosmetic surgery in order to polish the artist's image to prepare them for their expanded public career (Levy, S., 2004).

Hiring a media consultant is a judgment call for the label, and they may or may not feel that an artist should receive training. What a media consultant does is train artists to handle themselves in public interviews. Working with the artist, the consultant prepares the artist for interviews by taking the unfamiliar and making it familiar to them, teaching them to know what to expect, and giving them the basic tools to conduct themselves well in a media interview. The debate comes from critics who say media consultants go too far by preparing artists with suggested answers, and thereby create cookie-cutter interviewees with nothing new or interesting to say.

Publicists and others at the label should take care not to compromise the unique qualities an artist has by developing an image that isn't consistent with who the artist is. The values the artist has should be apparent in their music and public image in order to tighten the connection with record consumers who share those values. Ideally, the label and the artist manager will work with the artist to help them define who they are personally and creatively, and then coach them to an image that is commercial but not artificial.

For established artists, care must be taken not to radically change their image. Fans are quick to pick-up on efforts to radically re-tool an image, and the result can ultimately turn fans away from an artist.

Evaluation of publicity campaigns

Top management and artists themselves will seek feedback about the effectiveness of a media campaign on behalf of an album project. Clipping services are available that will search publications and television shows for copies of articles or video clips that will demonstrate the news item connected with the media. AristoMedia, an independent publicity company in Nashville, often uses Google as an efficient way to check the impact of its news releases a few days following dissemination (Lathrop, T., and Pettigrew, J, Jr., 1999). It is good for artist relations to present the artist and manager with a box of news clippings to indicate the successful efforts of the publicity department in securing media coverage.

Target marketing through the publicity plan

Label publicity, as in any part of the marketing process, must keep its focus on the target market for the recording. The starting point is to get a clear understanding of what that target market is and how publicity fits within the project's marketing plan, then to develop a plan to reach the market through publicity. The plan created by the publicity department is coordinated with the other departments within the label. For example, A&R will define the artist, the music and the genre. Marketing will provide the overall plan that will include sales objectives, the target market, the marketing communications plan, as well as an estimated release date for the single and the album. Radio promotion will provide its timeline to work the record to radio, and let publicity know if the music will be promoted to multiple formats. That helps the publicist know where to place their efforts. Sales will give its timetable so that publicity efforts will be timed to maximize sales at retail.

Reaching the target audience with press releases requires an understanding of the actual audience of the media. For example, Table 9.1 outlines the media type and its intended audience.

Trade publications, like those noted above, offer the benefit of setting up a recorded music project to garner the interest of radio and video channels. New artists especially require a major push by publicity to create that invaluable buzz among the electronic gatekeepers of radio and television. The trades must be worked before the release of the album, and radio should be worked prior to the release of the first single. Press releases can be sent to trade publications in the hopes of getting an article or at least a blurb about the upcoming release.

Table 9.1 Media and target examples	
Media	**Targets and examples**
Daily newspapers	General readers from a national, regional, local base
Lifestyle, entertainment magazines	Broad national readership with a certain editorial focus *People, Entertainment Weekly, Cosmopolitan*
News weeklies	Young, urban, sophisticated, culturally aware readers *Village Voice, SF Weekly, Boston Phoenix*
Music and pop culture magazines	Young, affluent adults with strong interest in . . . music *Rolling Stone, Spin, Vibe*
Genre-based music magazines	Customers of chain record stores *Country Weekly, The Source, Flipside*
Promotional magazines	Customers of chain record stores *Pulse, Request*
Magazines for music hobbyists and pros	Players of instruments, recording engineers *Guitar World, Mix, Musician*
Magazines for record collectors	Collectors and aficionados of oldies and rare discs *Goldmine, DISCoveries*
Trade publications	Music industry professionals *Billboard, Radio & Records, Variety*
Fanzines	Pre-teen and teenage fans of new artists *Tiger Beat*

Source: *This Business of Music Marketing and Promotion*, Lathrop and Pettigrew, 1999.

The savvy publicist researches the magazines, newspapers, fanzines, and so forth before submitting materials for publication. The types of stories, reviews and features that each publication prints should be noted. Then, only those that match the target market and that regularly feature the types of stories being pitched should be approached. Nothing aggravates a writer or editor more than someone pitching inappropriate material to their publication. It is generally easier to get placement in a niche magazine than a more general publication. Genre-based music magazines are more receptive to publishing suitable material, and there is less competition than one would find with a general-reading publication such as *People*.

The publicity plan

Peter Spellman says, "the first ingredient for a successful publicity plan is a clear idea of your market audience: who they are, what they read and listen to, and where they go. Each style of music is a subcultural world . . . your job is to understand this world" (Spellman, P., 2000a). The publicity plan is designed to coordinate all aspects of getting nonpaid press coverage, and is timed to maximize artist exposure and record sales. The plan is usually put into play several weeks before the release of an album. Spellman also says that "your publicity objectives can only be realized through successive 'waves' of media exposure'' (Spellman, P., 1996b) because of the short attention span of the public. Each wave needs a promotional angle: an album release, announcement of tour dates, or anything newsworthy.

The publicity plan starts with goals: awareness, motivation, sales, positioning, and so on, and priorities are established (Yale, D. R., 1993). The marketplace is then researched, and media vehicles targeted. The materials are developed and the pitching begins. Lead time is the amount of time in advance of the publication

Table 9.2 The publicity plan
The publicity plan
• Set publicity goals • Identify target market • Identify target media • Create materials • Set up timetable w/deadlines • Pitch to media • Provide materials to media • Evaluate
Source: *The Publicity Handbooks*, David R. Yale, 2001.

that a journalist or editor needs to prepare materials for inclusion in their publication. A schedule is created to ensure that materials are provided in time to make publication deadlines. Long-lead publications are particularly problematic for the publicist as they need to have materials prepared months in advance of the release date, and sometimes those materials are not yet available. If an artist suddenly breaks in the marketplace, it is too late to secure a last-minute cover photo on most monthly publications.

Pitch letters are then sent out to media requesting publication or other media exposure. The pitch letter is a carefully thought-out and crafted document specifically designed to grab the interest of a busy, often distracted journalist or TV producer. It is never mass-mailed, but is specifically tailored to each media outlet being contacted (D'Vari, M., 2003). The pitch letter should begin with a few words presenting the publicist's request, and then quickly point out why the media vehicle being contacted should be interested in the artist or press material. Prep sheets are also developed and sent to radio so that DJs can discuss the artist as they spin the record. As the publicity plan unfolds, it is necessary to evaluate the efforts through clipping services and search engines (see section on evaluation of publicity campaigns).

Budgets for money and time

A budget for the publicity campaign is developed based on the objectives of the project and the expectations of the label for the part publicity will play in stimulating interest in the artist's music. If this is the first album for the artist, the development of new support materials may be necessary, such as current photos and a bio. If it is an established artist, budgets could be considerably higher, in part because of the expectations of the artist to receive priority attention from the director of publicity.

Table 9.3 Example publicity timeline for a major label

Example publicity timeline for a major label

Time frame	Publicity task
Upon signing the artist	Schedule meetings with artist Press release announcing signing
During the recording	In-studio photos Invite key media people to studio
Also during this time period	Schedule media training if needed Select media photos Determine media message
When masters are ready	Hire bio writer Create advance copies for reviews Create visual promo items
When advance music is ready—Ideally four months out	Send advances to long-lead publications Send advances w/bio and photo to VIPs—magazines, TV bookers, and syndicators Begin pitch calls to secure month-of-release reviews Start servicing newsworthy bits on the artist on a weekly basis to all media
Advance Music—one month out	Send advances/press kits to key newspaper and TV outlets Begin pitch calls to guarantee week-of-release reviews
One week out	Service final packaged CDs to all media outlets Continue follow up calls and creative pitching
After release	Continue securing coverage and providing materials to all media outlets

Source: Amy Willis, Media Coordinator, Sony Music Nashville

Table 9.4 Preliminary publicity timeline for indie artist Prototype

Preliminary publicity timeline for indie artist *Prototype*

Day	Date	Event
Fri	12/6	Advance promo copies at The Rocket, UW Daily, Pandemonium, Seattle Times, and the Seattle Weekly
Mon	12/9	Arrange interviews on KCMU, Seattle and KUGS Bellingham
Fri	12/13	Deadline for completing database of print/broadcast mailing list
Fri	1/10	Single sent to College radio
Fri	1/17	*Calling It A Day* **CD release day**
Mon	1/20	Album mention in UW Daily/Seattle Times
Tues	1/21	Album mention in Seattle Weekly
Thur	1/24	Tour begins in Seattle
Thu	1/24	Album mention in CMJ/Hits
Tue	1/28	Album mention in the Rocket and Pandemonium
Tue	1/27	Album mention in Spokane and Tri-City Daily papers
Mon	2/2	Album mention in Cake and Fizz
Fri	2/6	Album mention in Flipside and Village Noize
Wed	2/15	Album mention in Spin
Wed	2/15	Album mention in Virtually Alternative
Fri	2/17	Album mention in Next
Fri	2/24	Album mention in Magnet
Mon	2/27	Interview on KUGS Bellingham
Mon	2/27	Interview on KCMU Seattle
Tue	2/28	Feature story in The Rocket

Source: Christopher Knab www.4frontmusic.com

Publicity costs include the expense of developing and reproducing materials such as press kits, photos, bios, and so forth, communication costs (postage and telephone bills, contact lists), and staffing costs. The minimum cost for an indie label would run about $8,000, with $3,000 of that for developing press kits and $2,800 for postage. Adding an outside consultant to the project would add another $1,500 or more per month. For major label projects, an outside publicist can be hired for six months to provide full support to a single, and album, and tour publicity for $25,000, which includes out-of-pocket costs such as postage, press kits, and other related expenses.

An equally important part of the plan is to budget adequate time to support the album based on when it will be released during the annual business cycle of the label. If the in-house staffing is adequate—given the timing of the project—the plan can be created. If, however, the publicity department is overloaded, the director may consider hiring an independent company to handle publicity for the project. This seemingly removes the burden from the director, but it adds oversight duties since the director must be sure the outside company is working the plan according to expectations. The ultimate success (and failure) is still the responsibility of the director of publicity.

Television appearances

News shows

Major entertainment television news shows, including syndicated news shows on major network affiliates, are most often interested in major acts. They see their viewers as people who want to know about the latest information on their favorite recording artists. Stars that are easily recognizable are those most often sought for their entertainment news stories. This creates a genuine challenge for the record label that is trying to publicize a new artist with consumers. A new act with a new single or album must have an interesting connection with consumers that goes beyond the music in order to compete with the superstars who will always get airtime. There are many more new artists looking for publicity than there are established artists, and it forces the best label publicists to be as creative as they can be on behalf of the few new acts.

For television interviews with new artists, a sit-down interview with the artist often is not enough. Television shows look for that added dimension to a new artist that makes them interesting to the viewers, and they often look for the nontraditional setting in which to present the story. Though at times it is overdone, connecting an artist with their charity work becomes an interesting angle for television.

The challenge to the label publicity department is to find those key personal differences that make their recording artists interesting beyond their music. Label publicists are sometimes criticized for citing regional radio airplay, chart position, or label financial support as the only positives that make their newest artists stand out. Those in the media say they look for that something special, different, and newsworthy that gives an angle for them to write about. In that light, it puts the responsibility on the label publicist to find several different angles to offer to different media outlets to generate the interest needed to get a story placed. Writers for major media want their own angle on an artist when possible because it demonstrates to media management that an independent standout story has been developed, making them different from their competition. Sometimes, though, the story angle about an artist is different enough that it stands on its own and most media will see the value it has for their audiences. Entertainment writers and producers are often self-described storytellers, and delivering that unique story to them is a continuing challenge to the successful label publicist (Pettigrew, J, Jr., 1997).

Talk-entertainment shows

Label publicists are often the facilitator of an artist's appearance on popular talk and entertainment shows. Among the most popular shows include those with hosts Jay Leno, David Letterman, Conan O'Brien, Oprah, and Regis Philbin. Other television shows that publicists seek as targets for the label's artists include the early morning shows like "Today," and "Saturday Night Live." Great performances on shows like these are expected of the artist, but Ashley Simpson's ill-fated 2004 appearance on "Saturday Night Live" will probably follow her forever. (Simpson had an embarrassing performance when tracks for the wrong song were cued up and played, revealing pre-recorded vocal tracks.)

Bookings to programs like these are handled by the publicist based upon their relationships with talent bookers on these shows. It is not uncommon for a publicist to precede a pitch for an artist to appear on one of these shows by sending a big fruit basket. However, the success of placing the label's new artist on one of these shows is also based on the ability of the publicist to build a compelling story for the artist that will interest the booker. Often the publicist will offer another major artist for a later appearance in exchange for accepting the new artist now.

Award shows

The value of having an artist perform on an award show is obvious—it sells records. These slots are coveted by all the record labels, and lobbying efforts may pay off in a big way.

Comparing publicity and record promotion

"The savvy labels recognize how important publicity is to the mix—it's almost as important as record promotion."

—**Phyllis Stark, Nashville Editor, Billboard Magazine**

The preceding chapter in this book looked at record promotion and its impact on the marketing of a recorded music project. The following chart in Table 9.5 is a look at the relationship that publicity has with its counterpart in the record promotion effort for an album.

Table 9.5 Comparison of promotion and publicity departments

Record promotion	Publicity department
Develops and maintains relationships with key radio programmers (gatekeepers).	Develops and maintains relationships with key writers, news program producers, and key talent bookers for network and cable channel TV shows.
Tells radio programmers that a new single or album is about to be released and to prepare for "add" date; sends promo singles and albums.	Prepares and sends a press kit to journalists announcing the new single or album project.
Schedules the new artist for tours of key radio stations for interviews and meet 'n greets with station personnel.	Schedules the artist and sometimes the album producer for interviews with both the trade press and consumer press.
Employs independent radio promotion people who have key relationships with important radio programmers.	Employs independent or freelance publicists who have key relationships with important media outlets.
Effectiveness of their work is measured by the number of "adds" they receive on the airplay charts of major trades.	Effectiveness is measured by the number of "gets" they receive, meaning the number of articles placed, number of TV news shows in which stories run, the number of talk/entertainment shows on which the artist performs (Phyllis Stark).
Gets local radio publicity and airplay for new artists based on the promise of an established artist making a local appearance sometime in the future.	Gets new artists booked on major talk/entertainmentshows based on the promise of making an established artist available to the show sometime in the future. Supports local press during touring.

Glossary

Bio – Short for biography. The brief description of an artist's life and/or music history that appears in a press kit.

Creative services department – A work unit at a record label that handles design, graphics, and imaging for a recorded music project.

Clippings – Stories cut from newspapers or magazine.

Cut-by-cut – This is a listing of comments made by an artist about each of the songs chosen to be included on an album project.

Discography – A bibliography of music recordings.

Independent publicist – Someone or a company that performs the work of a label publicist on a contract or retainer basis.

Lead time – Elapsed time between acquisition of a manuscript by an editor and its publication.

Media consultant – Trains artists to handle themselves in public interviews with the media.

Music pusher – A term used in the early 1900s for someone who was promoting music.

Press kit – An assemblage of information used to provide background information on an artist.

Press release – A formal printed announcement by a company about its activities that is written in the form of a news article and given to the media to generate or encourage publicity.

Slug – A short phrase or word that identifies an article as it goes through the production process; usually placed at the top corner of submitted copy.

Talent bookers – These are people who work for producers of television shows whose job it is to seek appropriate artists to perform on the program.

Tear sheets – A page of a publication featuring a particular advertisement or feature, and sent to the advertiser or PR firm for substantiation purposes.

A special thanks to Tom Baldrica, Bill Mayne, Jeff Walker, Phyllis Stark, Ed Benson, the folks at Starpolish.com, and the CMA for their assistance in providing added information and insight into this chapter.

Bibliography

D'Vari, M. (2003). http://www.publishingcentral.com/articles/20030301-17-6b33.html. How to Create a Pitch Letter.

Frascogna, X., and Hetherington, L. (2004). *This Business of Artist Management,* New York: Billboard Books.

Hyatt, A. (2004). http://arielpublicity.com. How to be Your Own Publicist.

Knab, C. (2001c). http://www.musicbizacademy.com/knab/articles/. Promo Kit Photos.

Knab, C. (2003a). http://www.musicbizacademy,com/knab/articles/pressrelease.htm. How to Write a Music-Related Press Release. (November, 2003).

Knab, C. (2003b). http://www.musicbizacademy.com/knab/articles.

Lathrop, T., and Pettigrew, J, Jr. (1999). *This Business of Music Marketing and Promotion,* New York: BPI Publications.

Lathrop, T., and Pettigrew, J Jr. (1999). *This Business of Music Marketing and Promotion,* New York: BPI Publications.

Levy, S. (2004). CMA's music business 101, *unpublished.*

Pettigrew, J, Jr. (1997). *The Billboard Guide to Music Publicity,* New York: Watson-Guptill Publications.

Spellman, P. (2000). http://www.harmony-central.com. /Bands/Articles/Self-Promoting__ Musician/chapter-14-1.html. Media Power: Creating a Music Publicity Plan That Works, Part 1 (March 24, 2000).

Spellman, P. (2000a). http://www.harmony-central.com/Bands/Articles/Self-Promoting__ Musician/chapter-14-1.html. Media Power: Creating a Music Publicity Plan That Works, Part 1. (March 24, 2000).

Spellman, P. (Spring 1996b). http://www.musicianassist.com/archive/newsletter/ MBSOLUT/files/mbiz-3.htm. Creative Marketing: Making Media Waves: Creating a Scheduled Publicity Plan. Music Business Insights (Issue I:3).

Yale, D. R. (1993). *The Publicity Handbook,* Chicago: NTC Business Books.

10 Advertising in the Recording Industry

Amy Macy, Thomas Hutchison

Basics of advertising

Advertising is a form of marketing communication. It is described by Bovée and Arens (1986) as "the nonpersonal communication of information usually paid for and usually persuasive in nature about products, services, or ideas by identified sponsors through the various media." The fact that advertising is paid for and persuasive (i.e., the seller controls the content and message) separates it from other forms of mass mediated communication such as publicity, news and features. Advertisements are usually directed toward a particular market segment, and that dictates which media and which vehicles are chosen to present the message. A *medium* refers to a class of communication carriers such as television, newspapers, magazines, outdoor, and so forth. A *vehicle* is a particular carrier within the group, such as *Rolling Stone* magazine or MTV network. Advertisers determine where to place their advertising budget based on the likelihood that the advertisements will create enough of an increase in sales to justify their expense. Advertisers must be familiar with their market and consumers' media consumption habits in order to be successful in reaching their customers as effectively as possible.

The most basic market segments for advertising are: (1) consumers, and (2) *trade*, which are people within the industry that make decisions affecting the success of your marketing efforts. The first market segment (consumers) is targeted through radio, television, billboards, direct mail, magazines, newspapers and the Internet. Consumer advertising is directed toward potential buyers to create a "pull" marketing effect (see Chapter 1). The second target market (the *trade*) consists of wholesalers, retailers, and other people who may be influenced by the advertisements and respond in a way that is favorable for the marketing goals. This creates a "push" marketing effect (again, see Chapter 1). In the recording industry, this would include radio program directors that may be influenced by an advertisement in *Radio & Records* to more favorably consider a particular advertised song for inclusion in the station's weekly playlist. *Trade advertising* is usually done through direct mail and through *trade publications* (usually magazines) aimed at people who work in the industry. Trade advertising is not common through television, radio and newspapers because they are too general in nature — not targeted enough to effectively reach the industry. Online publications for trade and consumers are becoming a factor in record label advertising.

Comparison of media options for advertising

The most complex issue facing advertisers involves decisions of where to place advertising. The expansion of media has increased the options and complicated the decision. The following chart in Table 10.1 represents a basic understanding of the advantages and disadvantages of the various media options.

Table 10.1 A comparison of media

Media	Advantages	Disadvantages
Television	• Reaches a wide audience, but can also target audiences through use of cable channels • Benefit of sight and sound • Captures viewers' attention • Can create emotional response • High information content	• Short life-span (30–60 seconds) • High cost • Clutter of too many other ads; consumers may avoid exposure • Can be expensive
Magazines	• High quality ads (compared to newspapers) • High information content • Long life-span • Can target audience through specialty magazines	• Long lead time • Position in magazine uncertain • No audio for product sampling (unless a CD is included at considerable expense)
Newspapers	• Good local coverage • Can place quickly (short lead) • Can group ads by product class (music in entertainment section) • Cost effective • Effective for dissemination of information, such as pricing	• Poor quality presentation • Short life-span • Poor attention-getting • No product sampling

Media	Advantages	Disadvantages
	Table 10.1 Continued	
Radio	• Is already music-oriented • Can sample product • Short lead, can place quickly • High frequency (repetition) • High quality audio presentation • Can segment geographically, demographically and by music tastes	• Audio only, no visuals • Short attention span • Avoidance of ads by listeners • Consumer may not remember product details
Billboards	• High exposure frequency • Low cost • Can segment geographically	• Message may be ignored • Brevity of message • Not targeted except geographically • Environmental blight
Direct Mail	• Best targeting • Large info content • Not competing with other advertising	• High cost per contact; must maintain accurate mailing lists. • Associated with junk mail
Internet	• Best targeting—can target based on consumer's interests • Potential for audio and video sampling; graphics and photos • Can be considered point-of-purchase if product available online	• Slow modem speeds limit quality and speed • Effectiveness of this new media still unknown • Doesn't reach entire market • Internet is vast and adequate coverage is elusive

Determining costs and effectiveness

Advertisers have many choices when it comes to how to spend the advertising budget to maximize effectiveness, and the options keep increasing. In the 1950s and 1960s, television and magazines were not as fragmented as they are today, and content was more general in nature. Media fragmentation is defined as the division of mass media into niche vehicles through specialization of content and segmentation of audiences. In the 1960s, advertisers sought to reach particular groups of potential customers, rather than paying for advertising in general media vehicles that would be wasted on viewers who were not interested in their product.

Magazines became the first medium to go through fragmentation. The popularity of general appeal magazines such as *Life*, *Look* and *Saturday Evening Post* gave way to the rise in specialty magazines that dealt with everything from fitness to alternative music. Advertisers would rather pay for ads in the particular publications that reach their market than pay the more expensive rates to reach the general population. Fragmentation spread to television with the expansion of cable TV. The Internet offers the ultimate in fragmented and specialized content or what is now being considered as "media personalization." As a result of this, advertisers can now reach people who are more likely to be potential consumers of the particular products being advertised.

Media planning

Media planning involves the decisions made in determining in which media to place advertisements. It consists of a series of decisions made to answer the following questions (Sissors, J., and Bumba, L., 1994):

1. How many potential consumers do I need to reach?
2. In what medium (or media) should I place ads?
3. When and how often should these ads run?
4. What vehicles and in which markets should they run?
5. What choices are most cost-effective?

The development of the media plan must incorporate the following considerations:

1. **Marketing objectives** – What are the goals of the entire marketing plan? To inform the buyer? To motivate the buyer? To change attitudes? To promote repeat purchases? To reinforce a previous purchase? To create a buzz? To emphasize price or value? To create sampling opportunities?
2. **Characteristics of the product** – Some products are more suitable for advertising in one medium over another. For example, perfume samples work well in magazines but not on radio. For music, audio is important but so are visuals.
3. **Pricing strategy** – The high costs associated with some media makes media buys impractical.
4. **Channels of distribution** – Media buys should be limited to areas where the product is available for sale and where they would support retailers.
5. **Promotion plans** – This involves determining the relative amount of effort needed for advertising, compared to other marketing aspects, and to complement and support them (Dunn, S. et al., 1990).

How advertising effectiveness is measured

The basic structure of measuring advertising effectiveness starts with information on reach, frequency, gross impressions, gross rating points (GRP), and cost per thousand (CPM). *Reach* is the number of different persons or households exposed to a particular media vehicle during the specified time period. *Frequency* is the number of times during that period that a prospect or a portion of the population is exposed to the message.

$$\text{Average frequency} = \frac{\text{total exposures}}{\text{reach}}$$

The concept used to measure these two notions together is *gross rating points*. Gross rating points (GRP) is a measure of the total weight of advertising that derives from a particular media buy; reach (expressed as a percentage of the market) times average frequency. It is the sum of the ratings for the individual announcements or programs. In television, GRP is the total weight of a media schedule against TV households. For example, a weekly schedule of five commercials with an average household rating of 20 would yield 100 GRPs.

GRP = reach × average frequency

Gross impressions (GI) are a measure of total media exposure. It is the sum of audiences of all vehicles used in a media plan. The number represents the message weight of a media plan. The purpose of GI analysis is to get a quick look at the total audience size of one or more media (Scissors, J. and Bumba, L., 1994). It is good for comparing with other media plans that may include a different group of media vehicles. GI can be calculated by adding the target audience sizes (audience size × number of exposures) delivered by each media vehicle in the plan.

When determining the effectiveness of an advertising plan, the costs must be factored in. This is done by determining the *cost per rating point* (for electronic media), and cost per thousand (a more general measure). Cost per rating point (CPP) is defined as determining what media programs are most cost-effective by dividing the cost of advertising by the show's expected rating.

CPP = Cost of ad schedule/GRP

Cost per thousand (CPM) is a term describing the cost of reaching 1,000 people in the medium's audience—the cost of delivering 1,000 gross impressions. It is used in comparing or evaluating the cost efficiency of various media vehicles. The formula varies depending on the media.

Table 10.2 Formulas for print and broadcast

Media	Formula
Print	$\dfrac{\text{Cost of ad} \times 1000 \text{ circulation}}{\text{circulation}}$
Broadcast	$\dfrac{\text{Cost of 1 unit of airtime} \times 1000}{\text{Number of households reached}}$

All other things being equal, you would expect advertisers to look for the lowest cost-per-thousand medium as being the most efficient. But this is seldom the case. Specialized products for narrowly targeted markets may require more scrutiny of each vehicle's ability to reach the target market and the effectiveness of the type of message being delivered. For example, a specialty music magazine such as *Down Beat*, designed to reach jazz fans, may cost more per each thousand people reached than local newspapers, but those who are reached may be more inclined to purchase a new jazz recording. Also, labels want to take advantage of the sampling capabilities of radio, television and the Internet, rather than rely on print media to sell music.

The role of advertising in marketing recorded music

Advertising is crucial for marketing recorded music just as it is for other products. The primary advertising vehicle in the recording industry is local print, done in conjunction with retail stores to promote pricing of new titles, and is referred to as co-op advertising. But the record industry also relies on magazine, radio, television, outdoor and Internet advertising. The impact of advertising is not easy to measure because much of its effect is cumulative or in conjunction with other promotional events such as live performances and radio airplay. SoundScan has improved the ability to judge the impact of advertising, but since marketing does not occur in a vacuum, the relative contribution of advertising to sales success remains somewhat of a mystery.

Advertising campaigns usually reflect the imaging of an artist's release. Consistency of artist photos, theme, and even the font of the text create indelible impressions that help consumers make the connection between the advertisement and the actual purchase of the release. Some ads may use related photos to the actual cover art of the CD, but where applicable, the actual CD cover will also be included.

Trade advertising

The recording industry uses trade publication advertising to set-up the release of a single to radio by advertising in publications often read by the radio industry, including *Radio & Records, Billboard, HITS magazine, Friday Morning Quarterback,* and *The Album Network*. Advertising in *Billboard* also reaches the retail industry and may influence retail buyers. Labels may also use direct mail flyers sent to retail accounts to help promote album releases.

These flyers may sometimes be sent through the distributors or one-stops. Postcard advertisements may also be sent to radio stations to generate interest in a single.

Coordinating with the publicity department

Any media campaign should be designed in conjunction with the publicity department. A coordinated media campaign is necessary because the vehicles targeted for advertising are the same vehicles targeted for publicity and some synergy may occur. For example, placing an ad in a particular publication may increase the likelihood of getting some editorial coverage. Or at least, if it has been determined that the vehicle reaches the targeted market, those consumers will be receptive to the editorial content of the publication, as well as the advertisement content. For example, if you wanted to sell hiking shoes, you would probably select a hiking, camping or backpacking magazine for your advertising, but you are also more likely to get a product review in those publications than in a parenting or sewing magazine.

Co-op advertising

Co-op advertising or *cooperative advertising* is generally defined as the joint promotion effort of two or more parties in the selling chain. It offers the potential for synergy between the two parties and is quite common in the recording industry. More specifically, co-op advertising is usually run by a local advertiser (retailer) in conjunction with a national advertiser (manufacturer). The national advertiser usually provides the copy and shares (or bears) the cost with the local retailer. Record labels team up with retail chains to create co-op advertising programs.

Co-op advertising offers a number of advantages to the manufacturer. It gains the retailer's support and endorsement, improves the relationship between the manufacturer and the retailer, helps get more products in the store and utilizes the retailer's knowledge of the local market.

Cooperative advertising programs must follow the guidelines and legal statues as monitored by the Federal Trade Commission to ensure fairness in trade practices. The Robinson-Patman Act of 1936 forbids price and payment (including payment for advertising) discrimination between suppliers and retailers. In other words, participating in a co-op promotion amounts to the retailer receiving a discount on goods purchased from the manufacturer. The discount comes in the form of a "rebate" paid for providing local advertisements.

If the plan is offered to one retail chain over another, it amounts to discriminatory promotional allowances in violation of the Robinson-Patman Act.

Although co-op advertising is usually targeted to consumers of a specific product, the ad dollars exchanged for the exposure will most likely gain added product placement with the retailer. For example, if a mini-album cover of a new release is placed in a retailer's Sunday circular, readers of the insert will then know about its release, what is looks like, the price, etc. But for placing the mini in the ad, the record label gains added exposure in-store as well, such as placement of the CD on the bestseller endcap, as well as extra product in the bin.

Co-op advertising can be perceived as a double-edged sword, meaning that consumers learn of new releases while the label gains additional exposure in the store. But understand that co-op advertising is usually part of the equation during the initial buy as well as continual support of the release through its lifecycle. Often times, co-op ad dollars offset margin, since a co-op advertising vehicles often feature a release at "sale" price. The retailer can afford to discount the consumer price of the release since co-op ad dollars are aiding the profitability of each unit sold. (Co-op advertising is discussed in greater detail in the retail chapter.)

Consumer advertising: the media buy

When considering a media buy, labels consider many factors:

- Budget
- Target market
- Media
- Timing
- Partners
- Artist Relations

Advertising budgets

Determining the budget sets the parameters and scope of any advertising campaign. Recognizing financial limitations can assist in creating a media buy that is the "best bang for the buck." Using the projected sales of the record plus a profit & loss analysis, labels can create a budget based on forecasted sales. To calculate a ballpark ad budget, published rate cards of advertising costs can be collected to assist in the budgeting process. The chapter regarding the profit & loss statement of a project discusses at length the implication of marketing dollars and advertising.

Target markets

Focusing on a target market increases the effectiveness of a media buy. Most products have a specific consumer to whom companies aim their promotions. This focus determines the type of media that is purchased. For example, many products created for parents are targeted in women's lifestyle magazines as well as daytime television, with research validating that stay-at-home mothers are consuming these media. When purchasing media, most consultants look at targets demographically and when these demographics are engaging the media. Women ages 25-54, and men ages 18-24 are examples of profiling the target market. Attention to dayparts when the target is engaged is also relevant. Specifying a particular time of day to increase chances of hitting the target is strategic in creating an effective advertising buy. Evaluating the media by target enables an advertising buyer to focus the purchase. Ratings of electronic media outlets are determined by Arbitron and Nielsen. As discussed in the chapter "How Radio Works," media outlets operate their businesses to attract a certain consumer, so that they can create revenue via the sale of advertising.

Media

Deciding which media to advertise through can be an agonizing process. Print advertising historically has the lowest cost per thousand (CPM) considering its wide readership. Newspaper advertising is known as the *shotgun* approach—a shotgun uses "shot" instead of a bullet. "Shot" will scatter and possibly hit the target, but will likely hit other areas too. A newspaper is consumed by many demographics, a small portion of which may be the target. Is this the best "bang for the advertising buck?" Depending on the product, it may be most effective. For music, newspaper advertising has an important role in alerting music buyers as to new releases. The Sunday circulars of various retailers often feature new releases that are streeting on the upcoming Tuesday. But some record label executives question, "Can a consumer hear a print ad?"

To hear the product, consumers would need to listen to radio, watch television, or engage the Internet to be able to "test drive" the product through advertising. Radio and television usually incur a higher CPM, but the audience is more concentrated, meaning that an advertisement on these outlets has a higher probability of hitting the target.

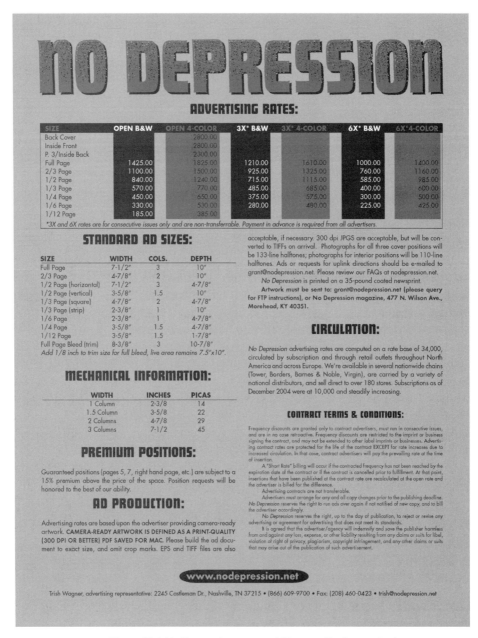

▲ *Figure 10.1 No Depression rate card (Source: No Depression)*

Timing

When to purchase the advertising "run" is a critical decision. Most advertising outlets have "lead" times in which they need to prepare for the actual advertisement. Note the above rate card information from "No Depression." To place a print ad for the January/February 2006 magazine, artwork has to be delivered to the magazine no later than November 17, 2005. Radio and television lead time tend to be shorter, since their advertising activities often occur in "real-time." But what does need time is the creation of the actual advertising spot. Both radio and television advertising require pre-recorded "spots" in which to air. Creation of these spots includes production issues such as musical content, voice-over ad copy, and in the case of television, visual images as well. Some labels will pre-produce spots with an eye towards the need to make a quick advertising decision.

In studying lifecycles of releases, most labels recognize the urgency of front-loading the advertising, since the bulk of sales occur early in the release of a record. Often, labels work diligently to alert consumers of a new release in order to increase the sales response at street date. But on occasion, a record may have a 2nd or 3rd single that is the reactionary song, giving the label an incentive to fuel the sales fire by generating an advertising buy.

Partners

Some advertising buys have a dual-purpose. Clearly, the primary objective is to alert consumers as to the availability of a release. But in many cases, the advertising purchase enables the label to leverage additional exposure by building in a promotional partner. When purchasing ad time, a record label may consider approaching a retailer as the destination store tag, such as "Available now at Tower Records." This activity would include logo placement of the retailer on both print and television advertisements. In turn, the retailer may purchase more records of the specified artist as well as prominently placing the release in stores, anticipating an increase in sales.

Artist relations

In an ideal situation, labels and their artists should look at their business relationship as a partnership. The artist delivers the product that the record label is to market and sell. An artist should expect from its label a level of commitment and promotional activity to give the release the best possible shot at success. Many efforts are put forth by record labels to ensure success, and one of these activities is advertising. An artist, with his or her manager, should review the promotional efforts being created and executed by the label, just as the label reviews the music of the artist.

Table 10.3 Sample TV advertising buy								
MARKET DAYPART	**PURCHASED**			**# of SPOTS**	**WM: 25-54**		**HOMES**	
	PNTS	**DOLLARS**	**CPP**		**IMPS**	**CPM**	**IMPS**	**CPM**
ATLANTA PRI	7.5	6000	800	1	75	80	115	52.17

	Market	Atlanta
PRI	Daypart	Primetime
PNTS	GRP	7.5
	Cost	$6,000.00
CPP	Cost/point	$800.00
	# of spots	1
IMPS	# of impressions of target Market—women 25-54	75,000
CPM	Cost/thousand on target	$80.00
IMPS	# of homes	115,000
CPM	Cost/thousand of homes	$52.17

This television advertising buy was bolstered by additional ad exposure at radio. These time buys were during the week leading up to the CMA Awards Show.

Table 10.4 Radio buy								
MARKET DAYPART	**PURCHASED**			**# of SPOTS**	**WM: 25-54**		**HOMES**	
	PNTS	**DOLLARS**	**CPP**		**IMPS**	**CPM**	**IMPS**	**CPM**
ATLANTA AMD	40	8000	200	27	440	18.18		
PMD	35	7000	200	23	375	18.66		
TOTAL	75	15000	200	50	815	18.4		

	Market	Atlanta
AMD	Daypart	AM Drive
PMD	Daypart:	PM Drive
PNTS	GRP	75
	Cost	$15,000.00
CPP	Cost/point	$200.00
	# of spots	50
IMPS	# of impressions of target Market—women 25-54	815,000
CPM	Cost/thousand	$18.40

The combined advertising gross rating point (GRP) for the Atlanta market was 82.5. Although $15,000 was spent at radio, the CPM was dramatically lower than at television. Additionally, many more women aged 25–54 were reached because of the focused advertising that radio and its specific targeted demographics can deliver. Generally, radio advertising does not include gross impressions of homes.

With any budget, the greater the GRP, the better the advertising "bang for the buck." An actual advertising campaign would be comprehensive in attempting to reach the target market while maintaining a budget. To do so, buys at many outlets would be secured. The optimal equation to increase the GRP includes varying the combinations of media, dayparts, and number of spots. Depending on the agenda as well as the budget, all advertising should help increase visibility and sales of a specific artist or project.

Glossary

Cooperative advertising (co-op) – An advertisement where the cost is shared by the manufacturer of the product and the local retail outlet.

Cost per thousand (CPM) – A dollar comparison that shows the relative cost of various media or vehicles; the figure indicates the dollar cost of advertising exposure to a thousand households or individuals.

Dayparts – Specific segments of the broadcast day; for example, midday, morning drive time, afternoon drive time, late night.

Frequency – the number of times the target audience will be exposed to a message.

Gross impressions – The total number of advertising impressions made during a schedule of commercials. GIs are calculated by multiplying the average persons reached in a specific time period by the number of spots in that period of time.

Gross rating point (GRP) – In broadcasting/cable, it means the size of the audience during two or more dayparts. GRPs are determined by multiplying the specific rating by the number of spots in that time period.

Media fragmentation – The division of mass media into niche vehicles through specialization of content and segmentation of audiences.

Media planning – Determining the proper use of advertising media to fulfill the marketing and promotional objectives for a specific product or advertiser.

Medium – a class of communication carriers such as television, newspapers, magazines, outdoor, and so on.

Rating – In TV, the percentage of households in a market that are viewing a station divided by the total number of households with TV in that market. In radio, the total number of people who are listening to a station divided by the total number of people in the market.

Rating point (a rating of 1%) – 1% of the potential audience; the sum of the ratings of multiple advertising insertions; for example, two advertisements with a rating of 10% each will total 20 rating points.

Reach – The total audience that a medium actually reaches; the size of the audience with which a vehicle communicates; the total number of people in an advertising media audience; the total percentage of the target group that is actually covered by an advertising campaign.

Trade advertising – Advertising aimed specifically for retailers and media gatekeepers.

Trade publication – A specialized publication for a specific profession, trade, or industry; another term for some business publications.

Vehicle – A particular carrier within the group, such as *Rolling Stone* magazine or MTV network.

Bibliography

Bovée, C. L., and Arens, W. F. (1986). *Contemporary Advertising (2nd edition),* Homewood: Irwin.

Dunn, S. et al. (1990). *Advertising: Its Role in Modern Marketing (7th edition),* Orlando: The Dreyden Press.

Sissors, J., and Bumba, L. (1994). *Advertising Media Planning (4th edition),* Lincolnwood: NTC Business Books.

11 Distribution

Amy Macy

Music distribution

Music distributors are a vital conduit in getting physical music product from record labels' creative hands into the brick-and-mortar retail environment. Recognize that as the marketplace shifts to a digital environment, distribution companies are re-evaluating their value in the food chain and continually developing sales models that reflect direct sales opportunities to music consumers. Whatever the sales channel, distribution companies think of themselves as extensions of the record labels that they represent.

Prior to the current business model, most individual records labels hired their own sales and distribution teams, with sales representatives calling on individual stores to sell music. It took many reps to cover retail, as Figure 11.1 shows. This model shows nine points of contact, where each label meets with each retailer. As retailers became chains, and as economics of the business evolved, record labels combined sales and distribution forces to take advantage of economies of scale, which eventually evolved into the current business model.

Figure 11.2 includes the distribution function, and shows the points of contact reduced to six. In today's business model, record label sales executives communicate with distribution as their primary conduit to the marketplace, but labels also have ongoing relationships with retailers. Depending on the importance of the retailer, the label rep will often visit the retailer with their distribution partner so that significant releases and marketing plans can be communicated directly

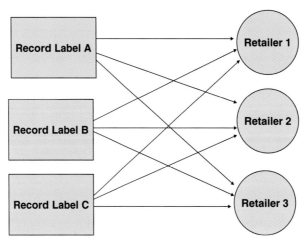

▲Figure 11.1 Direct contact concept

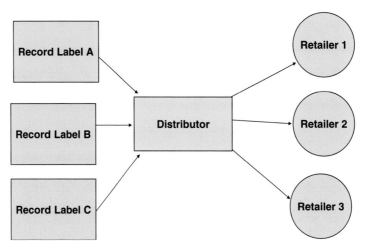

▲Figure 11.2 Distribution-centered concept

from the label to the retailer. And as deals are struck, both orders and marketing plans can then be implemented by the distributor.

Vertical integration

Three out of the Big 4 conglomerates (major labels) share profit centers that are vertically integrated, creating efficiencies in producing product for the marketplace. To take full advantage of being vertically integrated, labels looking for

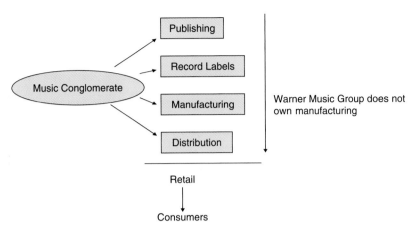

▲ *Figure 11.3 Vertical integration of conglomerates*

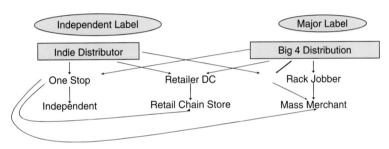

▲ *Figure 11.4 The distribution pathway*

songs would tap their "owned" publishing company. (Each of the major labels has a publishing company. If they only recorded songs that were published by their sister company, more of the money would stay in-house.) Once recorded, the records would be manufactured at the "owned" plant. In-turn, the pressed CDs would then be sold and distributed into retail—with the money being "paid" for each of the functions staying within the family of the music conglomerate.

In addition to the Big 4 companies, there are many independent music distributors that are contracted by independent labels to do the same job. Ideally, the distribution function is not only to place music into retailers, but also to assist in the sell-through of the product throughout its lifecycle.

Music supply to retailers

Once in the distributor's hands, music is then marketed and sold into retail. Varying retailers acquire their music from different sources. Most mass merchants are serviced by *rack jobbers*, who maintain the store's music department including inventory management, as well as marketing of music to consumers. Retail chain stores are usually their own buying entities, with company-managed purchasing offices and distribution centers. Many independent music stores are not large enough to open an account directly with the many distributors, but instead work from a *one-stop's* inventory as if it were their own. (One-stops are wholesalers who carry releases by a variety of labels for smaller retailers who, for one reason or another, do not deal directly with the major distributors.) Know that retail chain stores and mass merchants will on occasion use one-stops to do "fill-in" business, which is when a store runs out of a specific title and the one-stop supplies that inventory. (Retail is discussed in detail in another chapter.)

Role of distribution

Most distribution companies have three primary roles: The sale of the music, the physical distribution of the music, and the marketing of the music. Reflecting the marketplace, most distribution companies have similar business structures to that of their customers—the retailers. Often times, the national staff is separated into two divisions: Sales and marketing. The sales division is responsible for assisting labels in setting sales goals, determining and setting deal information, and soliciting and taking orders of the product from retail. Additionally, the sales administration department should provide and analyze sales data and trends, and readily share this information with the labels that they distribute.

The marketing division assists labels in the implementation of artists' marketing plans along with adding synergistic components that will enhance sales. For instance, the marketing plan for a holiday release may include a contest at the store level. Distribution marketing personnel would be charged with implementing this sort of activity. But the distribution company may be selling holiday releases from other labels that they represent. The distribution company may create a holiday product display that would feature all the records that fit the theme, adding to the exposure of the individual title.

The physical warehousing of a music product is a huge job. The major conglomerates have very sophisticated inventory management systems where music and its related products are stored. Once a retailer has placed an order, it is the distributor's job to pick, pack, and ship this product to its designated location. These sophisticated systems are automated so that manual picking of product is reduced, and that accuracy of the order placed is enhanced. Shipping is usually managed through third-party transportation companies.

As retailers manage their inventories, they can return music product for a credit. This process is tedious, not only making sure that the retailer receives accurate credit for product returned, but the music itself has to be retrofitted by removing stickers and price tags of the retailer, re-shrink wrapping, and then returning into inventory.

Conglomerate distribution company structure as it relates to national retail accounts

Although there are many variations and nuances to these structures as determined by each distribution company, the basic communication chart applies. At each level, the companies are communicating with each other. At the national level, very complex business transactions are being discussed, including terms of business as well as national sales and marketing strategies that would affect both entities company-wide. As mandated by the national staff, the

▲ *Figure 11.5 Conglomerate distribution company structure*

regional/branch level is to formulate marketing strategies, either as extensions of label/artist plans, or that of a distribution focus. At the local level, implementation of all these plans is the spotlight. But by design, these business structures are in place to create the best possible communication at every level, with an eye on maximizing sales.

National structure

To optimize communication along with service, distributors need to be close to retailers. Many of the major conglomerates have structured their companies nationally to accommodate the service element of their business. Most distributors have regional territories of management containing core offices and distribution centers. Each region contains satellite offices, getting one step closer to actual retail stores. As reflected in the structure chart in Figure 11.6, field personnel are on the front line, reporting to satellite offices, who then report to the regional core offices.

This map shows basic regions for a major distributor with satellite offices and distribution centers (DC). Core regional offices are located in Los Angeles,

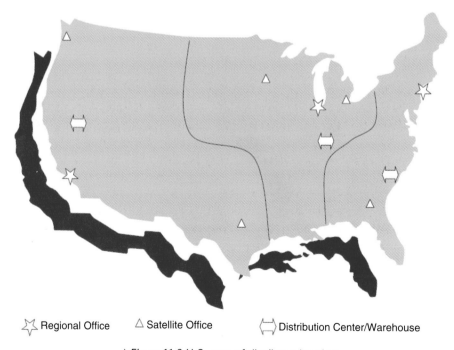

☆ Regional Office   △ Satellite Office   ⟨⟩ Distribution Center/Warehouse

▲ Figure 11.6 U.S. map of distributor locations

Chicago, and New York. Regions are naturally eastern, central, and western, with regional offices including Atlanta, Detroit, Minneapolis, Dallas, and Seattle. The distribution centers are centrally located within each region. In this example, the DCs are located in Sparks, NV, Indianapolis, IN, and Duncan, SC. No destination is more than a two-day drive from the DC, making product delivery timely.

Although the core regional offices are centrally located, these offices are primarily marketing teams, executing plans derived by both the labels that they represent and the distribution company. To be clear, over 80% of the music business is purchased and sold through ten retailers. The locations of these buying offices are key sales sites, and designated distribution personnel are placed near these retailers so that daily, personal interaction can occur. These locations and retailers are shown in Table 11.1.

Table 11.1 Retail-chain home office location	
Bentonville, AR	Wal-Mart
Minneapolis	Best Buy, Target, and Musicland
Detroit	Kmart
Albany	Transworld

Timeline

The communication regarding a new release begins months prior to the street date. Although there are varying deadlines within each distribution company, the ideal timeline is pivotal on the actual street date of a specific release. For most releases, street dates occur on Tuesdays. Working backwards in time, to have product on the shelves by a specific Tuesday, product has to ship to retailer's distribution centers approximately one week prior to street date. To process the orders generated by retail, distributors need the orders one week prior to shipping. The sales process of specific titles occurs during a period called *solicitation*. All titles streeting on a particular date are placed in a solicitation book, where details of the release are described. The solicitation page, also known as *Sales Book Copy*, usually includes the following information:

- Artist/Title
- Street Date
- File under category—where to place record in the store
- Information/History regarding the artist and release

- Marketing elements:

  - Single(s) and radio promotion plan
  - Video(s) and video promotion plan
  - Internet Marketing
  - Publicity Activities
  - Consumer Advertising
  - Tour and Promotional Dates

- Available POP
- Bar Code

This information is also available online on the business-to-business (B2B) sites established for the retail buyers.

Table 11.2 shows the actual timeline for labels to be included in the solicitation process.

Table 11.2 Sales timetable

Prior to SD	Activity	Example dates
8 weeks	Sales book copy due to distributor	September 27
6 weeks	Solicitation book mailed to retail buyers	October 11
4 weeks	Solicitation	October 25–November 5
2 weeks plus	Orders due	November 5
1 week	Orders shipped wholesalers/retail chain	November 16 orders received
5 days	Orders shipped to one stops	November 19 orders received
Street date		November 23

One sheet – the solicitation page

On the website, *MusicDish Industry e-journal*, Christopher Knab of ForeFront Media and Music describes a one sheet as:

"A Distributor One Sheet is a marketing document created by a record label to summarize, in marketing terms, the credentials of an artist or band. The One Sheet also summarizes the promotion and marketing plans and sales tactics that the label has developed to sell the record. It includes interesting facts about an act's fan base and target audience. The label uses it to help convince a distributor to carry and promote a new release" (Knab, C., 2001).

The one sheet typically includes the album logo and artwork, a description of the market, street date, contact info, track listings, accomplishments, and marketing points. The one sheet is designed to pitch to buyers at retail and distribution. The product bar code is also included to assist in the buy-plugging the actual release into the inventory code system.

▲ Figure 11.7 One sheet for Buck 65 (Source: V2 Records)

Consolidation and market share

As consolidation continues, the Big 4 just keep getting bigger. Looking at market share data generated by over-the-counter sales of SoundScan, one can view how large these entities have become over time. Note that the piece of the Big 4 pie fluctuates, with independents representing the remaining market share.

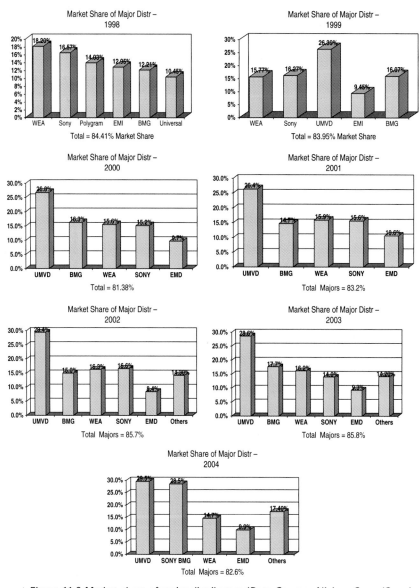

▲ Figure 11.8 Market share of major distributors (Data Source: Nielsen SoundScan)

Digital distribution

Since over 90% of music is still purchased through brick-and-mortar stores, the primary function for music distributors continues to be that of physical distribution of CDs into physical stores. But to reflect the burgeoning downloading trends, most of the major distributors have sites where consumers can buy tracks and albums directly from the conglomerate. These sites usually promote and sell music from their represented labels only, which makes it difficult for consumers to experience one-stop shopping.

In-turn, within most conglomerate families resides a department that licenses music to third-parties, which are legal downloading sites such as iTunes® and Napster™. These licensing departments are critical to the future of distribution, positioning them as gatekeepers for the growing downloading environment. At these third-party sites, consumers can purchase from the many sources of music, clearly beyond that of one conglomerate. Currently, the wholesale price of a digital track is approximately $0.68 a license.

Revenues from a 99-cent download

As consumership of music continues to evolve, so does that of distribution of music. Current trends reveal that younger consumers look to purchase tracks, making the classic album less attractive to this buyer. The digital download arena is where these consumers are satiating their needs. And yet, 90% of music is still purchased in the classic brick-and-mortar environment, making the need for physical distribution a continuing viable business entity.

Table 11.3 Revenues from a 99-cent download

Participant	Revenue
Label	47 cents
Distribution affiliate	10 cents
Artist	7 cents
Producer	3 cents
Music publisher	8 cents
Service provider	17 cents
Credit card fees	5 cents
Bandwidth costs	2 cents
	99 cents

Source: Brian Garrity.

Distribution value

Several distribution companies are exploring ways to add value to the conglomerate equation. Creating distribution-specific marketing campaigns with nonentertainment product lines helps validate distribution's existence, while hopefully enhancing the bottom line. Marketing efforts such as on-pack CDs with cereal, greeting card promotions, and ringtone services add to the branding of the participating artists, while increasing overall revenue through licensing and/or sales of primary items.

The ultimate value for today's distribution companies is that of consolidator. Distributors can consolidate labels to create leverage points within retail. To gain positioning in the retail environment, one must have marketing muscle, and by using the collective power of the label's talent, the entire company can raise its market share by coattailing on the larger releases in the family.

Glossary

Big 4 – These are the four music conglomerates that maintain a collective 85% market share of record sales. They are Universal, Sony/BMG, Warner, and EMI.

Economies of scale – Producing in large volume often generates economies of scale—the per-unit cost of something goes down with volume. Fixed-costs are spread over more units lowering the average cost per unit and offering a competitive price and margin advantage.

Fill-in – One-stop music distributors supply product to mass merchants and retailers who have run out of a specific title by "filling in" the hole of inventory for that release.

Mass merchants – Very large retail chains that sell a variety of goods and depend on volume sales. Wal-Mart and Target are examples.

One stop – A wholesaler who carries releases by a variety of labels for sale to small retail stores.

Rack jobbers – Companies that supply records, compact discs and other items to department stores, discount chains and other outlets and service (rack) their record departments with the right music mix.

Sales Book – Distribution companies compile all their releases for a specific street date into a "sales book," which contains one-sheets for each title that outlines the marketing efforts.

Sell-through – Once a title has been released, labels and distributors want to minimize returns and "sell-through" as much inventory as possible.

Solicitation period – The sales process of specific titles occurs during a period called *solicitation*. All titles streeting on a particular date are placed in a solicitation book, where details of the release are described.

Vertical integration – The expansion of a business by acquiring or developing businesses engaged in earlier or later stages of marketing a product.

Bibliography

Knab, C. (2001). http://www.musicdish.com/mag/index.php3?id=3357. The distributor one sheet. (3-25-2001).

Weatherson, Jim, President of Ventura Distribution and former Executive Vice President of Universal Music Distribution, from personal interview, July 2004.

Garrity, Brian. "Seeking Profits at 99¢", *Billboard*, July 12, 2003.

12 The Music Retail Environment

Amy Macy

Marketing and the music retail environment

The four P's of product, promotion, price, and placement converge within the music retail environment—this being the last stop prior to music being purchased. This environment is designed to aid consumers in making their purchasing decision. This decision can be influenced in a number of ways, depending on the consumer. For example, ''Does the store have hard-to-find releases?''; ''Do they have the lowest prices?''; ''Is it easy for consumers to find what they're looking for?''; ''Does the store have good customer service and knowledgeable employees?'' These questions should be answered, in one way or another, within the confines of the retail environment through the use of the four P's.

NARM

To assist music retailers in determining business strategies, companies look for current sales trends as well as educational and support networks. The National Association of Recording Merchandisers (NARM) is an organization conceived to be a central communicator of core business issues for the music retailing industry.

Founded in 1958, NARM is an industry trade group that serves the music retailing community in the areas of networking, advocacy, information, education and promotion. Members include brick-and-mortar, online and "click-and-mortar" retailers, wholesalers, distributors, content suppliers (primarily music labels, but also video and video game suppliers), suppliers of related products and services, artist managers, consultants, marketers, and educators in the music business field.

Retail members, who operate 7,000 storefronts that account for almost 85% of the music sold in the $12 billion U.S. music market, represent the big national and regional chains—from Anderson (Wal-Mart wholesaler), Best Buy, Borders, Circuit City, Handleman (Wal-Mart and Kmart wholesaler), Hastings Entertainment, Musicland, Newbury Comics, Target, Tower, TransWorld (FYE), Virgin—to the small, but influential "tastemaker" independent specialty stores. Additional members also include online retailers such as Amazon.com, eBay and iTunes.

Distributor and supplier members account for almost 90% of the music produced for the U.S. marketplace. This includes the four major music companies: Sony BMG, EMI, Universal Music and Video, and WEA, and many of the labels they represent. A few of the label members are Atlantic, Blue Note, Capitol, Capitol-Nashville, Columbia, Curb Records, DreamWorks Records, Elektra Entertainment, Epic Records, Geffen Records, Hollywood Records, Interscope Records, The Island Def Jam Music Group, J Records, Jive Records, Maverick, Mercury/MCA Nashville/Lost Highway, RCA Label Group, RCA Music Group, RCA Victor Group, Rhino, Sony Classical, Verve Music Group, Walt Disney Records, Warner Bros., Welk Music Group and Wind-up Records. This list extends even further to include influential independent labels and distributors. For more information regarding NARM and activities, visit www.narm.com.

Retail considerations

The four P's are applicable as an outline for retailer considerations in doing business:

#1 P—Product

How does a retailer learn about new releases? Distribution companies are basically extensions of the labels that they represent. To sell music well, a distribution company needs to be armed with key selling points. This critical information is usually outlined in the marketing plan that is created at the record label level. Record labels spend much time "educating" their distributors about their new releases so that they can sell and distribute their records effectively.

Distribution companies set up meetings with their *accounts*, meaning the retailers. At the retailer's office, the distribution company shares with the *buyer*—that is the person in charge of purchasing product for the retail company—the new releases for a specific release date, as well as the marketing strategies and events that will enhance consumer awareness and create sales. Depending on the size of the retailer, buyers are usually categorized by genre or product type, such as the R&B/hip-hop buyers, or the soundtrack buyer.

Purchasing music for the store

When making a purchase of product, the buyer will take into consideration several key marketing elements: radio airplay, media exposure, touring, cross-merchandising events, and most critical, previous sales history of an artist within the retailers' environment; or if a new artist, current trends within the genre and/or other similar artists. On occasion, the record label representative will accompany the distribution company sales rep with the hopes of enhancing the knowledge of the buyer of the new release. The ultimate goal is to increase the purchasing decision, while creating marketing events inside the retailer's environment.

Most record labels, along with their distributors, have agreed on a forecast for a specific release. This forecast, or number of records predicted to sell, is based on similar components that retail buyers consider when purchasing product. Many labels use the following benchmarks when determining forecast:

Initial orders or I.O.—This number is the initial shipment of music that will be on retailers' shelves or in their inventory at release date.

90-day forecast—Most releases sell the majority of their records within the first 90 days.

Lifetime—Depending on the release, some companies look to this number as when the fiscal year of the release ends, and the release will then rollover into a catalog title. But on occasion, a hit release will predicate that forecasting for that title continues, since sales are still brisk.

Inventory management

Larger music retailers have very sophisticated purchasing programs that profile their stores' sales strengths. Using the forecast, as well as an overall percentage of business specific to the label or genre, the retailer will determine how many units it believes it can sell. This decision is based on historical data of the artist and/or trends of the genre along with the other marketing components.

Keeping track of each release, along with the other products being sold within a store is called *inventory management*. Using point-of-sale (POS) data, the store's computer notes when a unit is sold using the Universal Product Code (**UPC**) **bar** code. Depending on the inventory management system, a store may have ten units on-hand, which is considered the ideal *maximum* number the store should carry. The ideal *minimum* number may be four units. If the store sells seven units, and drops below the ideal inventory number of four as set in the computer, the store's inventory management system will automatically generate a re-order for that title, up to the maximum number. This min/max inventory management system may then download the re-order through an electronic data interchange (EDI) to its supplier, and the product is then shipped to the retailer within a few days. To avoid waiting for the product to arrive, a retailer may opt to drop-ship product directly from the distribution company, avoiding the delay of processing at either the headquarters' distribution center (DC) or the one-stop supplier.

Turns and returns

Know that a store's success is based on the number of units it sells within a fiscal year. Clearly, the size of the store dictates how much product or inventory it can hold. On average, a 2,500 square-foot store may hold 20,000 units. According to NARM 2002 data, the average annual inventory *turn* for music is 4.4 times. As an industry standard, this store would sell: 20,000 units × 4.4 turns = 88,000 units in a fiscal year. This does not mean that every title sells 4.4 times, but rather the store averages 4.4 sales per year for every unit it is holding.

To manage the real estate within the store, the best-selling product should receive the most space. To keep a store performing well, the inventory management system should notice when certain titles are not selling. Music retailers have an advantage over traditional retailers in that if a product is not selling, they can send it back for a refund, called a *return*. The refund is usually in the form of a credit, and the amount credited is based on when the product is returned along with other considerations such as if a discount had been received. According to NARM 2002 data, the industry return average is 19.5%

for music. To put this statistic another way, for every 100 records in the marketplace, nearly 20 units are returned to the distributor.

Loss prevention

According to the National Security Retail Survey of 2002, the average retailer loses about 1.7% of product to shrinkage—or shoplifting. Other research notes that the music retailer can lose between 4–4.5% in product. But in any outlet, nearly 50% of the missing product is due to employee theft.

What do music retailers do to protect against shrinkage? Regarding employees, retailers use many screening techniques such as past employment records, criminal checks, multiple interviews and personal reference checks. As for consumers, there are various techniques to help protect against shoplifting. Loss prevention staff is also employed by many retailers.

Since thieves need privacy, training employees to serve the customer has a dual-purpose: To help secure a sale, and to let the would-be thief know that the "store" is watching. Reducing clutter and sighting blind spots help keep visibility high. The use of uniformed and plain-clothed detectives is common in larger format stores, with good results; but the use of technology has been a big aid in dissuading the shoplifter. *Source tagging* is the use of embedded security devices placed within the CD packaging. With the use of electronic article surveillance (EAS), stores place security panels at the exit. As a product is sold, the source-tag is deactivated. If a product that has not been purchased moves through these security panels, a warning siren occurs, notifying store personnel of the potential shoplifting event.

#2 P—Promotion

Survey after survey indicates that music consumers find out about new music primarily via the radio. So it would seem that music retailers would use radio as their primary advertising vehicle to promote their stores and product. But they do not. Cost considerations including reach and frequency of ads have caused music retailers to choose print advertising as their primary promotional activity. Consumers have been trained to look in Sunday circulars for sales and featured products on most any item. Major music retailers now use this advertising source as a way to announce new releases that are to street on the following Tuesday, along with other featured titles and sale product.

Once inside the store, promotional efforts to highlight different releases should aid consumers in purchasing decisions. These marketing devices often set the tone and culture of the store's environment. Whether it is listening stations near a coffee bar outlet, or oodles of posters hanging hodge-podge on the wall, a brief encounter within the store's confines will quickly identify the music that the store probably sells.

Featured titles within many retail environments are often dictated from the central buying office. As mentioned earlier, labels want and often do create marketing events that feature a specific title. This is coordinated via the retailer through an advertising vehicle called *cooperative advertising*. *Co-op advertising*, as it is known, is usually the exchange of money from the label to the retailer, so that a particular release will be featured. Following are examples of co-op advertising:

Pricing and positioning (P&P)—P&P is when a title is sale-priced and placed in a prominent area within the store.

End caps—Usually themed, this area is designated at the end of a row and features titles of a similar genre or idea.

Listening stations—Depending on the store, some releases are placed in an automatic digital feedback system where consumers can listen to almost any title within the store. Other listening stations may be less sophisticated, and may be as simple as using a free-standing CD player in a designated area. But all play back devices are giving consumers a chance to "test drive" the music before they buy it.

Point-of-purchase (POP) materials—Although many stores will say that they can use POP, including posters, flats, stand-ups, and so on, some retailers have advertising programs where labels can be guaranteed the use of such materials for a specific release.

Print advertising—A primary advertising vehicle, a label can secure a "mini" spot in a retailer's ad (a small picture of the CD cover art), which usually comes with sale pricing and featured positioning (P&P) in-store.

In-store event—Event marketing is a powerful tool in selling records. Creating an event where a hot artist is in-store and signing autographs of his or her newest release guarantees sales, while nurturing a strong relationship with the retailer.

The following grid is a sample forecasting and P&P planning tool used to predict I.O. and initial marketing campaign activities within a retailer's environment.

Rack jobbers such as Anderson and Handleman purchase nearly 30% of overall music. They supply music to mass merchants such as Wal-Mart and Kmart.

Alliance is the largest one-stop, whose primary business is supplying music to independent music stores.

Figures are based on a major distributor's overall business with these actual accounts. Note that ten accounts purchase nearly 80% of all records, which makes them very important in the marketing process. For Target to *not* purchase a title would mean that over 10% of sales would be lost, based on these numbers. This does not mean that a record would sell 10% less over the lifetime of the release, but it does mean that consumers will have to find that title in another store other than Target.

Artist Name
Title
Selection Number

Account	% of Business	Target	Account Advertising P&P	Cost
Anderson Merchandisers	14.6%	14,600		
Best Buy	14.0%	14,000		
Handleman Company	12.3%	12,300		
Target	11.8%	11,800		
Alliance Entertainment	5.3%	5,300		
Transworld	5.0%	5,000		
Musicland	4.7%	4,700		
Circuit City	4.5%	4,500		
AAFES	3.6%	3,600		
Borders	3.3%	3,300		
TOP 10 Accounts	79.1%	79,100		
All Others	20.9%	20,900		
Total	100.0%	100,000		

▲ *Figure 12.1 Sales forecasting grid*

Program	Program Includes	Mall Stores # of stores	Rural Stores # of stores	Urban Stores # of stores	Total Stores	Cost Per Title	Number of Titles
Best Seller	P&P on the Best Seller in-store position	500	250	75	825	$ 22,500	20
New Release	P&P on the New Release in-store position	500	250	75	825	$ 17,000	16
Music Video DVD	P&P on the Music DVD in-store position	500	250	75	825	$ 2,700	8
Fast Forward	Positioning on the FF in-store position	500	250	75	825	$ 11,000	12
Catalog Promotion	P&P on multipleprime in-store positions Supporting In-store signage Supporting Promotional Levers	500	250	75	825	To be negotiated	TBD
Premium Artist Package	P&P in all concepts TMG-TVin all concepts Insert (1 cut) (6.5 MM circulation) Magazine Direct Promotion Magazine RCC (monthly e-newsletter)	500	250	75	825	$ 40,000 $ 1,500 $ 3,500 $ 5,500 $ 4,500	3
TOTAL PREMIUM PACKAGE PRICE						$ 55,000	
GENRE ENDCAPS							
POP1/RK1	P&P on endcap	500	250	75	825	$ 13,500	12
POP1	P&P on endcap		250	75	325	$ 3,500	2
RK1	P&P on endcap		250	75	325	$ 3,500	2
POP2	P&P on endcap	350	200	75	725	$ 10,000	8
POP3	P&P on endcap	250	75	75	400	$ 7,000	8
RB1/RAP1	P&P on endcap	500	250	75	825	$ 8,500	12
RB1	P&P on endcap		100	75	175	$ 1,600	2
RAP1	P&P on endcap		250	75	325	$ 3,500	2
URBAN 1	P&P on endcap	250	75	75	400	$ 5,100	8
URBAN 2	P&P on endcap	150	41	76	267	$ 3,300	8
OPT1 (open endcap)	P&P on endcap	350	200	75	725	$ 5,000	8
OPT2 (open endcap)	P&P on endcap	250	75	75	400	$ 3,000	8
CNJ1 (Classical New Age Jazz)	P&P on endcap	133		76	209	$ 1,800	8

▲ *Figure 12.2 Sample retail chain P&P costs and programs*

Sample Rack Jobber P&P Programs and Costs

SUPER FEATURE: Sunday Circular
$60,000 (ONE MONTH P&P)

FEATURE
$40,000 (ONE MONTH)

GENRE FEATURES (ONE MONTH Endcap)
URBAN
$5,000
COUNTRY
$5,000
LATIN
$1,000

POP - SHOWBOARDS (ONE MONTH)
$24,000
URBAN – SHOWBOARDS (ONE MONTH)
$12,000

COUNTRY - SHOWBOARDS (ONE MONTH)
$20,000
CCM - SHOWBOARDS (ONE MONTH)
$5,000
GOSPEL – SHOWBOARDS (ONE MONTH)
$2,500
VARIOUS ARTISTS - SHOWBOARDS (ONE MONTH)
$10,000
SOUNDTRACK – SHOWBOARDS (ONE MONTH)
$10,000

RISING STARS DEVELOPING ARTIST (ONE MONTH)
$9,000

POD- Position out of Department (ONE MONTH endcap))
$5,000
$7,500
URBAN DIRECT ROTO (ONE WEEK)
$1500 SINGLE CUT
$3000 DOUBLE CUT
ROTO- Sunday Circular (ONE WEEK)
$10,000 per cut

▲ *Figure 12.3 Sample Rack Jobber P&P program and costs*

#3 P—Pricing

Although record labels set the suggested retail list price (SRLP) for a release, this is *not* what retailers are required to sell the product for. Most often, the SRLP sets the *wholesale price* or the cost to the retailer. In negotiating the order, the retailer may ask for a discount off the wholesale price. The retailer may also ask for additional *dating*, meaning that the retailer is asking for an extension on the payment due date. Each distributor has parameters in which this transaction may occur.

Generally, music product comes in box lots of 30 units. A retailer will receive a better price on product if it purchases in box lots. For example, a retailer wants to purchase 1,200 units of a new release with a 10% discount and 30 days dating. See Table 12.1.

Table 12.1 Box lot discount					
SRLP	$18.98	SRLP	$18.98		
Wholesale	$12.04	Wholesale	$12.04		
×	1200 units	−10% disc.	$1.20		
	$14,448			The 10% discount saved the retailer $1,440.	
		Total/unit	$10.84		
		×	1200 units		
			$13,008		

Normally, the money due for this purchase would be received at the end of the following month that the record was released. However, with an extra 30 days dating, the due date is extended, giving the retailer a longer timeframe in which to sell all the product. Adding extra dating is often a tactic of record labels that want retailers to take a chance on a new artist that may be slower to develop in the marketplace.

The price of product reflects the store's marketing strategy. The major electronic superstores look to music product as the magnet to get customers through their doors, which is why music prices in these environments are often lower than anywhere else. Often, these stores will sell music for less than they purchased it, called *loss-leader* pricing. But these stores will also raise the price after a short period of time, usually within first two weeks after the street date.

LPs Non-Returnable

SRLP	Box	Loose
698	4.20	4.33
798	4.70	4.85
898	5.26	5.42
998	4.50	4.50
998A	5.84	6.02
1098	5.00	5.00
1098A	6.70	6.70
1198	7.00	7.22
1298	6.00	6.00
1298A	7.62	7.86
1398	8.26	8.52
1498	8.00	8.00
1598	8.00	8.00
1598A	9.48	9.77
1698	9.00	9.00
1698A	10.48	10.80
1898	11.10	11.44
1998	11.68	12.04
2098	12.11	12.48
2498	18.00	18.00

Video

	SRLP	Box	Loose
	598	3.59	3.59
	898	5.39	5.39
	998	5.99	5.99
	1098	6.59	6.59
	1298	7.79	7.79
	1498	8.99	8.99
	1598	9.59	9.59
	1798	10.79	10.79
(2)	1998	10.99	10.99
	2198	12.99	12.99
	2498	14.99	14.99
	2998	17.99	17.99

DVD

	SRLP	Box	Loose
	798	4.62	4.62
	998	5.78	5.78
	1198	6.95	6.95
	1498	9.45	9.45
	1798	10.43	10.43
(3)	1998	11.59	11.59
	2098	12.58	12.58
	2498	15.75	15.75
	2998	17.39	17.39

Compact Disc

SRLP	Box	Loose
398	2.32	2.40
498	3.00	3.09
598	3.60	3.70
698	4.40	4.54
798	4.80	4.95
898	5.39	5.56
998	6.60	6.80
999	6.00	6.18
1098	6.92	7.13
1179	5.99	5.99
1198	7.88	8.12
1298	8.40	8.66
1398	9.14	9.42
1498	9.77	10.07
1598	10.37	10.69
1698	10.79	11.12
1798	11.41	11.76
1898	12.04	12.40
1998	12.99	13.40
2098	13.48	13.90
2198	14.23	14.67
2298	14.88	15.34
2398	15.52	15.99
2498	16.16	16.66
2998	19.39	19.99
3198	20.67	21.31
3498	22.47	23.16
3598	22.97	23.68
3998	25.83	26.63
4498	28.97	29.87
4798	30.71	31.66
4998	31.97	32.96
5498	35.49	36.55
5998	36.10	37.22
7998	52.72	52.72
12998	82.71	85.19

DVD Audio

SRLP	Box	Loose
1798	11.69	12.05
1998	12.98	13.38
2298	14.94	15.39
2498	16.23	16.72

Super Audio CD

SRLP	Box	Loose
2198	13.95	14.38

Cassette Singles

SRLP	Box	Loose
349	1.93	1.93

CD Single/CD-5

SRLP	Box	Loose
199	1.24	1.24
298	1.65	1.65
349	2.06	2.06
398	2.65	2.65
498	2.94	2.94
598	3.48	3.59
649	3.80	3.80
698	4.14	4.14
749	4.45	4.45

12" Vinyl/Maxi Cass Returnable

SRLP	Box	Loose
498	2.75	2.84
598	3.37	3.37
649	3.64	3.64
698	3.91	3.91
898	5.03	5.03
998	5.62	5.62
1198	6.00	6.00
1298	6.50	6.50

Cassette

SRLP	Box	Loose
398	2.35	2.42
598	3.48	3.59
779	3.99	3.99
798	4.70	4.85
898	5.25	5.41
998	5.84	6.02
1098	6.48	6.68
1198	7.00	7.22
1298	7.62	7.86
1598	9.48	9.77
1998	11.68	12.04

7" Singles - Non-Returnable

SRLP	Box	Loose
198	1.28	1.28
259	1.19	1.19
298	1.98	1.98
348	2.28	2.28
398	2.58	2.58

▲ *Figure 12.6 Sample distribution pricing schedule*

Actual pricing of product

Margin and markup

Margin and markup are both calculated using the wholesale purchase price of the product. Percent margin uses the selling price as the denominator, whereas percent markup uses the purchasing (wholesale) price as the denominator for calculating.

Margin percentage on product is determined with the following calculation:

$$\text{Formula:} \quad \text{Margin\%} = \frac{\text{Dollar Markup}}{\text{Retail}}$$

If an SRLP CD of $18.98 is purchased wholesale for $12.04 and the store wants to sell it for $13.98, the margin percentage is $13.98 − $12.04/$13.98 = 13.9%.

Although there are variable ways to calculate margin, most stores use this retail markup calculation since it takes into consideration differing price lines, product extensions and customer demands in retail value.

Markup uses a similar calculation, but divides the dollar markup by the wholesale cost.

$$\text{Formula:} \quad \text{Markup\%} = \frac{\text{Dollar Markup}}{\text{Wholesale}}$$

If an SRLP CD of $18.98 is purchased wholesale for $12.04 and the store wants to sell it for $13.98, the markup percentage is $13.98 − $12.04/$12.04 = 16.1%

There is always an arithmetical relationship between gross margin and markup.

A gross margin of 40% requires a markup of 66.67% calculated as 40 ÷ (100 − 40).

A gross margin of 60% requires a markup of 150% calculated as 60 ÷ (100 − 60).

To achieve a target gross margin of 13.9% on the previous example, based on the purchase cost, the calculations are as follows:

A gross margin of 13.9% requires a markup of 16.1% calculated as 13.9 ÷ (100 − 13.9) = 16.1%.

	$
Wholesale CD cost	12.04
Margin 13.9%	× .161 = 1.94
Total	13.98

When setting prices, retailers think about markup since they start with costs and work upwards. When thinking about profitability, retailers think about margin, since these are the funds leftover to cover expenses as well as account for profit. Importantly, negotiating for the best discount off the wholesale price improves both markup and margin.

Where the money goes

When the consumer plunks down their money at the cash register to purchase a CD, Figure 12.4 shows how that money is divided between all the invested parties.

$	18.98	SRLP	
$	12.04	Wholesale	
		Label	$ 5.00
		Distr	$ 1.80
		Design/Manu	$ 1.00
		Artist Roy	$ 1.50
		Mech Roy	$ 0.80
		Rec Cost	$ 1.00
		Mkt/Promotion	$ 0.90
		Retail Profit	$ 3.00

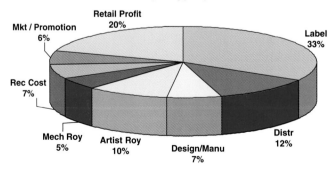

▲ *Figure 12.4 Where the money goes*

If the retailer receives a 10% discount off the wholesale price, then the label gets a reduction of 3% and the distributor a reduction of 2%, while the retailer enjoys a 25% *increase* in margin, from 20% of retail price to 25%, as shown in Figure 12.5.

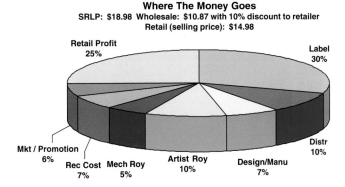

Where The Money Goes
SRLP: $18.98 Wholesale: $10.87 with 10% discount to retailer
Retail (selling price): $14.98

▲ *Figure 12.5 Where the money goes with 10% discount to retailer*

Free goods

Many labels use free goods to fund co-op advertising. If a retailer has a co-op advertising program that costs $120, instead of exchanging dollars, the label may fund the advertising with product. Wholesale for a specific CD may average $12.00, so the label would send ten CDs to the retailer—being the same dollar value of the advertising program.

This kind of exchange is a win-win scenario for both players. The retailer can then sell those ten CDs for $14.00/unit, garnering $140 in revenues, which is more than the actual advertising program cost. And, the label value of the CDs on the accounting books is not that of $12.00/unit, but rather $6.00/unit—meaning that in cost to the label, the co-op advertising program was purchased for $60 total.

#4 P—Place

Although promotional activities within the store include the *placement* of product in designated areas, the choosing of a location or place of the actual store is paramount to the success of the business. Factors to consider are:

Location and visibility—Having built-in traffic helps a store attract more customers. Being in a mall or a high traffic area is ideal for the store selling mass-appeal product. Additionally, being visible and easy to find and access can help a store succeed.

Competition—Depending on the store's marketing strategy, having competition nearby can detract from a store's success. Knowledge of area businesses is key to success.

Rent—Besides inventory and employee costs, rent is a big expense that can impact profitability. Again, depending on store strategy, some retailers build their own storefront, while others rent from existing outlets (Hall, C., 2000).

Use clauses—A store must be sensitive to non-compete use clauses that can restrict business. Some malls restrict the number of record retail outlets and/or electronic store outlets to help ensure success of existing tenants.

Image—Store image reflects corporate culture along with product for sale.

Floor plans—The store's culture and marketing strategy is best observed in the design and layout of the floor plan. Market research observes that the longer a person is in a store, the higher the probability that the person will purchase an item, based on a marketing concept called *time spent shopping (TSS)*. On that premise, some stores draw in their customers, creating interesting and interactive displays further back in the store. Sale items may be placed in the back, or in some cases, there may be multiple floors, all with an eye to keep the shopper in the store longer. Other stores may entice browsers with new releases right inside the front door. Others include the addition of coffee bars to boost TSS.

Genre placement, along with related items, helps define a store. Display placement, including interactive kiosks, listening stations, featured titles, and top-selling charts, helps consumers make purchasing decisions. Larger stores with broader product offerings couple music product with related items. The placement of these elements and traffic flow design can aid consumers and increase sales.

Store target market

A music store's target market or consumer generally dictates what kind of retailer it will be. To attract consumers interested in independent music, or to attract folks who are always looking for a bargain, determines the parameters

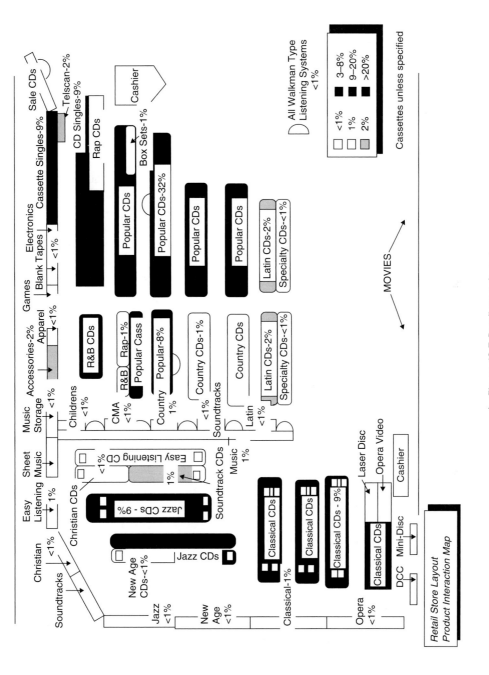

▲ *Figure 12.7 Retail store layout*

in which a store operates. Music retailers have traditionally been segmented into the following profiles:

Independent music retailers cater to a consumer looking for a specific genre or lifestyle of music. Generally, these types of stores get their music from one-stops. Independent stores are locally or regionally owned and operated, with one or just a few stores under one ownership.

The **Mom & Pop** retailer is usually a one-store operation that is owned and operated by the same person. This owner is involved with every aspect of running the business and tends to be very passionate about the particular style of music that the store sells. This passion can be interpreted as being an expert in the knowledge of the genre and can be a unique resource for the consumer looking for the obscure release. Mom & Pop storeowners tend to have a personal relationship with their customer base, knowing musical preferences and keeping the customers informed about upcoming releases and events.

Alternative music stores profile very similarly to Mom & Pop stores, but with the exception that they tend to be lifestyle-oriented. An electronica music retailer may have many hard-to-find releases along with hardware offerings such as turntable and mixing boards.

Chain stores tend to attract music purchasers who are looking for deep selection of releases along with assistance from employees who have strong product knowledge. These stores are often in malls and cater to a broad spectrum of purchasers. These stores have been studied and replicated so that entering any store with the same name in any location feels very similar. Often, they have the new major releases upfront with many related items for sale, such as blank media and entertainment magazines. Chain stores traditionally buy their music inventory directly from music distributors, with warehousing and price stickering occurring in a central location. Some examples of chain stores are FYE, Hastings, and Musicland/Sam Goody.

Electronic superstores do not make the bulk of their profits from the sale of music. But rather, use music as an attraction to bring consumers into their store environments. By using loss leader pricing strategies, these stores often sell new releases for less than they purchased it, but for a limited time. Meanwhile, they have created traffic to the store in order to make money from the sale of all the other items offered such as electronics, computers, televisions, and so forth. Electronic Superstores also buy their music inventory directly from music distributors, with warehousing and price stickering occurring in a central location. Some examples of electronic superstores are Best Buy and Circuit City.

Mass merchants use the sale of music as event marketing for their stores. Each week, a new release brings customers back to their aisles with the notion that they will purchase something else while there. There is little profit in the sale of music for the mass merchant, but the offering of music is looked at as a service to customers. Often, mass merchants use rack jobbers to supply and maintain music for their stores. It is the rack jobber who initially purchases the music for the mass merchant environment. Some examples of mass merchants are Wal-Mart and Kmart

Sources of music

The following figure shows the basic flow of music as it reaches the consumer level. Recognize that one stops' primary business is servicing independent records stores, but that they also do what is called *fill-in* business for all music retailers.

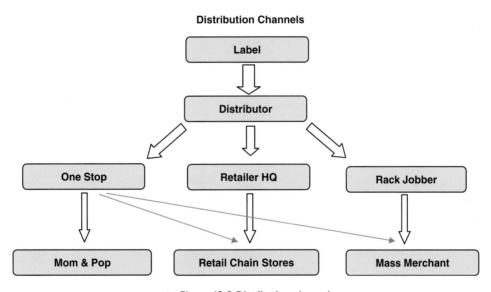

▲ *Figure 12.8 Distribution channels*

Internet marketing and sales

Until recently, it was the retailers who had the brand identities that were winning the Internet sales wars. Consumers went to their favorite retail store websites to browse and purchase music, meaning the actual CD that was to be delivered to the consumer's door. These well-known retailer sites left many start-up websites with unknown names with little traffic. As noted in the sales overview, downloading and file-swapping has become big "business," but not perceived as a potentially profitable business ... until now. Downloading activities have hurt not only the labels and their distributors, but the retailers as well. With the aggressive prosecution and education campaigns to alert downloaders as to their illegal practices, consumers are beginning to use legal downloading sites to purchase music. Sites such as iTunes, Napster, MusicMatch, Rhapsody, and yes, Wal-Mart are all experiencing a high volume of downloads, with more sites coming online soon. Well-known online sites such as Amazon and CDNow have also had an impact on retailing. For more on Internet sales and marketing, refer to Chapter 14 on Internet Marketing.

Glossary

Bricks and clicks – The term given to a retailer who has physical stores and an online retail presence.

Brick and mortar – The description given to physical store locations when compared to online shopping.

Box lot – Purchases made in increments of what comes in full, sealed boxes receive a lower price. (For CDs with normal packaging, usually 30).

Buyers – Agents of retail chains who decide what products to purchase from the suppliers.

Chain stores – A group of retail stores under one ownership and selling the same lines of merchandise. Because they purchase product in large quantities from centralized distribution centers, they can command big discounts from record manufacturers (compared to indie stores).

Computerized ordering process – An inventory management system that tracks the sale of product and automatically reorders when inventories fall below a preset level. Reordering is done through an electronic data interchange (EDI) connected to the supplier.

Co-op advertising – A co-operative advertising effort by two or more companies sharing in the costs and responsibilities. A common example is where a record label and a record retailer work together to run ads in local newspapers touting the availability of new releases at the retailer's locations.

Discount and dating – The manufacturer offers a discount on orders and allows for delayed payment. It is used as an incentive to increase orders.

Distribution – A company that distributes products to retailers. This can be an independent distributor handling products for indie labels or a major record company that distributes its own products and that of others through its branch system.

Drop ship – Shipping product quickly and directly to a retail store without going through the normal distribution system.

Electronic data interchange (EDI) – The inter-firm computer-to-computer transfer of information, as between or among retailers, wholesalers, and manufacturers. Used for automated re-ordering.

Electronic superstores – Large chain stores such as Circuit City and Best Buy that sell recorded music and videos, in addition to electronic hardware.

End cap – In retail merchandising, a display rack or shelf at the end of a store aisle; a prime store location for stocking product.

Floor designs – A store layout designed to facilitate store traffic to increase the amount of time spent shopping.

Free goods – Saleable goods offered to retailers at no cost as an incentive to purchase additional products.

Indie stores – Business entities of a single proprietorship or partnership servicing a smaller music consumer base of usually one or two stores. (Sometimes known as *mom & pop stores.*)

Inventory management – The process of acquiring and maintaining a proper assortment of merchandise while keeping ordering, shipping, handling, and other related costs in check.

Listening station – A device in retail stores allowing the customer to sample music for sale in the store. Usually the devices have headphones and may be free-standing or grouped together in a designated section of the store.

Loose – The pricing scheme for product sold individually or in increments smaller than a sealed box.

Loss leader pricing – The featuring of items priced below cost or at relatively low prices to attract customers to the seller's place of business.

Margin – The percentage of revenues leftover to cover expenses as well as account for profitability.

Markup – The percentage of increase from wholesale price to retail price.

Mass merchants – Large discount chain stores that sell a variety of products in all categories, for example, Wal-Mart and Target.

Min/max systems – A store may have ten units on-hand, which is considered the ideal *maximum* number the store should carry. The ideal *minimum* number may be four units. If the store sells seven units and drops below the ideal inventory number of four, as set in the computer, the store's inventory management system will automatically generate a reorder for that title, up to the maximum number.

National Association of Recording Merchandiser (NARM) – The organization of record retailers, wholesalers, distributors and labels.

One-Stop – A record wholesaler that stocks product from many different labels and distributors for resale to retailers, rack jobbers and juke box operators. The prime source of product for small mom & pop retailers.

Point-of-purchase (POP) – A marketing technique used to stimulate impulse sales in the store. POP materials are visually positioned to attract customer attention and may include displays, posters, bin cards, banners, window displays, and so forth.

Point-of-sale (POS) – Where the sale is entered into registers. Origination of information for tracking sales, and so on.

Price and positioning (P&P) – When a title is sale priced and placed in a prominent area within the store.

Pricing strategies – A key element in marketing, whereby, the price of a product is set in order to generate the most sales at optimum profits.

Rack jobber – A company that supplies records, cassettes, and compact discs to department stores, discount chains and other outlets and services (racks) their record departments with the right music mix.

Returns – Products that do not sell within a reasonable amount of time and are returned to the manufacturer for a refund or credit.

Sales forecast – An estimate of the dollar or unit sales for a specified future period under a proposed marketing plan or program.

Shrinkage – The loss of inventory through shoplifting and employee theft.

Source tagging – The process of using electronic security tags embedded in a product's packaging.

Theft protection – Systems in place to reduce shoplifting and employee theft in retail stores. These systems may include electronic surveillance.

Time spent shopping (TSS) – A measure of how long a customer spends in the store.

Turn – The rate that inventory is sold through, usually expressed in number of units sold per year/inventory capacity on the floor.

Universal Product Code (UPC) – The bar codes that are used in inventory management and are scanned when product is sold.

Wholesale – The price paid by the retailer to purchase goods.

Bibliography

Hall, Charles W. and Taylor, Frederick J. "Marketing in the Music Industry", Pearson Custom Publishing, 2000.
Gloor, Storm. From personal interview, March 2004.
National Association of Recording Merchandizers (NARM), www.narm.com.

13 Grassroots Marketing

Paul Allen

Grassroots marketing

Consumers are bombarded everyday with more commercial messages from more sources than ever before. The competition for our attention ranges from a subtle product placement in your favorite television show to an in-your-face ad from a car salesman on the radio. Ads pop-up at us while we are online, they precede theater movies, they're on seat backs at stadiums, and they're even in public facilities where graffiti once was posted. It is this competition for the attention of the consumer that has driven marketers to find alternative methods to connect products and services with billfolds and credit cards. This chapter presents some of those less traditional marketing tools and how they may be used to encourage the consumer to buy recorded music.

The power of word of mouth

Very few concepts in business are as universally accepted as being axiomatic as is the power of word of mouth (WOM) in marketing. Everett Rogers, in the book, *Diffusion of Innovations*, talks about the role of "opinion leaders" in the diffusion process (see Chapter 1) and how these trendsetters can be used in facilitating word-of-mouth communication messages about a new product or other innovation (Rogers, E. M., 1995). Word of mouth is more effective at closing a deal with a consumer than any pitch from any paid medium. Word of mouth is simply someone you know whose opinion you trust saying to you, "You gotta try this!" And the companies most effective with this kind of marketing are those who give their consumers reasons to talk about products to others, and then facilitate that communication.

From the company's standpoint, there are several basics to developing a plan to reach new consumers through word of mouth. They are:

- Educating people about your product or services
- Identifying people most likely to share their opinions
- Providing tools that make it easier to share information
- Studying how, where, and when those opinions are being shared
- Looking and listening for those who are detractors, and being prepared to respond (Word of Mouth 101, 2005)

Among the most popular ways to effectively use word-of-mouth marketing and promotion in the recording industry is through the artist fan club. The club is a network of people who have a passion for the music of the artist and who are willing to be actively involved in promoting the career of the artist. Members of the club are regular chat room visitors, bloggers, video bloggers, message board respondents, and hosts of discussion groups.

This discussion of word-of-mouth marketing is not intended to be a short course on fan club development, but the elements of it as a catalyst for "spreading the word" are among the best central strategies a label can use to draw from the strength of the idea. Word of mouth is often compared to the work of the evangelist, and fan club members easily fit the definition. Energizing them and arming them with available tools has the potential to spread positive information about the artist in geometric proportions.

Word-of-mouth marketing takes several forms. Among those are:

Table 13.1 Word-of-mouth style

Word-of-mouth style	Elements of the style
Buzz Marketing	Using high profile people or journalists to talk about the artist.
Viral Marketing	Creating interesting, entertaining, or informative messages electronically by email and passed along to targets.
Community Marketing	Forming or supporting fan clubs or similar organizations to share their interest in the artist; provide access to music, videos, tour schedules, etc.
Street Team Marketing	Organizing and motivating volunteers to work in personal outreach locally.
Evangelist Marketing	Cultivating key opinion leaders and tastemakers to spread the word about the artist.
Product Seeding	Giving product samples and information about the artist to key groups and individuals.
Influencer Marketing	Identifying other artist fan clubs and key members of those clubs to help spread the word about the artist.
Cause Marketing	Working with charities and causes that the artist supports.
Conversation Creation	Creating catch phrases, highlighting lyric lines, creating emails, or designing fun activities to start the word-of-mouth activity.
Referral Programs	Creating tools that let help fans refer the artist to their friends.

Adapted from www.womma.org.

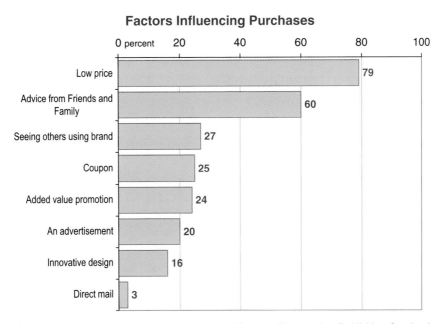

Factors Influencing Purchases

▲ *Figure 13.1 Factors influencing purchases (Source: Haynmarket Publishing Services)*

The importance of word-of-mouth marketing was demonstrated by a U.K. trend-watcher in its study in 2002 that shows the primary motivators for people to try a new product or a new brand. Translating this to apply to a new artist or new music for a record label, you can see by the chart in Figure 13.1 the importance of word of mouth.

Seeing the strength of "advice from friends and family" (60%), and "seeing others using the brand" (27%), suggests that word of mouth can have a powerful influence on the decision of a consumer to try and to buy new music.

Word-of-mouth promotion as defined by Malcolm Gladwell in his book, *The Tipping Point*, shows the marketer that this seemingly simple strategy can actually have a deep-rooted sophistication. He suggests there are three rules to its effectiveness. First, there is the *law of the few people who are connected* to many others, and who can spread the word about a product in ways that have epidemic proportions. Second, the *law of stickiness* suggests that the idea shared by the connector has to be memorable and must be able to move people to action. And third, the *power of context* means that people "are a lot more sensitive to their environment than they may seem" (Gladwell, M., 2000a) and the effectiveness of word of mouth has a lot to do with the "conditions of, and circumstances of, the times and places" (Gladwell, M., 2000b) where it happens.

A great deal of care must be taken to set up a word-of-mouth promotion to make it effective.

The World Wide Web has quickly become one of the most effective media for word-of-mouth marketing, and we have dedicated a chapter in this book to the use of the Web in marketing plans. But briefly, in its online white paper, *Consumer Generated Media*, Intelliseek writers Pete Blackshaw and Mike Nazzaro point to Internet word-of-mouth vehicles they see as the fastest-growing and most viable. They include, "consumer-to-consumer e-mail, postings on Internet discussion boards and forums, consumer ratings websites for forums, blogs (short for weblogs, or digital diaries), moblogs (sites where users post digital images/photos/movies), video blogs, social networking sites, and individual websites" (Intelliseek, 2005). The impact of consumer generated media, or word-of-mouth marketing, is estimated to total over 2 billion web postings by the end of 2006. These reviews, ratings, and postings are building into a massive archive of consumer opinions, nearly doubling the 2004 estimate (Presentation at Ad-Tech, 2004).

Sometimes an aggressive and competitive spirit shows itself in word-of-mouth marketing as fans or consumers are deceived. Spamming is occasionally used, and some use automated software to post to message boards online. Marketers may try to hide the knowledge that they are behind a big word-of-mouth program. And there are occasions when people represent that they are fans, when they are actually the hired help. One fan club president lost her job when it was learned that the "Gulf War wives" requesting a certain song on the radio were actually just fans of the artist. The credibility of the sources of word-of-mouth campaigns will ultimately determine whether the effort was effective, so it is important to keep participants reminded about the ethics of this type of promotion.

While word-of-mouth marketing has traditionally been the tool of independent and private record labels, major label EMI has joined in a big way. The music giant has created an alliance with Procter & Gamble to test their new music through P&G's American network of 200,000 teens and young adults. The record company plans to use the young people to help it decide which singles to release by sending early copies of new releases, and then they plan to use the network for word-of-mouth promotion to others (Rees, J., 2004).

For the record label, coordinating the energy and passion for artists through their fans can generate sales. Dave Balter, owner of BzzAgent, a word-of-mouth marketing firm, says, "The key is about harnessing something that's already occurring. We tap into people's passions and help them become product evangelists'' (Strahinich, John, 2005). And that is the essence of the word-of-mouth marketing strategy as it is applied to the recording industry.

Street teams

The idea of street teams is an adaptation of the strategy of politicians everywhere: Energize groups of volunteers to promote you by building crowds, creating local buzz, posting signs, and rallying voters. There isn't a lot of difference between political volunteer groups and artist street teams.

For the artist, they are volunteer ambassadors who quite literally promote personal appearances and music. For the record label, the ultimate rallying point is getting consumers to purchase the artist's music. While many artists use street teams today, they were originally formed in order to promote music that was not radio-friendly. Without radio airplay, creators of alternative music sought other ways to connect consumers with their music, and employing street teams became an effective way to do that.

Street teams are local groups of people who use networking on behalf of the artist in order to reach their target market. These team members make up the core of the fan base of the artist and often have the deepest passion for the music and message of the artist. Often they are friends and fans of the artist. Some refer to street team members as ''marketing representatives'' who promote music at events and locations where the target market can be found. Often these places are tied to the lifestyle of the

---

Case study: Columbia Records and Switchfoot

"UBI was enlisted to help introduce the band Switchfoot, and their latest album, "The Beautiful Letdown", to the masses. Switchfoot had found success in the Christian Rock format, but their label was interested in creating crossover appeal by targeting college campuses, lifestyle/community locations, clubs/bars, and record stores in 10 major cities across the United States. The goal was to increase awareness of the band, to bring attention to their live performances and to drive sales of their new release.

"Universal Buzz Intelligence mobilized their corps of highly trained operatives for an intense 9 week promotional campaign in 10 major markets. The campaign focused on the distribution of promotional materials and other collateral marketing materials throughout college campuses and the communities that surround them. The guerilla marketing effort focused heavily on driving retail store sales for "The Beautiful Letdown" album, as well as promoting the band's live tour dates.

"By all accounts the Switchfoot release "The Beautiful Letdown" far exceeded label expectations with over one million copies sold to date (the band sold an aggregate of 100,000 albums for their previous three releases). Of the 10 markets covered by UBI, the band realized 7 sold out performances. In addition, six of their top ten selling markets were markets covered by UBI (which had not previously been strong markets for the band), with the remaining 4 cities in the UBI campaign falling within the bands top 25 selling markets. The bands crossover popularity has brought them video exposure on MTV and The Fuse Network."

Source:
http://www.universalbuzzintelligence.com/case/switch.htm

▲ *Figure 13.2 Universal Buzz case study (Source: Universal Buzz Intelligence)*

target market such as at specialty clothing stores, coffee houses, and at concerts of similar acts. A key to effective street teams is for members to understand where to find the target market of the artist and to provide tools and guidance on how to communicate to the target (Tiwary, V. J., 2002).

Street teams originally were formed as a way to reach consumers by companies who either did not have the resources for mass-media marketing, or to reach segments of the market who were not as responsive to mass mediated messages as they are to peer influence (Holzman, K., 2005). Now, major marketing and advertising agencies have started to realize that street teams are an effective form of youth marketing.

Among the tools that may be provided to the street team members are postcards and flyers, email addresses to contact local fans, small prizes for local contests, Web addresses for music samples, actual CD samplers to give away at appropriate events, advance information about tour appearances, and release dates for new music. Street team members may engage in *sniping* — the posting of handbills in areas where the target market is known to congregate. Some labels economize on printing by emailing flyers and posters to local street team coordinators, and asking them to arrange for printing.

Street teams require servicing. The most important thing a street team coordinator can do is to find ways to say "thanks" to the street members. Volunteers for causes require a measure of recognition for their effort in order to keep them energized, and it is no different with those working for free on behalf of an artist. The first thing the label coordinator must do is to regularly communicate with core members of the team. Keep them current on the planned activities of the artist, and make them feel they are important. Provide incentives to the extent the budget will allow. Incentives can be CDs, tee shirts, meet-and-greets with the artist, and free tickets.

The hierarchy of street team management often begins at the label with a coordinator, who recruits regional street team captains who then coordinate at the local level. Data captured by artists through their website or via links from the label's website becomes an important building block for street teams.

While major labels often have in-house grassroots or "new media" departments to run street teams, there are several independent or outside marketing companies that specialize in street team marketing. The advantage of using one of these companies is that they often promote other lifestyle products and the opportunities for cross-marketing are increased. As mentioned earlier, EMI Group teamed up with Tremor, a street team division of Proctor and

Gamble, to not only promote new music, but to coordinate music promotion with P&G's products.

Some of the outside marketing companies used for street marketing are Street Attack, Buzz Intelligence, and Streetwise Concepts and Culture. Street Attack boasts on their website that their understanding of street psychology and Generation Y mindset makes them specialists in building street teams (www.streetattack.com, 2005). Buzz Intelligence boasts that V2, Virgin, Arista, Interscope, Sony, Universal, Geffen, Atlantic, Blue Note and Warner Bros. are among their clients. Streetwise claims A&M, Atlantic, Capitol, Columbia, Epic, Epitaph, Geffen, Interscope, Island, Maverick, Razor & Tie, RCA, Reprise, Sanctuary and Universal have used their street marketing services (www.streetwise.com, 2005).

Guerilla marketing

The subject of guerilla marketing must begin with Conrad Levinson. He is the author of the best-selling book, *Guerilla Marketing*, first published in 1984. Levinson is credited with coining the term, which generally means using nontraditional marketing tools and ideas on a limited budget to reach a target market. In Levinson's words, guerilla marketing is "achieving conventional goals, such as profits and joy, with unconventional methods, such as investing energy instead of money" (www.gmarketing.com, 2005).

In the recording industry, much of the activity of street and e-teams is basic guerilla marketing. Postings in chat rooms, handing out music samplers, and giving promotional buttons and bumper stickers at the competition's concerts are examples of low-cost, but effective guerilla marketing. Figure 13.3 includes a list of guerilla marketing tactics. While they may not directly apply to promoting music, many of them can be developed and adapted for specific uses by a label to help market recorded music.

Guerilla tactics in the marketplace aren't limited to smaller labels with limited resources. When Mike Kraski was vice president of marketing at Sony Nashville, he wanted to find a way to encourage Wal-Mart store associates to become familiar with their new music so that callers to the store could be told the item was available. He announced to the stores he was beginning a promotional campaign where someone from Sony Nashville would occasionally call Wal-Mart record departments and ask about Sony new releases. To those associates who were able to tell the caller about the new music, he gave a limited number of big screen Sony television sets as prizes.

1. Marketing plan	51. Neatness
2. Marketing calendar	52. Referral program
3. Niche/positioning	53. Sharing with peers
4. Name of company	54. Guarantee
5. Identity	55. Telemarketing
6. Logo	56. Gift certificates
7. Theme	57. Brochures
8. Stationery	58. Electronic brochures
9. Business card	59. Location
10. Signs inside	60. Advertising
11. Signs outside	61. Sales training
12. Hours of operation	62. Networking
13. Days of operation	63. Quality
14. Window display	64. Reprints and blow-ups
15. Flexibility	65. Flipcharts
16. Word-of-mouth	66. Opportunities to upgrade
17. Community involvement	67. Contests/sweepstakes
18. Barter	68. Online marketing
19. Club/Association memberships	69. Classified advertising
20. Partial payment plans	70. Newspaper ads
21. Cause-related marketing	71. Magazine ads
22. Telephone demeanor	72. Radio spots
23. Toll-Free phone number	73. TV spots
24. Free consultations	74. Infomercials
25. Free seminars and clinics	75. Movie ads
26. Free demonstrations	76. Direct mail letters
27. Free samples	77. Direct mail postcards
28. Giver vs taker stance	78. Postcard decks
29. Fusion marketing	79. Posters
30. Marketing on telephone hold	80. Fax-on-demand
31. Success stories	81. Special events
32. Employee attire	82. Show display
33. Service	83. Audio-visual aids
34. Follow-up	84. Spare time
35. Yourself and your employees	85. Prospect mailing lists
36. Gifts and ad specialties	86. Research studies
37. Catalog	87. Competitive advantages
38. Yellow Pages ads	88. Marketing insight
39. Column in a publication	89. Speed
40. Article in a publication	90. Testimonials
41. Speaker at any club	91. Reputation
42. Newsletter	92. Enthusiasm & passion
43. All your audiences	93. Credibility
44. Benefits list	94. Spying on yourself and others
45. Computer	95. Being easy to do business with
46. Selection	96. Brand name awareness
47. Contact time with customer	97. Designated guerrilla
48. How you say hello/goodbye	98. Customer mailing list
49. Public relations	99. Competitiveness
50. Media contacts	100. Satisfied customer

▲ *Figure 13.3 One-hundred marketing weapons*
Source: Jay Conrad Levinson, www.gmarketing.com/articles.

Basic grassroots marketing, whether you call it word of mouth, street teams, e-teams, peer-to-peer, guerilla marketing, viral marketing, or the latest term *du jour*, can create an environment that can set a record label's marketing plan and results apart from those of competing artists. One of the biggest challenges of a label-marketing department is to find new and unique ways to present its product to consumers. Including an element of grassroots in the marketing plan can create a plan that steps away from overused templates to add a unique element.

Glossary

Bloggers – Short for weblogs, and refers to those who write digital diaries on the Internet.

Grassroots marketing – A marketing approach using nontraditional methods to reach target consumers.

Guerilla marketing – Using nontraditional marketing tools and ideas on a limited budget to reach a target market.

Moblogs – Websites where individuals post still images and videos.

Marketing representatives – Another term sometimes used for members of a street team.

Sniping – The posting of handbills in areas where the target market is known to congregate.

Street teams – Local groups of people who use networking on behalf of the artist in order to reach their target market.

Video bloggers – Also known as vloggers, they are the video counterparts to bloggers except the content contains audio and video.

Bibliography

Gladwell, M. (2000a). *Tipping Point,* New York: Little, Brown, and Company, pp. 29.
Gladwell, M. (2000b). *Tipping Point,* New York: Little, Brown, and Company, pp. 139.
Holzman, K. (February 20, 2005). Effective Use of Street Teams. http://www.indiemusician.com/2005/02/effective__use__o.html. Music Dish Network.

Intelliseek (2005). *Word-of Mouth in the Age of the Web-Fortified Consumer.* www.intelli-seek.com.

Presentation at Ad-Tech (2004). Measuring Word of Mouth, *Ad-Tech. New York, November 8, 2004, Word of Mouth Marketing Association.*

Rees, J. (Nov 14, 2004). EMI Clinches Research Deal, *Associated Newspapers. (LexisNexis Acacdemic).*

Rogers, E. M. (1995). *Diffusion of Innovations (4th edition),* New York: The Free Press.

Strahinich, John (January 23, 2005). What's all the buzz; Word-of mouth advertising goes mainstream, *The Boston Herald. (LexisNexis Academic).*

Tiwary, V. J. (2002). Starting and Running A Marketing/Street Team.

http://www.starpolish.com /advice/article.asp?id=31.

Word of Mouth 101 (2005). Word of Mouth Marketing Association www.womma.org.

www.gmarketing.com (2005). www.gmarketing.com/what__is__gm.html.

www.streetattack.com (2005). http://www.streetattack.com /Default.aspx.

www.streetwise.com (2005). http://www.streetwise.com/indexnew.php.

14 Internet Marketing

Thomas Hutchison

The Internet

The Internet has become a powerful force in marketing and commerce and now accounts for an ever-increasing percentage of retail sales. Forrester Research reports that e-commerce in the United States is expected to grow from $95.7 billion in 2003, to $230 billion in 2008. According to Jupiter Communications, 25% of online adults had purchased physical CDs over the Internet by 2003. Legal digital music distribution is projected to grow from 4% of online music spending in 2003, to 27% by 2008.

In the recording industry, the web has become useful to create awareness and demand for artists and their recordings. But the web is also capable of electronically delivering the product to the customer—a claim that food, beverage and clothing companies will never be able to make. The Internet provides two opportunities to the music industry: (1) to market and promote artists and their products, and (2) actual distribution of recorded music.

Web marketing should be a part of any marketing plan, but should not be the only aspect of the plan. Even though the Internet has become a great tool for selling music, the traditional methods of live performance, radio airplay, advertising, and publicity are still a significant part of marketing and should not be neglected.

> **Rule #1: Don't make the Internet your entire marketing strategy. Internet marketing should not be a substitute for traditional promotion. The two strategies should work together, creating synergy.**

In the music business, it is necessary to build brand awareness. Whether it's with the artist or the record label, you need to create a sense of familiarity in the consumer's mind. This can be done through many tactics available both on and off the Internet, and should be designed to lure customers to the artist's website, or drive them into the stores to buy the artist's records.

The web is a great tool to reach a large number of people, with minimum expenses involved. But the web is so vast that unless potential customers are looking for a particular artist, they are unlikely to stumble across an artist's site by chance. Also, many customers are more responsive to traditional marketing methods such as radio airplay and retail store displays. An artist's website is a great place for their customers to learn more about their products, such as the artist's live shows and recordings, but it is not necessarily the best way to introduce new customers to the artist's products.

> **Rule #2: Build a good website, but don't expect customers to automatically find it on their own.**

A solid marketing plan incorporates the company website into every aspect of marketing and promotion, and while the website will be the cornerstone of Internet marketing efforts, it is just the start. Building a good website is crucial to marketing success, but the old adage, "If you build it, they will come," does not apply to the Internet.

Every marketing angle should tie in to the web presence. Internet marketing is more effective when it is conducted in conjunction with other aspects of the plan. Every piece of promotional material should contain the website address to help direct the potential customer to the website. Any posters, flyers, postcards, press releases, and so on, should have the artist's website address prominently displayed. And it goes without saying that CD tray cards and cassette J-cards should also contain the address. One band had hand-stamps printed up at the local office supply store, and the stamps had the band's web address on them. The band requested that the bouncers use them to stamp customer's hands as they entered the venue where the act was performing.

> **Rule #3: Incorporate the artist's website address into everything you do online and offline.**

This chapter will discuss how Internet marketing and web commerce can become an integral part of the overall marketing plan. Many record labels have an emerging department known as the *new media* department, which is in charge of Internet marketing. We will be looking at the following aspects of Internet marketing:

- The artist's website
- Promoting the artist's website on the Internet
- Promoting the artist's website off the Internet
- Promoting the artist's brand on the Internet other than the artist's website

The website as your home base

The domain name

The first aspect of building a website involves registering a domain name. The uniform resource locator (URL) is the means of identifying the website location on the Internet. An Internet address (for example, *http://www.yourname.com*) usually consists of the access protocol (*http*), and the domain name (*www.yourname.com*). The URL will also contain the directory path and file name. The general default-loading file is named *index.html* or *index.htm*. This file name should be the name of your homepage, since this is the page that will load when your customers enter in the address. The file, index.html, is served by default if a URL is requested that corresponds to a directory on the server your website resides on.

It is important for the URL, or Internet address, to be simple and easy to remember. A long URL will confuse customers and prevent them from finding the website. There is also more possibility of error when the customer enters in a long name such as: http://www.cheapdomainprovider.com/~yourcustomername/personalweb/bandname/index.html.

There are many services available on the Internet that will register the artist's domain name. A quick online check with these providers can determine if the artist's name or band's name is available. However, it is not necessary to actually possess a physical site with the address http://www.bandname.com. This name can be used as the URL and then visitors can be redirected, or forwarded to the actual website, which may have a longer address. In other words, the actual physical site for an artist may be http://www.recordlabelname.com/artists/the__artist/index.html but fans only need to type in http://www.bandname.com.

For example, my work website address is http://www.mtsu.edu/~thutchis. Web surfers looking for my personal site can type in http://www.tomhutchison.com and automatically be redirected to the proper site. It is also important to register derivations of the artist's name, just to cover those visitors who may not know the correct spelling.

Website developer Dr. Tandem (www.drtandem.com, 2004) suggests the following qualities for a good domain name:

Make it easy to remember
Keep it short
Make it descriptive of the site
Use the "dot com" domain, if available
Use a keyword in the name, if possible

Web hosting

The host is the physical server in which a website is stored. If a company has its own server, they can just redirect or forward the artist's URL to their server. If they do not have a host provider, there are many commercial companies that offer "space" to host a website. The free hosting services should be avoided because they tack on banner advertisements and pop-up ads to the site. These advertisements discourage your customers from visiting the site. It is worth the money to pay for a cleaner site, without all the ads.

It is important to ensure the hosting site has enough data storage space for all the functions a music-based site needs. Thirty megabytes (MB) of space is probably not enough to hold graphics, photos, and sound files—one MP3 file can take up to 4 MB. It is necessary to have enough space for expansion. An artist may only need to feature three audio files from the current album, but when subsequent albums are released, the demands on the site will also increase. Up to 5 MB are needed for graphics and web documents. A thorough site should include: a bio, photos, band news, tour itinerary, sign-up for a mailing list, press and media coverage (reprints), discography, lyrics, audio files, contests and giveaways, merchandise, links to other favorite sites, video, and contact information. It is vital to include some way for the fans to purchase the artist's recordings—either directly from the website, or by directing them to an online retailer. All this can add up to quite a bit of storage space. It is recommended that at least 50 MB of storage space and sufficient *bandwidth* be allocated. Bandwidth is the amount of data that can be transmitted in a fixed amount of time. If a server is accessed by a lot of users simultaneously, it requires sufficient bandwidth to transmit the information from the server to the users in a timely fashion.

Before a website design can begin, it is necessary to outline the goals for the website. These goals can include creating brand awareness, creating a demand for the products and fulfilling that demand through e-commerce, and creating a sense of community among the artist's fans. The site should also be designed to create repeat traffic.

Branding is defined as creating a distinct personality for the product (in this case the artist, not the label) and telling the world about it. Artists who create a well-known brand can parlay that into endorsement deals, an acting career, or becoming a spokesperson for a worthy cause. Artists Madonna and Missy Elliott appeared in TV ads for The Gap; Queen Latifah has moved into films; and Bono of the rock group U2 has become the spokesperson for solving the problems of third-world debt and global trade. So branding should be an important part of any website. Creating an easily identifiable logo and maintaining consistency in style and design can help support the brand.

The website should create a demand for the artist's products. The primary products are recorded music and tickets to live shows. Secondary products include tee shirts and other souvenirs. Enticing the web visitor to purchase, or creating a desire to purchase the recordings, should be an important aspect of the site. This can be achieved through offering music samples on the site as well as information about the music and the recording. Links to online retailers will encourage the visitor to follow through on purchase intentions and may create impulse buys. A demand for concert tickets can be created through tour schedules, concert photos, samples of live recordings and links to sites that sell tickets. The website is also a great place for the merchandising of tee shirts and other souvenirs. Photos of these items are important too.

Artist websites are a good place for like-minded people to connect with each other. These fans probably have much in common—especially their interest in the artist. A sense of community can be created through message boards, online chats with or without the artist, and by including photos of fans at concerts. These devoted fans can become opinion leaders (the virtual street team), influencing their friends to become fans of the artist, and encouraging their friends to visit the website.

The website should also provide information for journalists looking for information on the artist, and contact information for both journalists and booking agents.

Website design

A good website is one that is attractive, uncluttered, quick to load and easy to navigate. The site must offer something of value to the consumer—information, products, and freebies. It also helps to make the website fun, refresh content, and give people a reason to return. Effective websites avoid large flash programs, flashing text, animation, and large graphic files. Most Internet users do not yet have cable modems and a website that is slow to load is sure to fail.

Some simple rules to follow include:

1. Don't make the pages too long, so that users have to scroll a lot
2. Don't load a lot of photos on one page. It will slow the loading time
3. Clearly specify the purpose of the website, and why the user should visit regularly
4. Don't make your background too cluttered or the text too difficult to read
5. Balance the design elements—don't go too heavy on either text or graphics

Website navigation is made easy with the inclusion of a navigation bar (nav bar). The bar usually has prominent placement in the same location on every page. Usually the bar is located across the top, under the banner, or in a column on the left side of the page. The nav bar allows the user to be one-click away from any place they want to go within the site.

A *splash page* can be used to enhance the image of the website. This is the first page a web surfer is directed to when they access the site. Usually, a splash page will have the logo and elaborate graphical layout, but not much information. The visitor is encouraged to enter the site by clicking on the page. Automatic splash pages direct the visitor to the homepage after a few moments. Do not overload the visitor with large flash programs that take a long time to load. An effective splash page has just a graphic image and the brand logo. Many web designers discourage the use of a splash page because it is just one more step or click required of the visitor before they get to the true substance of the website. If you must use one, Client Help Desk (www.clienthelpdesk.com, 2004) advises one to "consider dropping a cookie into each site visitor's computer that automatically skips the splash screen on subsequent visits. Even people with the patience to deal with a splash screen once will be tested upon seeing it repeated each time they return to the site."

The homepage identifies your company or brand and extols the benefits of the site and its products. This page should be updated frequently to reflect changes and notify the visitor of new and interesting developments. Announcements are frequently found on the homepage, although a "What's New" page can also contain news briefs and updates. The homepage should also contain links and navigational tools for the rest of the site. A good web page should balance text with graphics or images and sport a layout that is inviting and interesting.

An artist's website should contain the following basic elements:

1. A description and biography of the artist.
2. Photos: Promotional photos, concert photos, and other pictures of interest. This can include shots of the artist that capture everyday life, photos of the fans at concerts, and other photos that reflect the artist's hobbies or interests.
3. News of the artist: press releases, news of upcoming tour dates, record releases, and milestones such as awards.
4. Discography and liner notes from albums.
5. Song lyrics, and perhaps chord charts.
6. Audio files: these may be located on the purchase page.
7. Membership or fan club sign-up page. Allow visitors to sign up for your newsletter, or for access to more exclusive areas of the site.
8. Tour information: tour dates, set lists, driving directions, touring equipment list.
9. E-store: merchandise page for selling records, tee shirts and other swag.
10. Contest or giveaways.
11. Links to other favorite sites, including links to purchase product or concert tickets, venue information, the artist's personal favorites, e-zines, and other music sites. Ensure that all your off-site links open in a new window, so the visitor can easily return to your site.
12. Contact information for booking agencies, club managers, and the press.
13. Message board or chat rooms: This allows the fans to communicate with one another to create a sense of community. This can be an area restricted to members only.
14. Blogs: Recently blogs have become popular on the Internet. A blog is simply a journal, usually in chronological order, of an event or person's experiences. Maintaining a blog of the touring experience is one way to keep fans coming back to the website to read the most recent updates to the journal. It also gives fans a sense of intimacy with the artist.

E-Commerce: the electronic storefront

One major decision facing web designers is whether to host an electronic storefront, or simply create links to one of the Internet's many online stores. Setting up a storefront can be complicated since it involves creating a security encryption through a secure sockets layer (SSL) and the ability to process credit card transactions. It also means having arrangements in place for product fulfillment: The ability to ship product directly to the customer. Processing credit cards is the most efficient way to handle monetary transactions, but it can be costly. Credit card services usually charge a transaction fee for each order. In addition, the site will need to contain a shopping cart page, wish list database, and order processing and tracking information, all of which can add significantly to website maintenance.

New to the scene is a company called PayPal (owned by eBay) that has become very popular. PayPal offers the ability to accept credit card payment online, software to setup the shopping cart page, and assistance with managing shipping.

For handling e-commerce on the website, it is necessary to follow these recommendations:

1. Provide thorough product descriptions, including graphics.
2. Prominently display the product name and price. If several formats are available, clearly identify the format.
3. Make it easy for the customer to purchase; the fewer clicks, the better. The "buy" button should be obvious.
4. Ensure the customer knows when the order is completed.
5. Once the order is placed, send an immediate email confirmation.
6. Ensure the orders go out as quickly as possible.
7. Ensure the customer knows how long it will take to receive the order.
8. Ensure you have the inventory to fill the orders. If you are out of stock, modify the storefront page immediately to reflect this fact.

The alternative to the do-it-yourself arrangement is to provide links to one of the many online retail stores. For the music business, this includes CDBaby, The Orchard, CDNow, MP3.com, and Amazon.com. Use of these services does mean giving up a percentage of each product sold, but requires little effort on the part of the webmaster. For selling merchandise such as bumper stickers, hats and T-shirts, www.cafepress.com offers a low-cost fulfillment service linked directly from your website. Links to these e-tailers should direct the shopper to the exact product page and not the homepage of the retail company.

Website promotion on the site

There are ways to promote a website and entice visitors to spend more time and visit the site more often. (The next section will discuss ways to promote your website in general on the Internet.) The most obvious way to have people return to your site is to offer them something of value. This can be as simple as information: Tour information, release dates, general industry information, or great links. You can add value to the site by offering such things as screen savers, shareware programs, electronic greeting cards with appealing photographs, jokes, recipes, links, and active message boards.

Contests also lure visitors to return to the site, especially if new contests are offered each month. Successful contests are setup to encourage the visitor to return to or further explore the website. One example would be to include trivia questions with answers that are easily found on other pages within the website. It is not wise to offer the main product as contest prizes. If 200 visitors sign up for the contest, they may hesitate on purchasing the product in the hope that they may be one of the winners. It is advisable to offer some other product as prizes.

On the homepage, be sure to remind visitors to bookmark the site so they can easily find it again. In 2002, 46% of website traffic came through bookmarks and direct navigation.

Tell-a-friend

The "tell-a-friend" or "recommend-it" function is a form of *viral marketing*. Viral marketing is "any strategy that encourages people to pass along a marketing message to others, creating the potential for exponential growth in the message's exposure and influence" (Wilson, R., 2000). Visitors to the website can be encouraged to spread the word and tell others. This is made easier by providing them with a feature to forward the web address to their friends (see Figure 14.1).

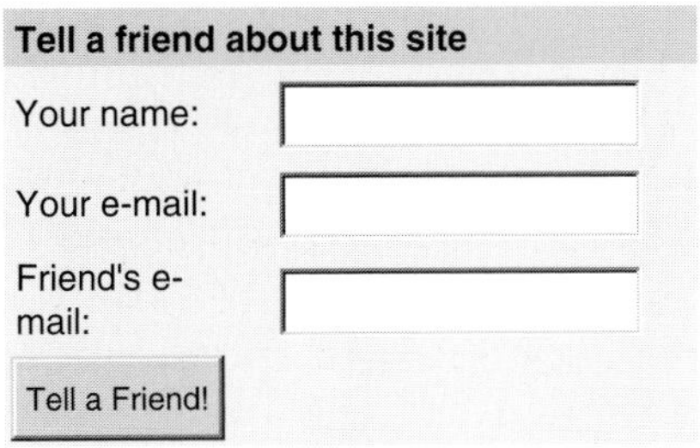

Tell a friend about this site	
Your name:	
Your e-mail:	
Friend's e-mail:	
Tell a Friend!	

▲ *Figure 14.1 Tell-a-friend about this site*

This type of marketing is very successful because word of mouth is one of the most credible forms of promotion. Visitors to the site may not be personally interested in what the site has to offer, but they may know someone who is. This makes it very successful in targeting the desired market. This feature can also be used to forward online newsletters, photos, and specific web pages through the "email to a friend" feature that usually contains a subject line such as "(your friend) thought you might be interested in this."

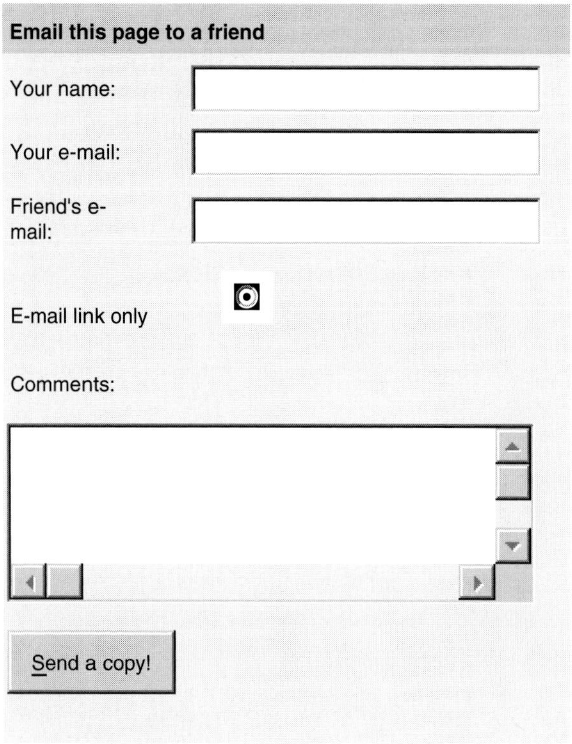

▲ *Figure 14.2 Email this page to a friend*

Posting pictures of fans (especially at live performances) is a good way to increase traffic. Those fans that appear in the photos will have an incentive to direct their friends to the site.

Visitor registration

The most valuable piece of information that can be obtained from consumers who visit the website is their email address and permission to add them to the "list." A list of email addresses, and permission to contact them, is necessary in this new age of *spamming*. Spamming is the activity of sending out unsolicited commercial emails. It is the online equivalent of telemarketing (more on spamming later). An effective up-to-date email list is a valuable marketing tool, and allows for e-newsletters to be sent to fans who have shown enough interest to sign up. Websites should post their privacy policies to avoid any confusion or legal complications if visitors end up on a mailing list.

When recruiting visitors to sign up, it is more attractive to present this as either a guest book or free membership to the artist's fan club, rather than just signing up to receive emails. Usually, websites include a visitor sign-in, registration, or "join now" button leading to a short online form.

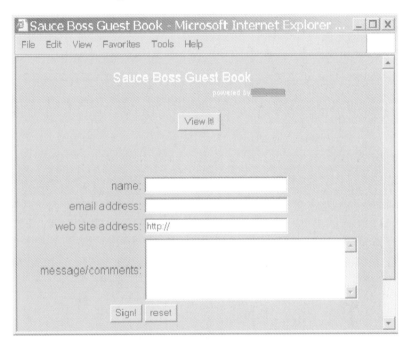

▲ Figure 14.3 Sauce Boss guest book (Source: www.Sauceboss.com)

When visitors are asked to sign a guest book or register to enter the site, there are two strategies for adding these visitors to the email list: "opt-in" and "opt-out." With opt-in, the visitor selects a blank check box to be added to the email list. With opt-out, the button default is in the checked position and the visitor must uncheck the box to avoid being included in the email list.

"Opt-in" means that visitors choose to join a site's mailing list—one that is generally aimed at notifying the visitor of new developments. Some marketers set the default setting to the "opt-out" approach by automatically checking the "Yes, sign me up" box. By default, visitors overlook the box "giving permission" to contact them.

Marketers advise to use the opt-in method only. This avoids annoying visitors who did not intend to sign-up for emails and simply overlooked the "uncheck" function. It also reduces the amount of follow-up work that must be done by the webmaster in honoring unsubscribe requests.

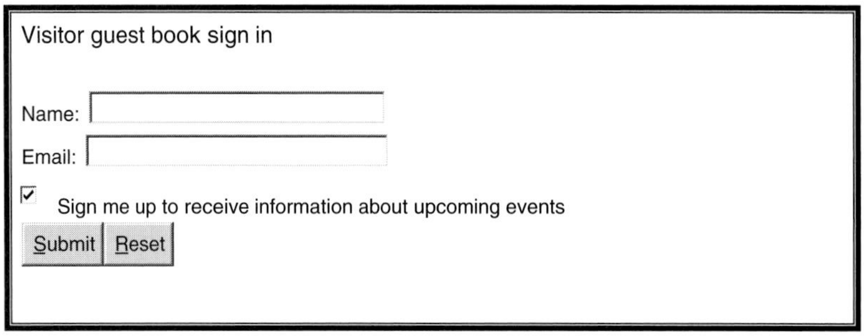

▲ *Figure 14.4 Visitor guest book sign in*

The Client Help Desk (www.clienthelpdesk.com) reports that almost 70% of Internet users say they unknowingly signed up for email distribution lists. Almost 75% of those who received unsolicited email took action to be removed from the sender's list. With opt-outs, the website can claim a larger number of subscribers (the willing and the unknowing), whereas with opt-ins, the site can get a better understanding of how many people want to receive the emails or e-newsletters sent out from the list.

Promoting the website on the internet

Building a website is not enough. With all the clutter on the Internet, it is necessary to promote the website by reaching out to web surfers and encouraging them to visit the website. It is a commonly held belief among Internet marketers that most of the website traffic will come from word of mouth, or word of mouse in this case.

Search engines

Once a website is completed, the URL is submitted to the various search engines that scour the Internet when users type in search terms. Search engines can be of some assistance if used correctly, but with so many websites now on the Internet, it is becoming harder to stand out from the crowd. According to StatMarket (2003) in 2003, search sites accounted for more than 13.4 percent of global referrals, up from 7.1 percent the previous year. The 7 most popular search engines on the web are:

1. Google 54.7%
2. Yahoo 22.1%
3. MSN Search 9.5%
4. AOL Search 3.7%
5. Terra Lycos 2.8%
6. Altavista 2.5%
7. Askjeeves 1.5%
   Source: (OneStat.com, 2004)

Search engines vary in how they find and list sites that match the subjects or search terms entered in by the web user. Some search engines automatically evaluate HTML documents, looking for particular information used to describe the site (meta tags), while others use humans who scour through applications to determine how to rank and categorize submissions.

Many webmasters customize their website description and keywords to maximize exposure and listing on search engine results. To ensure adequate listing on these search engines, webmasters will typically submit the website to the search engine either directly or through services that offer a directory listing. To get listed in Google, for example, the site must be submitted to Open Directory at www.dmoz.org. All submissions are evaluated by editors, making this a human-driven search engine. Others search engines such as Excite and GoTo use spiders or "bots." These are automated software programs that constantly roam the Internet, cataloging new web pages they find. They search the site to determine how to categorize information that was submitted to the search engine. Savvy web designers use meta tags to help the spider program define the site and which pages to list.

Meta tags

Meta tags are author-generated HTML commands that are placed in the head section of an HTML document. These tags specify which search terms should be used to list the site on search engines. Popular meta tags can affect search engine rankings and are generally listed in the sections "Meta Keywords" and "Meta Description." A meta tag can be generated automatically by the site www.submitcorner.com. The following example tag will let the "bot" know to categorize this website under blues music, slide guitar, and by the artist's other endeavor, Louisiana-style cooking.

< META NAME="KEYWORDS" content="blues music hot sauce gumbo slide guitar Bill Wharton datil pepper habanero Liquid Summer recipe contest Sauce Boss Jimmy Buffett Parrothead">
< META NAME="DESCRIPTION" content="Hot Sauce and Scorching Slide-Florida Bluesman Bill Wharton makes Liquid Summer Hot Sauce and has made gumbo for over 80,000 people during his high energy concerts with his band-The Ingredients">

(http://www.sauceboss.com)

The web page title is also very significant and should reflect the nature of the site. The <TITLE> tag is the caption that appears on the title bar of your browser and is the name on the clickable link listed in the search engine results. (Example: <title> Sauce Boss </title>.)

Music directories

Standard directories are defined by the I-search Digest (www.led-digest.com) as "a server or a collection of servers dedicated to indexing Internet web pages and returning lists of pages which match particular queries. Directories (also known as *indexes*) are normally compiled manually, by user submission (such as at www.whatsnew.com), and often involve an editorial selection and/or categorization process." Directories are arranged hierarchically, from general to specific. Since these directories are not compiled automatically by spiders, a site's webmaster must submit the site to the directory. Then their editors will visit the site to determine if it is appropriate to list in the categories specified in the request.

Music directories are very helpful at directing web surfers to new sites of interest. Some of the more popular directories are: the ultimate band list (IUMA), sonicswitchblade.com, FarmClub, music-sites.net, music crawl, Eclef, Listen. com, and StarPolish. It is also important to target major traffic sites with advertising, such as AOL, and Yahoo! Launch.

Banner exchange/link exchange/reciprocal linking

Exchanging links or banners with another similar website is a great way to expand traffic to a website. Exchanges occur when two websites agree to link to one another, sharing their visitors with the other site. A banner is described as a graphical web-advertising unit with a hot link to the advertiser's website. Banner exchange occurs when two or more websites agree to display banners promoting the partners' sites. There are banner exchange services available that help coordinate banner exchanges. Some experts advise against participating in banner exchanges because they may distract visitors and direct them away from the initial site before they have had a chance to explore. Nonetheless, trading links with other websites that target the same market has been a proven success in increasing website traffic. The advantages of trading links, according to Insite Media (www.insite-media.com) are:

1. They bring in a highly targeted audience because the visitor was on a similar website.
2. Links from other sites create "link popularity" that improves positioning on search engines.
3. It is a more stable form of creating traffic, since once these links are setup, it is unlikely that they will change or disappear. As the link partners increase traffic to their own sites, they send more traffic to your site.

Webrings

In the article "Music and the Internet," webrings are defined by Sara Gibson (2003) as "simply a conglomerate of sites that are linked up together because they share a common interest or goal." This is mostly a grassroots phenomenon and can be created by any willing webmaster. The ring creator is responsible for attracting and approving applicants who then become part of the web ring navigational structure. Members are required to provide navigation buttons on their website for moving among member sites, primarily using the "next" and "previous" links, although other options such as "random site" and "complete list" often exist. The banner on the ring hub or complete list page may look like the item below.

▲ *Figure 14.5 Webring menu*

Members of the webring are required to include a navigation bar like the example in Figure 14.6 on their homepage.

▲ *Figure 14.6 Webring banner (Source: www.Sauceboss.com)*

Viral marketing

Publicityadvisor.com describes viral marketing as a new buzzword for the oldest form of marketing in the world: referral, or word of mouth, which they update to "word of mouse." Viral marketing is any strategy that encourages individuals to pass on a marketing message to others, creating an ever-expanding nexus of Internet users "spreading the word." On the web, this is done through several strategies, including the "tell-a-friend" button mentioned earlier. Incentives can be added by entering each person who "tells a friend" into a contest. You can also add viral tell-a-friend buttons to many of the pages on your website.

The most common form of viral marketing is through email signature attachments. Hotmail.com has successfully employed this technique by appending their message to the bottom of every email generated by its users. Hotmail offers free email service to users but also places a viral tag at the end of each message that its users generate. If that message is passed along, the advertising tag goes with it.

Ralph F. Wilson, in his article, "Six Simple Principles of Viral Marketing," says there are six elements of viral marketing:

1. Give away something of value – Giveaways can attract attention. By giving something away up front, the marketer can hope to generate revenues from future transactions. For example, the tag can say "Win free music at www.yourartist.com."
2. Provide effortless transfer to others – "The medium that carries your marketing message must be easy to transfer and replicate."
3. Expand exponentially – Scalability must be built-in so that one message transmitted to 10 people gets passed on to 10 each, for 100 new messages.
4. Exploit common motivations and behaviors – Success relies on the basic urge to communicate and share experiences and knowledge with others.
5. Utilize existing communication networks – Learn to place viral messages into existing communications and the message will "rapidly multiply in its dispersion."
6. Take advantage of others' resources – A news release reprinted elsewhere will include the viral message and perhaps the link.

Newsgroups/discussion groups/bulletin board/chat rooms

The Internet is full of discussion groups, Usenet groups, and special interest websites that foster the exchange of information and ideas among its members. These topic-oriented websites usually provide chat rooms and bulletin boards for discussions. Chat rooms allow for real-time interaction between members, while bulletin boards allow individuals to post messages for others to read and respond to. Since these members or users have a mutual interest in the site topic, user groups are an excellent way to locate members of the target market. Music-related Usenet groups are often defined by a common interest in one artist or a genre of music. Internet marketers will often define the target market for their unknown artist by identifying artists with similar appeal. Then marketers will visit sites that cater to those fans.

These sites usually have specific rules about the posting of commercial messages and harvesting email addresses, which is universally frowned upon by group moderators and system operators. Many groups, however, welcome brief messages from industry insiders notifying interested members of new products. These messages often are embedded with hot links to the site actually promoting or selling the product. Some user groups require that you register before participating in their group discussions. Marketing professionals as well as web surfers often engage in "lurking" behavior when first introduced to a new users group. Lurking involves observing quietly—invisibly watching and reading before actually participating and making yourself known. Often, user groups have their own style and "netiquette" (Internet etiquette), and it's best to learn these before jumping in. Lists of these groups can be found at http://groups.google.com/ and http://www.liszt.com/.

▲ *Figure 14.7 List directory (Source: Big Bend Blues Directory)*

Marketers have also been known to join in chat discussions. Again, targeting the right chat rooms in the right discussion group is important. In street marketing, marketing representatives who are part of the target market pose as fans of the artist and engage in discussions on the merits of the artist and his or her works (see section on fan-based promotions). Marketing through online chats and message boards is the online counterpart of street teams.

Fan websites

It is not uncommon for fans of an artist to create unofficial websites or web pages on their personal website featuring their favorite artists. These sites can be used to spread the word. Most of these dedicated fans welcome receiving press releases and digital photos and will gladly display these on their sites. They usually are also willing to create a link to the artist's official website. Sometimes these sites are difficult to find because they are not always listed with search engines, but many of the more popular ones can be located through the creative use of search engines or directories.

Some grassroots marketers encourage their street teams to create their own "fan" websites. These members of the target market add information about the artist to their personal website and generate word-of-mouth promotion from the fan's point of view.

Email addresses, spam, and e-zines

Spam is defined as unsolicited commercial email messages—the equivalent of junk mail or telemarketers. Because of the negativity and problems associated with spam, successful marketers have adopted a code of conduct. The U.S. government and several foreign countries have passed legislation regulating or outlawing the act of spamming. Many special interest sites (discussion groups) strongly discourage *harvesting* (collecting from posted messages) email addresses to be used for spamming. The general rules for sending commercial emails are:

1. Don't send unsolicited messages, period. Instead, create email lists by encouraging website visitors to leave their email address so they can receive valuable updates and information.
2. Send the right messages to the right people. Touring information is only important to fans in the area of the performances. Leave the comprehensive touring schedule to the website, and use email for only those fans within driving distance to the venue.

3. Only send emails when there is something new to report. Make the message specific and make the best use of the subject field.
4. Use the email as a teaser, providing enough information to entice the reader into visiting the website for the rest of the story. (Be sure to include a hot link or the URL address to your site in the email.) Have a great opening, use short sentences, and focus on the recipient's self-interest.
5. You can link various parts of the email message to specific pages on your website.
6. Don't send out large files full of graphics and attachments. Instead, rely on the website to provide the images.
7. If you can't send out individual emails, use the blind courtesy copy (BCC) function so that recipients cannot view, and thus harvest, other email addresses from your list.

Email newsletters have become an effective way to stay in touch with the market on a regular basis. It is the cheapest and most effective way to generate repeat traffic to your website. The email newsletter (sometimes called an *e-zine*) lets the marketer control which messages are sent out to fans, and how often.

Directing customers to online retailers

The goal of an online newsletter is to either, (1) get the fan to revisit the website, or (2) get the fan to purchase the new recording. The fan doesn't necessarily have to revisit the website to purchase a new release. An effective newsletter announcing the new release should include a direct link to a merchandising web page, whether it is part of the artist's website or the product page of an online retailer. The idea is to make it easy for the customer to buy. If the fan would prefer to bypass the artist website and just purchase the new release, embedded links in the newsletter should facilitate this. Companies such as Amazon.com make it easy to link directly to the product page so the customer doesn't have to sift through storefront pages and menus to find the product. For example, the album *The Best of the Sauce Boss* is available at Amazon.com on the dedicated page at http://www.amazon.com/exec/obidos/tg/detail/-/B00004SW5Z/qid=1069273304/sr=1-1/ref=sr__1__1/102-6269284-5258519?v=glance&s=music. Instead of dumping customers at the "front door" of www.amazon.com and having them use the search feature, they are directed to the specific page for that release.

Some fans may prefer to shop at a reputable online retailer, rather than purchase from the artist's website. The fan may already have an account with a particular online retailer, making the transaction easier. Creating a link to each of these sites increases the likelihood that the shopper would choose their favorite retailer and complete the purchase. Links to online retailers should also be included on the artist's website for the same reasons.

Offline retailers, or those with "click and mortar" stores are often included in the favorite links area of the artist's website, thus supporting traditional retailers who often feel overlooked in the world of cyber-promotion. The e-zine, or newsletter should also mention that the album is available in stores, if indeed it is. Some fans just can't wait for a mail-order copy to arrive. This brief mention in the newsletter may send anxious fans to a physical record store. As mentioned earlier, effective marketing plans integrate offline and online marketing components.

Fan-based promotions

Fan-based, or "perceived" fan-based web promotions are very effective and contain the "word of mouth" or "street credibility" that other forms of marketing lack. Marketers have begun to exploit this technique by hiring *street teams*, who pose as avid fans of the artist and engage in peer-to-peer promotions. While some of these tactics border on deception, marketers try to include street team members who actually are fans and part of the target market, thus increasing the credibility of their efforts. Nonetheless, these hired e-team members masquerade as fans, and engage in activities much like those of the avid fans they purport to be.

These hired fans typically will find discussion groups and join in the discussions. They will subtly introduce the artist they are promoting and coax participants into visiting an artist website or previewing the artist's music. They may seek out members of a discussion group most likely to respond to the artist being promoted and send them personal instant messages and emails. They may also post messages on the group message board inviting members to check out a particular website promoting the artist. Often, these messages take on a tone of fan discovery, such as, "Hey all, I've just discovered a great new artist and wanted to share her (him) with you." Online street teams should be coached or given talking points so that the appropriate messages will be disseminated.

There was a commercial for Subway™ sandwiches that ran recently on television. In the commercial, a young man gets on a crowded bus and sees a friend on the bus who works at Subway and is wearing the official uniform. The young man asks his friend what's new at Subway. When the friend describes the new menu offerings, other riders on the bus become interested and head for the nearby Subway restaurant when the bus stops. The two young men then congratulate each other on the effectiveness of their pre-planned tactic. This scheme also is being used in chat rooms to promote artists. One co-conspirator may strike up an online conversation with their accomplice who then displays an interest in the information on the new artist or release. The accomplice will then begin asking questions designed to elicit more promotional information from the initiator. While the ethics of this process may be questionable, the effects are not.

Hired fans also write fan reviews for websites that post reviews, such as on Amazon.com. Often, artists will ask friends and family members to write a fan review for them and post it on Amazon.com or CD-Now, or participate in the "rate this item" feature. It is not considered unethical for an artist to send out an appeal in the newsletter for fans to log on and rate their music, or vote for them on some other website.

Integrating offline and online promotions

To reiterate the theme at the beginning of this chapter, successful marketing plans are those that integrate online marketing with offline promotions. The two strategies working together create synergy. This starts by incorporating the artist's web address into every aspect of offline marketing. There are three main areas of offline promotion that should be exploited for creating online traffic:

1. Concerts, live performances
2. The press
3. Retail products

Concerts

The artist's URL should be prominently displayed at all live performances. Announcements from the stage and handouts such as decals, bumper stickers, buttons, and postcards (and even temporary tattoos) can help remind fans of the web address. However, promotion of the website extends beyond merely providing the URL. Concertgoers should be given a reason to visit the site, including contests, photos and music from the performance, product availability, and updates on the next local event.

In addition to supplying fans with the URL, email addresses should be collected at the venue to add to the e-zine list. Fans can be encouraged to signup through contests and the opportunity to be informed of upcoming events and new releases.

The press kit

The press kit should contain numerous references to the website, and the fact that additional materials and updates are available there. All press releases should contain the URL. A special section of the website can be dedicated to supplying materials normally used by the media to cover the artist, such as print-quality promotional photos, artist history, discography, music samples, and so forth. The press should also be encouraged to include the artist's web address in all articles published.

Retail products

All music-related merchandise should reference the artist's website, perhaps encouraging visitation by offering value-added features such as bonus tracks or remixes available only to buyers. The interior liner notes can contain key information for accessing restricted portions of the website. Again, contests can be geared toward fans that have purchased the music. Fans can also be encouraged to leave feedback or complete a survey. Retail coupons can be offered with arrangements already made for redemption at brick-and-mortar retailers.

Conclusion

Internet marketing is a cost-effective way to promote music and artists. It offers the ability to provide fans with music samples and artist information. It has the potential to generate new fans, create a sense of community, and transcend geographical boundaries. The Internet is a vast resource. The goals should be to carve out a niche, identify a market and cultivate current and potential consumers in that market.

The website serves as the home base for both online and offline promotions. Fans should always know where to look to find updates on their favorite artists. The website is a dynamic marketing tool and it should not be left unattended. While the ultimate marketing goal is to increase the sales base for recorded music and concert tickets, the Internet can be also useful for creating brand awareness and nurturing existing customers.

Glossary

Banner – A graphical web-advertising unit with a hot link to the advertiser's website.

Bandwidth – The amount of data that can be transmitted in a fixed amount of time. Bandwidth is usually expressed in bits per second (bps) or bytes per second.

Banner exchange – Two or more websites agree to display banner advertisements promoting the partners' sites.

Blog – A journal that is available on the web, typically updated daily or weekly. Postings are usually arranged in chronological order.

Bots – Automated software programs that constantly roam the Internet cataloging new web pages they find.

Branding – Creating a distinct personality for a product.

Chat – An online discussion in real time in which two or more people engage in a written conversation.

Cookie – The name for files stored on your hard drive by your web browser that hold information about your browsing habits, such as what sites you have visited, which newsgroups you have read, etc.

Directories – A server dedicated to indexing Internet web pages and returning lists of pages and links that match particular queries.

Discussion groups – A group of Internet users who meet online in a virtual meeting place to discussion a topic of mutual interest. They can use message boards or real-time chats.

Domain name – The name given to a host computer or site on the Internet.

E-store – A virtual online store capable of taking orders, collecting money, and shipping product.

E-zine – An electronic magazine, or email organized in a magazine format and sent via email to interested Internet users or made available on a website.

Harvesting – collecting email addresses by visiting user groups and copying email addresses from their message boards.

Homepage – The document that is first accessed when visiting a website, also known as index page.

Host provider – The physical place where your website resides.

Hot links – A highlighted or underlined word or a graphic that is linked to another page or website. The user clicks on that word or image to jump to the new location.

HyperText Markup Language (HTML) – The language used to create links, format documents and communicate with web browsers to emulate the look intended by the creator.

Lurking – The act of observing or viewing a chat room, newsgroup, usenet and forums without contributing.

Message boards – A section on some websites where users post public messages. Users often respond to each other's postings.

Meta tag – Author-generated HTML commands that are placed in the head section of an HTML document. These identification tags help search engines identify the content of the document.

Navigation (nav bar) – The set of buttons on a website with hot links that, when clicked on, take you to other sections of the site. The nav bar is usually visible on every page and allows the user to jump to other sections of the website by using the bar.

Opt-in/Opt-out – The act of explicitly requesting an email distribution by checking a box. For example, when you sign up to receive an email newsletter, you are "opting" to receive it. Opt-out occurs when the check box is auto-matically checked as the default position. The user must then "unselect" the box to avoid being added to the email list.

Reciprocal linking – Trading links with websites that target the same audience as your own.

Search engine – A program located on a website that acts as a library card catalog for the Internet. Google is an example of a search engine.

Secure Sockets Layer (SSL) – A method of encrypting data as it is transferred between a browser and Internet server. Commonly used to encrypt credit card information for online payments.

Spamming – The activity of sending out unsolicited commercial emails. The online equivalent of telemarketing or junk mail.

Splash page – An introductory first page or front page that you see on some websites, usually containing a click-through logo or message or a fancy graphic.

Uniform resource locator (URL) – An address that identifies the location of any type of Internet resource.

Usenet – The collection of the thousands of bulletin boards residing on the Internet.

Viral marketing – Any strategy that encourages people to pass along a marketing message to others, creating the potential for exponential growth in the message's exposure and influence.

Webring – A conglomeration of sites that are linked up together because they share a common interest or goal.

Bibliography

clienthelpdesk.com (2004). www.clienthelpdesk.com. The Client Help Desk (no longer available).

drtandem.com (2004). www.drtandem.com. Website Design.

Gibson, S., (2003). Music and the Internet: cybermarketing and promotion, *www.starpolish.com.*

Insite Media (2004). www.insite-media.com.

I-search Digest (2004). www.led-digest.com.

OneStat.com (2004). Number one real-time intelligence web analytics. www.onestat.com.

StatMarket, (2003). The pulse of the Internet. www.statmarket.com.

Wilson, R. (February 1, 2000). The six simple principles of viral marketing. *Web Marketing Today,* **70**.

15 Music Videos

Paul Allen

Video production and promotions

Music videos can play a key part in the success of the well-crafted marketing plan of a record label. They can add another dimension to the song and its lyrics by adding that visual element to the musical performance of the artist. It can create strong images to pique the passion of the video viewers and get them more involved in the music. Ultimately, the marketer's objective is to direct that passion into the need to possess the music in the form of a purchased CD, or download from a music retailer.

From the point of the record label, the three functions of a music video are:

- **Programming** for video channels
- **Promotion** for records
- **Products** to sell in record stores (*personal interview with Amy Macy*)

This chapter is dedicated to gaining a better understanding of music videos, how they fit into a marketing plan, and how they are used strategically to stimulate retail sales of a recording.

History of the music video

The earliest form of video promotion of music began in 1890. From then on and through the next 25 years, "illustrated songs" became the rage of thousands of large and small theaters across America, and were credited with selling millions of copies of sheet music. In these theaters, vocalists accompanied by bands or small orchestras would perform songs, while hand-painted glass slides were sequenced and projected on a screen depicting the story of the lyrics. Often, a vocalist would then lead the audience in singing the song as the lyrics were projected (PBSKids, 2004).

When radio was introduced in the 1920s, music promotion changed. One of the earliest of radio deejays was an announcer named Al Jarvis, who worked in the 1930s at a radio station in Los Angeles. He created a radio show featuring recorded music mixed with chatter. It was also Al Jarvis who went on television in early 1950s with a program that featured recordings, guests, and information which made him one of the original video deejays. His program was replicated in numerous other television markets, and became a staple of many major broadcast companies. By the mid-1950s, video deejay programs were prominent in all top 50 media markets.

Some of these early television shows featured what was known as a *soundie*. A soundie was a video creation by the Mills Novelty Company that featured performances by bands, and was used in coin-operated video machines. In the early 1940s, Mills Novelty created over 2,000 promotional soundies. In the early '50s, producer Louis Snader created 750 "visual records," and Screen Gems and United Artists began creating their versions of early music videos.

The granddaddy of video deejay television shows was *Bandstand*, which began airing in Philadelphia in 1952 with host Bob Horn. In 1956, Dick Clark replaced Horn, and the show soon became *American Bandstand* on the ABC Television Network.

Television in the 1960s and early '70s featured many shows that included musical performances, but it wasn't until the mid-'70s that artists like Queen and Rod Stewart created videos designed specifically to assist in the promotion of their recorded music projects. But the link of the music video to record promotion caught fire when cable television began to flourish in the early 1980s. Among the first to begin using music videos was a program called *Night Flight* on the USA Network. Since then, scores of cable channels and hundreds of regional and local shows have continued to regularly feature music videos as part of their entertainment programming (McCourt, T. and Zuberi, N., 2005).

Deciding to produce the music video

Investing in a music video means that the label views it as an important promotional piece for the recording. It is a very strong promotional device designed to promote the label's recording, but places the image of the artist in the homes of concert ticket buyers, too. There are instances where the music and its images are inseparable, and a video is a necessity. At other times—for whatever reason—the artist may not be camera-friendly, and the decision to create a video will include images that do not feature or focus on the image of the artist.

The general kinds of music videos created can be classified as shown in Table 15.1.

Table 15.1 Elements of music videos

Elements of music videos	
Scratch videos	These videos use pieces of film that are synchronized with the tempo of the music.
Performance videos	This is a concert view of the artist performing and singing.
Abstract images	These videos incorporate surrealistic special effects and sometimes bizarre images.
Storyline video	The video follows the storyline of the song lyrics.

From the record company's perspective, a number of considerations come into the decision matrix to determine whether a video makes business sense for a specific recorded music project. Those considerations are detailed in Table 15.2.

Table 15.2 Considerations to making a music video

Considerations to making a music video	
Is there a story that a video can tell?	Creative services, the song producer, the publicist, the artist, and marketing may all help determine if the song lyrics and the artist can enhance the impact of the recording through a video creation.
Is there a budget available for a video?	Reasonable video production costs have ranges of $125,000–$150,000 for a new artist; $250,000–$500,000 for a mid-level artist; and $500,000–$750,000 for major acts (Passman, D., 2003). The budget must be adequate to make the video competitive with others that are currently airing on target video programs.
How much of the production cost is the artist expected to pay?	Artist contracts typically include a provision that the artist pays for half of the cost of video production, though some require the artist to pay for all costs above $100,000. The artist must be willing to accept these recoupable costs.
Is there a viable programming opportunity for the video?	The video promotion department must confirm that there are video programming opportunities for the message and concept of the video, and an adequate number of outlets must be identified that are likely to present it to their audiences.

Producing the music video

Creating a music video has all of the standard financial considerations that are applied to a label's recorded music project. But, it comes down to an answer to the question: "Will the video help us sell more music than it will cost to create?"

Creating a budget that stays within available funds is the first step in creating the music video. The criteria for the quality of the production are determined by looking at the video channels themselves. Any music video that the label produces must at least be comparable to production elements used by those that actually make it to the air on the video channels. We have heard about the low-budget music video that gets airplay, but instances like that are anomalies; the label must create a music video for programming that meets the level of creative and technical quality that matches or exceeds videos currently airing on the target network or program. Often this reaches $250,000, and can at times exceed a million dollars (Tiwary, V. J., 2002).

When the inclusion of a music video becomes part of the promotional plan for a recorded music project, the label will typically request proposals from video production houses for storyboards and ask them to include their suggested concepts for the video. Replies by video companies will also include the budget necessary to produce the video based on the suggested concept. The label will either accept the idea and its budget, or it may negotiate for a larger or smaller production and accompanying budget.

The Wall Street world of financial accountability has pushed some record companies towards more of a conservative approach to music videos by delaying the production decision until it is clear that the single is finding airplay chart success at radio. In the past, this strategy carried risks because music video production required six to eight weeks to take the video from concept to completion. By that time, it might have been too late to get a music video promoted to video channels to help the single. This was especially true with rock and urban stations, where time on an airplay chart can be fewer than ten weeks.

However, it is possible today to request proposals for concepts and then keep the proposals on file until the single achieves chart success. If the video is required, video production companies can use current technology to provide rough cuts of the produced video in a week, and have the final version ready for promotion in less than two weeks.

From this point, the challenge is in the hands of those responsible for promoting the new video to convince video programmers to find a place to add it into the playlist.

Uses for the music video

Promotion

In a perfect world it would be a simple matter to create the video and be able to determine the impact it has had in supporting the sale of a recorded music project. Certainly a bump in sales figures can often be seen in SoundScan data when a music video has been added into the rotation of a video channel or local cable program, but there are other opportunities for use of the video to support the project. Often, it is difficult to measure the impact of these other uses for the video, but following are some of those additional opportunities the music video offers the record label.

The publicity department and the independent publicist rely on a video to help them garner media attention for the artist. When a new artist is being pitched to a journalist for a story, the video is helpful by associating faces and performances with the music. The publicist also uses the video when vying for a performance opportunity on early morning and late night network television shows, including *Saturday Night Live, Today Show,* or *Oprah*. This added tool gives the gatekeepers of these television shows a view of what they might expect when the artist appears.

The radio promotion staff may find the video useful, especially if the artist is new. Radio programmers are mailed a copy of the music video, or they might be emailed an mpeg or an .avi file to make an impression that goes beyond the music. An example of the power of the video is the debut single of country group Trick Pony. Some radio programmers increased spins on their stations based on the strength of their video for "Pour Me."

For labels with new media departments, the music video can create a presence on the Internet. Yahoo and other sites have set aside an amount of dedicated space to feature music videos. The label's website often includes a link to the music video, and the artist's website will have an associated link with the music video. Artist management companies that maintain an active web presence will also include links to either the label's or the artist's site to give web surfers the experience of the music video.

Sales

The marketing department also has specialized uses for the music video by including it as part of a "value added" feature of the release of the CD. It is not uncommon for CD jewel cases to display an added bonus for the consumer in the form of video performances either as a separate DVD, or as a single disc with the music CD on one side, and a DVD on the other side (dual-disc). Packaging music video performances as part of the experience the consumer purchases with a music CD adds value to the product, and positions it to better compete with movie DVDs, which regularly include added features.

Based on RIAA shipment data, shipments of VHS music videos peaked in 1998. The DVD format has grown significantly since its introduction in '98 to command a majority of the music video sales market. Even with the introduction of the DVD format, the music video market had not recovered to 1998 levels until 2004, according to the RIAA. For 2004, even VHS music video shipments were up slightly. Based on RIAA shipment figures, cost per unit on VHS music videos slipped from $12.65 in 2003, to $12.49 in 2004. The average cost of a DVD music video in 2004 was $19.34.

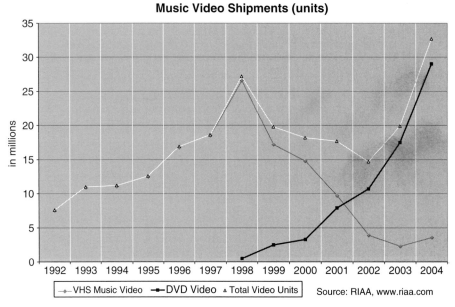

▲ Figure 15.1 Music video shipments

The SoundScan numbers tell a slightly different story, with sales for 2001 at only 8.9 million units. The RIAA shows more units of VHS were shipped in 2001 than DVD, whereas SoundScan shows more DVD units sold in 2001 than VHS units. Both show DVD sales increasing exponentially from 2001 to 2004, and the overall video market growing with it—more than tripling in just two years. VHS, on the other hand, is in the decline stage according to SoundScan, accounting for just a fraction of sales by 2004.

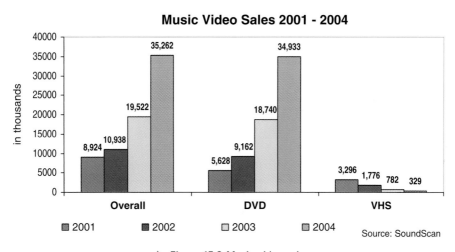

Music Video Sales 2001 - 2004

Source: SoundScan

▲ *Figure 15.2 Music video sales*

Label staffing

Labels do not do their own music video production, preferring to hire outside firms that are specialists in the field. However, all labels have someone on staff who coordinates the creation of music videos as one of their job responsibilities. Depending on the label, video coordination may be a stand-alone job, it could be part of creative services, or it may be a part of the publicity department. Likewise, many labels hire out video promotion.

Coordination of the creation of a music video involves working with A&R, publicity, marketing, radio promotion, video producers, and creative services to find a common vision for the video. The coordinator seeks to find a story or concept that can enhance the music, and then find a consensus among the various responsibility centers in the label.

Independent video promotion

Contracting with an outside music video promotion company can be efficient for many record labels because the video promotion function is not one that is done on a full-time basis at most labels.

Video promotion takes on many of the features of radio promotion of a single. Independent music video promotion company AristoMedia has a 12-week program for its music video promotion clients. The company:

- Takes new music videos to cable channels and pitches them to the programming staff;
- Puts videos on a compilation DVD, called a video pool, and services local, regional, and national video outlets;
- Follows up with video programmers;
- Creates tracking sheets so they know the frequency the video appears on each of the outlets to which they promote;
- Follows the charts of each of the music video outlets;
- Always promotes for the heaviest rotation possible.

Among the services offered by some independent video promoters is a semi-monthly creation of genre-specific compilation discs, sometimes referred to as video pools, which are sent to video channels. A promotion like this is fee-based, but it is an efficient method to distribute music videos to hundreds of music video outlets. Some of the businesses that use reels from video pools include:

- Nightclubs and DJ services
- Music retail outlets like Best Buy, Tower, and Sam Goody
- Retails outlets like Sears, Nordstrom, and Footlocker
- Restaurants like Colton's Steak House & Grill
- Bowling alleys
- Theme parks and cruise lines such as Pleasure Island
- U.S. Military entertainment complexes
- Health clubs such as Gold's Gym
- Airline entertainment reels (AristoMedia, 2005)

The video channels

The penetration of music videos into American households through cable and satellite channels is growing each year. The data shown in Table 15.3 indicates a considerable presence in homes, underscoring the important part they play in getting music in front of consumers.

Top music video channels		
Cable channel	**Audience**	**Service**
BET	79 million households	Entertainment and informational programming targeting African-American audiences.
MTV2	60 million households	Continuous music videos.
VH1	87 million households	Music videos, events, and series.
VH1 classic	33 million households	A combination of current and vintage music videos, musical performances.
CMT	77 million households	Music videos, original programming, mostly targeting consumers of country music.
GAC	34 million households	Music videos and related programming targeting consumers of country music.

Table 15.3 Top music video channels

Source: PR Newswire Association LLC

Submitting a video

Most video channels have their own protocol for acquiring videos for consideration. The video networks associated with MTV, which include VH1, CMT, and Nickelodeon, have a detailed submission form available from any of the network-affiliated channels. It is a four-page contract that gives the MTV networks certain rights for the use of the label's music videos. Among elements of the agreement between the label and MTV are:

- The network has the right to use the video in all available technologies that deliver MTV programming
- The video will be used by MTV at no cost to the network
- The label will pay for all expenses of providing copies to the network and must meet network specifications on the format of the copies
- Videos will be submitted with the inclusion of closed-captioning
- The network has the privilege to use the video for ten years
- The network must be able to use the video, or parts of the video, to promote itself and its programming in other media (MTV, 2005)

After the video has been accepted for consideration, the video channels follow their own weekly schedule for programming decisions. An example of how these decisions are made is by looking at one of the MTV network channels, CMT. Their weekly calendar for programming music videos is like this:

Wednesday—The programming staff reviews airplay information from BDS and MediaBase 24/7 to determine how often radio is playing certain songs. They also review SoundScan data released every Wednesday, which indicates how many units of sales the singles and albums have had during the most recent reporting period.

Thursday—Members of the staff will hear pitches from video promoters who are trying to establish airplay for a particular single.

Friday—The programming staff decides which videos will be added and which ones will be taken off of the current video playlist. At this meeting, the staff will circulate viewer feedback from the telephone request line and from the channel's website.

Monday—The new video rotation begins.

The future of the video

The impact of the music video will continue to increase as cable and satellite television services penetrate international markets that currently are not serviced. The resulting opportunity is for the presentation of American music in non-English speaking countries with visual images to help interpret the lyrics and connect the viewer with the music. And even the definition of the traditional "video" will continue to be revised as the programming for cable stations and consumer dual-discs include visual content to add entertainment dimensions beyond the standard music video.

Technology has combined the CD with the DVD onto a single disc, placing recordings into a position to compete with the added features found on movie DVDs. However, it is unclear if the ability to download music videos to cell phones will become as big of an opportunity as music ringtones have become.

Some of the answers to questions about the music video's future will become apparent as new generations of 3G cell phones, PSPs, PS3s, and video iPods are developed, and when satellite radio broadcasters and others add portable video offerings to their services. The acceleration of broadband deployment into homes has made video broadcasts via the Internet a way to present quality video music performances to web users, and offers new opportunities for content creators.

Glossary

AVI - A computer file format that compresses the size of the video to make it easier to view on the Internet or to send via email.

Broadband – This is high speed Internet access by means other than a 56k modem.

Illustrated songs – An early form of promoting music in theaters featuring live music and projected slides depicting the lyric story line.

MPEG – A computer file format that compresses the size of the video to make it easier to view on the Internet or to send via email.

Soundie – Black and white short films of live performances presented in video jukeboxes in the 1940s.

Video deejay – A television program that features an announcer, information, and music videos.

Video pool – A compilation of music videos that are distributed on a regular basis (semi-monthly and monthly) by independent video promoters to local, cable, and satellite video channels which use this kind of programming.

Visual records – Video film productions in the 1950s created to promote sound recordings.

A special thanks to Jeff Walker for his input and help with this chapter.

Bibliography

AristoMedia (2005). Aristomedia.com and Jeff Walker.

McCourt, T. and Zuberi, N. (2005). http://www.museum.tv/archives/etv/M/htmlM/musicontele/musicontele.htm.

MTV (2005). MTV Networks Videoclip Submission Specifications form http://www.mtv.com/onair/mpeg_us/release.jhtml.

Passman, D. *All You Need to Know About the Music Business,* Simon & Schuster, NY p. 156

PBSKids (2004). *Music Video 1900 Style*.http://pbskids.org/wayback/tech1900/music/index.

PR Newswire Association LLC, 2004, Lexis-Nexis.com, various 2004 client news releases

Tiwary, V. J. (2002). http://www.starpolish.com/advice/print.asp?id=35.

16 The International Recording Industry

Thomas Hutchison

International recording industry

The sale of recorded music overseas offers challenging, yet potentially rewarding opportunities for future growth of the record business. Countries with well-developed music markets are considered to be mature markets, characterized by slow growth and saturation. Market saturation is when the quantity of products in use in the market place is close to or at its maximum. Newer markets are considered emerging markets and offer the potential for rapid expansion. This opportunity is not without complications, the most critical being music piracy.

In this chapter, we will examine how markets are analyzed internationally and what opportunities and challenges the international recording industry faces.

MIDEM conference

Each year, Reed MIDEM holds an international music conference in Cannes, France. It is an opportunity for global networking among music industry professionals. The event includes a large trade show and seminars on international issues in the recording industry. It is considered by many in the U.S. record business to be the most important international conference.

The IFPI

The predominant organization for monitoring and supporting the recording industry internationally is the International Federation of Phonographic Industries (IFPI). The goals of the IFPI include fighting international piracy, fostering the development of music industries globally, and compiling and reporting economic information about the recording industries.

According to the IFPI, global sales of recorded music fell in 2003 for the fourth straight year, a four-year total decline of 16.3%. As a result, the major record labels have further consolidated and trimmed their payrolls to reduce overhead. Global sales were considered flat for 2004, with the growth of DVD music videos and digital downloads offset by a slight reduction in physical audio sales.

The nature of oligopolies

The recording industry is considered a mature industry. Mature industries are characterized by the presence of a structure called an *oligopoly*. An oligopoly is defined as a situation in an industry where the market is dominated by just a few companies who control most of the market share. By contrast, new, emerging industries are characterized by extreme competition among many small players who are all competing to increase their market share as the industry grows and expands.

The consolidation process

Usually at the early stage of marketplace development, the barriers to entry are low, and companies already in the marketplace are not successful in preventing startups from encroaching on their markets. Eventually, the industry is flooded with suppliers and this leads to a shakeout, with some of the more successful small players extending their market share through acquisitions, mergers and driving out the competition. The motivations for expanding and acquiring other companies include: (1) an increase in market share, (2) reduction of costs through economies of scale, (3) more control over the market including protecting pricing and preventing the entry of upstarts. Eventually, these "chains" or medium-sized companies drive out most of the small entrepreneurs or force them into specialized niche markets. At that point, the medium-sized firms begin to buy up or merge with each other until ultimately, just a few major firms control most of the marketplace—an oligopoly. Examples include the tobacco, automotive and soft drink industries, which are all dominated by two or three companies controlling over three-fourths of the market share.

Oligopolies

Economists differ in their evaluation of whether an oligopolistic marketplace stifles product diversity or enhances it. Some argue that innovation will more often occur in an oligopolistic market structure because firms have better ability to finance research and development (Vaknin, S., 2001). With less competition and larger market share, these companies can better afford the costs and risks associated with developing and introducing new products (new artists in the case of the recording industry) than could a market situation with many smaller players.

Others believe in a negative relationship between industry concentration and product diversity. Their research suggests that complacency and status quo are the norm—that is, as long as these few companies control the marketplace, there is no incentive to invest in research and development (Burnett, R., 1996). Several scholars have noted that periods of high industry concentration in the recording industry (fewer labels controlling more of the market share) have been accompanied by periods of low product diversity (fewer top-selling titles). This trend started to change in the 1990s when major record labels realized that product development was more effective with smaller business units, such as independent labels. Major labels made great attempts to forge partnerships and partial ownership of the most successful independent labels without completely absorbing them into the larger company. Joint ventures, pressing and distribution deals, vanity labels and imprints became the norm, thus preserving the innovative edge of the indie label environment.

Pricing by oligopolies

It is also believed by some scholars that consolidation leads to higher pricing as the few remaining firms have a lock on the marketplace. This notion has been refuted by other economists who have found that mergers do not always drive prices higher. Companies fear losing market share to new entrants and existing competitors. This is especially true in a market where entry by new players is easy and cheap. In the recorded music industry, technology has enabled startup companies to record top quality products at a fraction of the traditional costs, and use the Internet as an inexpensive way to market and distribute those products. In a report commissioned by UNESCO, David Throsby (2002) states "... [the majors'] capacity to continue to exert the market power that they have enjoyed in the past is seriously threatened by the fluidity and universal accessibility of the internet."

Universal Music, the largest of the major music companies (when publishing is included), slashed its retail prices in 2002 in an effort to recapture customers and improve market share. However, as a condition of the new pricing strategy, the company did demand that retailers provide them with 32% of their shelf space and promotional considerations. So far, none of the other major labels has followed.

The four major labels

As a result of consolidation in the industry throughout the 1980s, 1990s and into the new millennium, the music industry is now dominated by four major companies who control about 70–80% of the world music market. These companies are Warner Music Group, Universal Music Group, Sony BMG, and EMI.

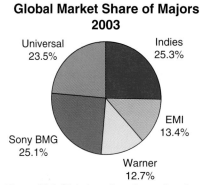

Global Market Share of Majors 2003

▲ *Figure 16.1 Global market share of majors (Source: IFPI)*

Universal Music Group (UMG) was the largest of the major labels until the Sony-BMG merger, coupled with a drop in market share, caused them to slip into second place, according to some analysts. The company controls about 25 record labels, including Interscope, Island, Def Jam, Mercury, Motown, Geffen and A&M. UMG is a subsidiary of the French telecommunications giant Vivendi Universal, which includes Universal Music Publishing, the world's third largest music publisher. Universal Music Group has operations in 71 countries. Universal's market share is strongest in the United States and Europe and weaker than other majors in Japan and Latin America.

EMI, with a 13.4% market share in 2003, did rank second behind Universal until the Sony BMG merger (IFPI, 2004a). Home to labels such as Blue Note, Capitol and Virgin, EMI has the smallest market share of the majors in the United States, but has done remarkably well in Africa, Europe and Australasia. EMI owns the world's largest music publishing company, with rights to over a million songs, but has no other communications or entertainment holdings. In 2003, EMI entered into acquisition talks with Warner Music Group, but lost out to former Seagram head Edgar Bronfman, Jr., who purchased Warner Music Group for $2.6 billion.

Warner Music Group (WMG) was purchased from AOL-Time/Warner by a group of private investors including Edgar Bronfman, Jr. and went public in May 2005. With labels such as Atlantic, Elektra, Reprise, and Rhino, WMG has a strong catalog but continues to lose market share with just 12.7% in 2003. Warner's share is strongest in North America and Latin America, and weak in Africa and Japan.

Sony Music Entertainment (SME) recently merged (in a joint venture that includes only the record label divisions) with BMG to rival UMG in market share. Before the merger, SME had a global market share of 13.2%. Sony has a strong presence in Latin America and, among the major labels, the strongest presence in Japan—home of the parent company. Sony produces consumer electronics products including the successful Sony PlayStation® and cell phones through its Sony/Ericcson joint venture. Labels include Columbia and Epic, and music sales account for about 7% of total revenue. The merger with BMG is designed as a 50-50 joint venture limited to the label division, and is called *Sony BMG*.

Interesting! CR is distributed by Columbia in Japan - doing great there.

★NSync in Germany

Bertelsmann Music Group (BMG), now a partner with Sony, is owned by the German media giant Bertelsmann. Bertelsmann has publishing, music, and broadcasting operations in nearly 60 countries. It owns the top book publisher Random House, several major newspapers, and has a 75% stake in magazine publisher Gruner + Jahr, which publishes *Family Circle* and *Inc*. Best known record labels include Arista and RCA, J Records and LaFace.

D.Cook   *NSync*   *Idol works w/ SonyBMG*

In 2002 and again in 2003, the independent label sector accounted for just over 25% of global sales. The indie market share is lowest in North America with 16%, and quite low in Europe and Australasia with 20%. The indie sector is highest in Japan, where several large regional labels contribute to the indie market share of 55.3%. *Woah!*

Markets

The IFPI reports industry statistics for regions as well as markets in individual countries. Calculations based on regions include only the most influential markets in each region.

Table 16.1 Countries included in regional analysis (Source: IFPI)	
Region	**Countries included in regional analysis**
North America	Canada, U.S.
Europe	Austria, Belgium, Czech Republic, Denmark, Finland, France, Germany, Greece, Hungary, Ireland, Italy, Norway, Netherlands, Poland, Portugal, Spain, Sweden, Switzerland and the U.K.
Latin America	Argentina, Brazil, Chile, Colombia and Mexico
Asia	Hong Kong, Indonesia, Japan, Malaysia, Philippines, Singapore, South Korea, Taiwan and Thailand
Africa	South Africa

How markets are evaluated

Before one can examine individual geographic markets, it is necessary to understand how the IFPI evaluates national and regional markets. Recorded music sales include an estimate of legitimate retail sales including singles, LPs, CDs, SACD, DVD Audio, MiniDisc (MD) and others. Final retail value is estimated and converted to U.S. currency values. Price splits are examined — the relative market share of topline product (full), midline product, and budget product. Repertoire origin is examined, which evaluates the relative market share of music originating within the country (domestic) to that which is produced elsewhere (international). World ranking and piracy level are evaluated for each market. The IFPI also looks at sales by genre and retail channels.

The proliferation of consumer electronic devices is seen as an indicator of market potential, and penetration levels of CD players, DVD players, optical recorders, digital music devices, mobile phones and Internet connections are evaluated by the IFPI for each of the 70 member countries. It is also an indicator of the maturity level of the market. Mature geographic markets are those that exhibit a high penetration level of consumer electronic devices, a higher level of domestic repertoire, a higher proportion of sales in newer formats, lower piracy rates, and higher per capita sales. There is less potential for growth in mature markets, which, by their very nature, are saturated — meaning that consumers are not likely to increase their music purchases from one year to the next.

A comparison of emerging markets and mature markets

Developing (or emerging) markets differ from mature markets on several factors. Emerging markets are commonly found in countries that are still growing their industrial sector, moving from predominantly agricultural products to

Table 16.2 IFPI measures (Source: IFPI)

IFPI measure	Description
Sales of recorded music	Sales are reported by format, by variable US$ retail value, by fixed US$ retail value, by local currency value, and by annual percentage of change in units. Fixed US$ retail value is a report based on historical local currency values re-stated at the exchange rate at the time the statistics are reported.
Format analysis	Unit sales are reported for singles, LPs, MCs, CDs, DVD, and VHS. Market share (in value) and percent change in value from the previous year are reported.
CD price splits	Based on units sold through retail channels and includes full—based upon full retail price, mid—product priced between 50-75% of full value, and budget—product priced between 35-50% of full price.
Sales by genre	The list of genres varies from country to country and is not included for all countries.
Repertoire origin	Compares domestic and international market share. Sometimes classical or regional are included as additional categories.
Electronic hardware penetration	As an indicator of the market for technology and entertainment software, the total ownership of hardware units is divided by number of households to yield penetration statistics for CD players, DVD players, optical recorders, mobile phones, and digital music players (most likely MP3).
Internet penetration	Digital music sales are not yet compiled by the IFPI, but in anticipation of this, the potential market for Internet delivery of digital music is measured through proportion of households with Internet and broadband connections.
Domestic piracy level	Domestic piracy is a measure of consumption of pirated goods, regardless of the origin of such product. It is measured as a percentage of total (legitimate and pirate) sales.

manufacturing. As this happens, workers have more disposable income and more time for recreation and entertainment. Emerging markets have more potential for growth as consumers are still purchasing their first recorded music players. Differences are also evident in the record industries. As emerging markets grow, the ability to nurture and develop local musical talent increases, ultimately increasing the amount of domestic repertoire. However, Throsby argues that the proportion of domestic repertoire is subject to decrease in growth markets. He claims that as international music becomes more available in these markets, and as incomes rise and consumer tastes change, consumer demand for international repertoire increases and the proportion of domestically produced product declines.

Table 16.3 Characteristics of emerging vs. mature markets

Characteristics	Mature markets	Emerging markets
Hardware penetration	High level	Low level
CDs vs. cassettes	Higher proportion CDs	Strong presence of cassettes
Repertoire origin	Domestic rate high	Domestic rate lower
Piracy	Low levels	High levels
Per capita sales	Higher	Lower
Industry concentration	Oligopoly	Strong Indie presence
Market growth	Low or no growth	Higher growth

A comparison of top ten markets and developing markets

The top ten markets all show characteristics of mature markets with the following exceptions. The market share of Indies in Japan is over 50%, due mainly to strong local labels that provide much of Japan's domestic repertoire. Despite the statistical figure for market share of indies, the market in Japan can be characterized as an oligopoly. Domestic repertoire is low for Australia and Canada which import music from other English-speaking countries, most notably the United States and United Kingdom. It is also low for the Netherlands, which imports much of their music from the rest of Europe. The piracy rate is higher in Italy, due to the presence of pirate networks and an influx of popular pirated music from nearby San Marino and other places.

Table 16.4 Characteristics of Top Ten markets

Characteristics of Top Ten markets 2003				
Country	Piracy	Domestic repertoire	CD player penetration	Market share of Indies
U.S.	<10%	93%	262%	16.1%
Japan	<10%	72%	122%	55.3%
U.K.	<10%	47%	163%	20.6%
France	<10%	60%	128%	16.6%
Germany	<10%	48%	124%	20.9%
Canada	<10%	22%*	112%	14.5%
Australia	<10%	26%*	93%	20.7%
Italy	25–50%	48%	65%	N/A
Spain	10–25%	46%	111%	35.8%
Netherlands	10–25%	19%*	138%	23.8%

Source: IFPI

For purposes of comparison to the top ten markets, emerging markets were selected from the IFPI world-ranking list and includes a sample of markets in the lower half of the chart. Attempts were made to include countries from differing geographic regions. Piracy levels are substantially higher and domestic repertoire is lower in most cases. Information on market share of labels and hardware penetration is not available for these smaller markets.

Table 16.5 Characteristics of emerging markets

Characteristics of emerging markets 2003			
Country	Piracy	Domestic repertoire	CD player penetration
Pakistan	>50%	n/a*	48%
Egypt	>50%	n/a*	52%
Slovakia	25-50%	35%	54%
Lebanon	>50%	46%*	58%
Bulgaria	>50%	61%	60%
Venezuela	>50%	19%	61%
Uruguay	>50%	40%	65%
		*Repertoire is listed as regional	
			Source: IFPI

The major markets and trends

The top ten markets for recorded music account for 85% of world sales in dollar value. Mexico and Brazil have recently fallen from the top ten with reduced sales resulting from an increase in piracy. Only one market in the top ten — Australia — experienced growth in the year 2003 (Broward Daily Business Review, 2004). Australia has recently reversed its ban on parallel imports, which led to a unit increase of 2.5% for 2000, but a value decrease of −5.2%, indicating an overall drop in retail prices. In 2001, Australia experienced a 14.4% increase in units, with only an 8.0% increase in value — further evidence of price reduction.

In 2004, volume growth of 2.8% in the U.S. and 4.5% in the U.K. helped stabilize the global market. Together, these two countries make up 47% of the world's value in sales (IFPI press release, March 22, 2005). The sale of digital downloads increased more than tenfold in 2004, to over 200 million tracks in the four major digital music markets of the United States, United Kingdom, France and Germany. Elsewhere, overall markets were mixed for 2004: down for Germany and Japan, up for Latin America, leaving the industry flat.

Table 16.6 Top Ten music markets

Country	Top Ten music markets retail value 2003		
	In millions USD	% Change from 2002	Per capita sales (units)
U.S.	11,848	−6.0%	2.7
Japan	4,910	−9.2%	2.0
U.K.	3,216	0.1%	4.3
France	2,115	−14.4%	2.3
Germany	2,041	−17.9%	2.2
Canada	676	−2.9%	1.8
Australia	674	5.9%	3.1
Italy	645	−4.4%	0.7
Spain	596	−9.4%	1.4
Netherlands	499	−5.1%	1.9

The 6.6% drop in global recorded music sales in 2003 marked the fourth consecutive year of decline. The IFPI attributes the decline to three factors: (1) illegal downloading and burning, (2) competition from other entertainment sectors, and (3) economic downturn.

Issues and challenges to marketing recordings globally

Piracy

The primary threat to the international recording industry is piracy. In 2003, 40% of all recordings sold worldwide were pirated copies. Over two-thirds of pirated product sold in 2003 was on discs. In 2003, disc piracy increased by 45 million units. The format of pirated product varies from market-to-market. Pressed discs dominate the pirate market in Asia and Russia, while CD-R accounts for the majority of pirate product in Latin America, North America and Europe. The IFPI examines piracy from two perspectives: (1) the production of pirated goods, and (2) the consumption of pirated goods. The consumption aspect is termed domestic music piracy, while most production issues deal with "pressing capacity."

Production

Pirated recordings include factory pressed CDs, where music albums are copied on blank discs, and pirated cassettes. The market for CD-R has grown in recent years, especially in more mature markets.

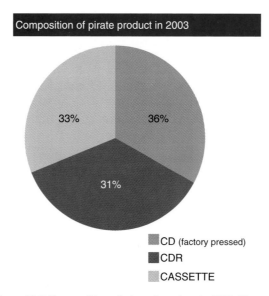

▲ *Figure 16.2 Composition of pirated product in 2003 (Source: IFPI)*

In markets where production of factory-pressed pirate CDs is high, the IFPI examines pressing capacity. This is an examination of all the CD factories' manufacturing capacity and a comparison with the legitimate demand for legal discs (of all software) in markets served by those manufacturers. For instance, Taiwan has factories capable of producing 7.9 billion units per year, but there is only a local demand for 270 million units in areas served by these producers. The IFPI claims that overcapacity is a key factor in the spread of disc piracy, including recorded music, movies and computer software. Various measures are used to monitor production in factories suspected of producing illegal copies (see section on anti-piracy efforts). The Asia-Pacific region is the primary location for factory-pressed discs, feeding illegal music markets around the world.

Piracy operations have recently been linked with organized crime as optical discs become a commodity linked with drug trafficking, illegal firearms, money laundering and even the funding of terrorist activities. Some recent raids on illegal music production facilities have netted caches of other contraband.

Consumption

The IFPI also looks at consumption of pirated product by analyzing domestic piracy levels for each member country. Countries that show a high level of domestic piracy are targeted in anti-piracy efforts including crackdown on distribution and importation.

Domestic music piracy levels in 2003 (units)			
< 10%	10% - 24%	25% - 50%	> 50%
Australia	Bahrain	Chile	Argentina
Austria	Belgium	Costa Rica	Brazil
Canada	Finland	Croatia	Bulgaria
Denmark	Hong Kong	Cyprus	China
France	Italy	Czech Republic	Colombia
Germany	Netherlands	Greece	Ecuador
Iceland	New Zealand	Hungary	Egypt
Ireland	Oman	India	Estonia
Japan	Qatar	Israel	Indonesia
Norway	Singapore	Philippines	Kuwait
Sweden	Slovenia	Poland	Latvia
Switzerland	South Korea	Portugal	Lebanon
UK	Spain	Saudi Arabia	Lithuania
USA	Turkey	Slovakia	Malaysia
	UAE	South Africa	Mexico
	Zimbabwe	Taiwan	Pakistan
		Thailand	Paraguay
			Peru
			Romania
			Russia
			Ukraine
			Uruguay
			Venezuela

Source: IFPI, National Group

▲ *Figure 16.3 Domestic music piracy levels (Source: IFPI)*

Anti-piracy efforts

The IFPI has been working with government organizations to reduce piracy through the following measures:

1. Encouraging copyright legislation and adequate enforcement, including monitoring borders and educating local law enforcement officials.
2. Regulation of optical disc manufacturing facilities, including implementing identification coding, and plant inspections.
3. Consumer awareness: Effective prosecution and deterrent penalties. Education and consumer awareness programs. Making local governments aware of the adverse effects on the development of local artistry.

Parallel imports

Parallel imports are defined as imported genuine copyright goods for resale by an agent other than the local authorized distributor. In other words, the importing of copyright goods into a country that were lawfully made in a foreign country, without obtaining permission of the holder of the copyright for those goods in the recipient country. In the recording industry, this occurs when a particular title is imported from another country when a domestic version of the same product is released in that country. Parallel imports are considered a problem when differential pricing policies are in effect. Recorded music is priced based upon what the market will bear, and that varies from country to country. As a result, prices are significantly lower in some countries. Problems occur when manufacturers attempt to maintain pricing policies that are not competitive with the imported version of the same product.

This is currently an issue in the prescription drug industry. Consumers want the ability to import cheaper prescription drugs from Canada. Drug manufacturers are attempting to restrict this practice through government legislation, therefore making the importation practice illegal. This allows the drug companies to maintain higher prices in the United States for the same drugs sold for lower prices in Canada.

Currently, the United States has no such restrictions for the importation of licensed recorded music. As long as the music is a legally licensed copy, it can be imported into the United States and sold alongside the version released for sale in the United States. Until recently, Australia had laws against the importation of recorded music. As a result, Australian consumers paid more for CDs than consumers in other comparable countries. Consumer groups and retailers petitioned the government and the restrictions were lifted. Also, policies in the European Union call for the gradual abolishment of tariffs and trade restrictions, allowing for parallel imports among member countries. While this practice is good for consumers and retailers, it causes problems for manufacturers.

The primary problem with parallel imports for manufacturers is that it undermines differential pricing policies by leveling the playing field, usually to the lowest available retail price. Parallel imports also create problems with customer support, as local agents are sometimes called upon to offer warrantee service on products purchased elsewhere. The imports may also cause problems for local agents who pay for marketing expenses, only to see the consumer demand they have created met with purchases elsewhere. Licensing also becomes less profitable as local agents may hesitate to pay licensing fees for fear that they may not be able to recoup the costs of licensing, manufacturing and marketing.

Parallel imports are also subject to erratic variations in exchange rates. In the 1990s, when the British pound was low compared to the German mark, German retailers were ordering their CDs from British suppliers, thus forcing TVG-WD, the top German music rackjobber, into bankruptcy. Several years later, when the British pound rose against the mark, these retailers could no longer afford to order CDs from the United Kingdom, and found the distribution infrastructure in Germany was now lacking. Now that the European Union has adopted a standard currency, this will no longer be a problem.

The DVD industry has avoided problems with parallel imports by selling systems in Asia that are not compatible with systems sold in the United States. Therefore, DVD movie titles sold in Asia will not work on DVD players owned by U.S. consumers.

Licensing

It is still a common practice in the recording industry among the nonmajors to license recordings to foreign labels to manufacture and sell in their own territory. Under this arrangement, the original record label grants a foreign record label a license to issue an artist's recording in its local or regional territory, as specified by the terms of the agreement. The licensor (original label) provides the licensee (foreign label) a master recording and artwork for the licensee to reproduce, distribute and market. The licensor is paid a percentage of the local published price to dealer (PPD), which is usually between 8–17% of the retail price (Lathrop, T. and Pettigrew, J. Jr., 1999).

In the absence of parallel imports and other forms of transshipment, licensing can be quite profitable for the licensee. However, in the spirit of global trading, many nations have reduced barriers to parallel imports, thus undermining the value of licensing. Also, Internet marketing and commercial music downloading services have reduced the effectiveness of territorial licensing by allowing the original record label to distribute globally without relying on other labels internationally to make recordings available on the local level. Licensing is still effective when a physical presence in local retail stores is important.

Currency exchange rates

Every nation that has an independent currency usually has one that fluctuates in value relative to the currencies of other nations. Although a few countries officially fix their exchange value to a key currency, the exchange rates between most currencies is primarily determined by market forces and can be erratic when engaging in international commerce. Currency rates are usually cyclical. If the U.S. dollar is strong, American consumers can easily afford to purchase imported goods. But products produced in the United States can become unaffordable overseas. As a result, the demand for U.S. goods is reduced, thus reducing the demand for U.S. dollars. According to the *purchasing power parity theory*, exchange rates will tend to adjust over time so that purchasing power will return to a status quo (The Foreign Exchange Market, 2004).

Not all countries allow their currency values to fluctuate on the open market. Some countries officially fix or peg their currency to that of another country. For example, China currently has their currency pegged to the U.S. dollar. Hong Kong, Saudi Arabia and Egypt also have their currencies pegged to the U.S. dollar. The main advantage of a fixed-rate system is that it removes the risks associated with unpredictable fluctuations in exchange rates over time. The disadvantage is that it ignores market demands for currency and requires countries to maintain large reserves of foreign currency to supply the market when there is an excess demand for them at the official exchange rate.

Fluctuations in currency value can interfere with sales expectations and are beyond the control of the marketing efforts of a company (see section on parallel imports). They can discourage many people from making foreign investments. To overcome this, the European Union has adopted a standard currency, the Euro, in an effort to stabilize pricing and encourage commerce among its member nations.

Cultural restrictions

Global cultural industries that include movies, music, television, books and magazines, and fashions are considered a threat to cultural diversity around the world. Since cultural goods are often mass-produced, there is homogeneity among these goods. As a result, consumers everywhere may wear the same styles and listen to the same music. The underlying assumption is that this will lead to cultural homogenization and a loss of the local culture. The term *cultural imperialism* is used to describe a system where a universal homogenized culture (usually U.S. culture) replaces the genuine local culture. To protect local culture, governments may seek to reduce the influence of outside cultural "imperialism" in several ways: (1) providing an alternative, (2) subsidizing local cultural products, (3) banning outside cultural products, and (4) setting quotas.

Examples of this cultural protectionism can be seen in France and China, where government regulation has restricted the importation of American music and films. Canada is an example of a country that has chosen to not only restrict U.S. media content, but to also offer cultural subsidies to local artisans, including musicians, to develop their craft and products. The United Kingdom is an example of a country that has chosen to offer a government-sponsored alternative to "Hollywood" media by subsidizing the British Broadcasting Corporation (BBC). Their stated purpose is "to enrich people's lives with great programmes and services that inform, educate and entertain" (BBC Website, 2004). The most common forms of cultural protectionism are quotas and censorship.

Quotas

Many nations have imposed quotas or trade barriers on the amount of cultural product imported into their country. Some examples are included in Table 16.7. All are done to preserve the local culture and local cultural industries.

Table 16.7 Quota action

Country	Action
Canada	The Canadian government currently requires that 30% of broadcast material on cable and radio be of Canadian origin.
European Union	The 1989 Broadcast Directive requires that a majority of "entertainment broadcast transmission time be reserved for European origin programming.
France	Extended the EU Broadcast Directive to require that 40% of programming be of French origin, including music on private and public radio.
Italy	Half of the European quota must be dedicated to Italian films (excludes TV films).
South Korea	Limits broadcasting of imported programs to 20% of airtime. Cable TV must limit imported programming to 50% for nonfiction (documentary) programs and 30% for movies.

Censorship

Censorship is imposed to protect the moral and cultural values of a society. Censorship may include imported goods, but may also apply to entertainment produced locally. There are examples of censorship in the United States, stemming from public protests over broadcast material deemed immoral or indecent by public standards. Public outcry over the 2004 Super Bowl halftime show, in which Janet Jackson had a breast exposed, lead to an increase in fines imposed by the Federal Communications Commission for indecency on the public airwaves.

The United States is considered generally more liberal than some other countries in allowing for free expression of ideas. In many parts of the Middle East, North Africa and China, censorship of music is common (Cloonan, M and Reebee G., 2003). In 1996, officials in Iran arrested 28 teenagers for possessing "obscene" CDs and cassettes. In 1997, South Korean radio station KBS banned teen pop music because of the clothing styles worn by the entertainers. In 2003, heavy metal fans in Morocco were jailed for "acts capable of undermining the faith of a Muslim" and "possessing objects, which infringe morals" (Index Online, March 7, 2003).

OMG!

China is one of the most heavily regulated countries with regard to music censorship (Orban, M., 2003). Nothing can be produced or distributed without permission from Chinese government authorities. Restrictions are placed on all materials that "... are against the basic principles of the Constitution ... that advocate obscenity, superstition or play-up violence or impair social morals and cultural traditions, and those that insult or defame others" (Brenneman, E. S., 2003).

Opportunities

Opportunities for growth in the international recording industry are coming from two areas: New markets and new technologies. This growth will be fueled by a reduction in trade barriers, improved economic conditions, the availability of new distribution channels, and continued adoption of next-generation technologies (Pricewaterhouse Coopers, 2004). Bessman claims that "by 2006 the 'experimental phase' of digital music services development will be over, with the subscription models in place ..." (Bessman, J., 2002).

Informa Media Group predicts a return to growth in the music industry in 2005 and expects that by 2008, the value of global sales will rise to U.S. $32 billion (Press Release, 2005).

Geographic growth areas

Asia/Pacific is emerging as the key area of growth in the entertainment and media industries, fueled mainly by China and India, both of which are investing in new communication technologies and media infrastructure as well as opening up their markets to international goods. China, with its population of 1.29 billion, and India, with its population of over one billion, offer the potential for growth and both have a low penetration of media hardware at this point in time. Efforts to stem piracy combined with heavy investments in media industries will drive growth in the region. Compact disc hardware penetration in India was reported at 9% for 2003, with no measured penetration in China. Mobile phone penetration was at 17% in China, and 1% in India in 2003.

New technologies

The recording industry will experience a major shift in the way music is distributed, especially in new markets that do not possess a current infrastructure for the distribution of physical CDs. Growth in these areas is expected to be fueled by the expansion of broadband Internet access and wireless communications. Between 2003 and 2008, the number of broadband households is expected to grow at 31.3% compound annual growth rate.

Music-to-mobile services are already big business in Japan, South Korea and Taiwan, and parts of Europe. Music-to-mobile is defined as the downloading of music—from ringtones to full master recordings—into handheld mobile devices. These devices are third-generation cell phones (3G) capable of providing entertainment in addition to phone services. In South Korea, music-to-mobile sales generated U.S. $29 million for record companies in 2003. As bandwidth increases, more companies will be offering mobile jukebox services. The United States and China represent the largest potential markets. Since the United States dominates in global music sales, growth is anticipated when wireless infrastructure catches up to the European and Asian levels. China represents the potential for growth due to its massive population, increasing disposable income and a vast young consumer base (IFPI, 2004b).

Trade agreements

GATT

From 1948 through 1994, the General Agreement on Tariffs and Trade (GATT) provided the rules for much of world trade and handled issues dealing with international commerce. Members of GATT pledged to work together to reduce tariffs and other barriers to international trade, and to eliminate discriminatory treatment in international commerce. The GATT was designed as an interim measure pending the creation of a specialized institution to handle the trade side of international economic cooperation. The first attempt, the International Trade Organization (ITO), failed to materialize when the U.S. Congress refused to ratify the treaty. However, the GATT remained in effect and was the only multilateral instrument governing international trade from 1948 until the World Trade Organization was established in 1995.

WTO

The World Trade Organization (WTO) was established as a result of the final round of the (GATT) negotiations, called the Uruguay Round. The WTO is responsible for monitoring international trade policies, handling trade disputes, and enforcing the GATT agreements (www.encyclopedia.com, 2004). The WTO website describes their mission as:

> "At the heart of the system—known as the multilateral trading system— are the WTO's agreements, negotiated and signed by a large majority of the world's trading nations, and ratified in their parliaments. These agreements are the legal ground-rules for international commerce. Essentially, they are contracts, guaranteeing member countries important trade rights. They also bind governments to keep their trade policies within agreed limits to everybody's benefit" (World Trade Organization, 2004).

TRIPS

Even before the WTO was formally operational, the Uruguay Round of the GATT negotiations produced the Agreement on Trade-Related Aspects of Intellectual Property Rights (TRIPS), negotiated from 1986-94. This agreement introduced intellectual property rules into the multilateral trading system for the first time, and addressed protection of copyright ownership of recorded music. The agreement establishes a minimum amount of copyright protection for creative works and specifies how countries should protect and enforce these rights.

Marketing products overseas

There are several avenues for marketing recorded music overseas: (1) the major labels tap their affiliates in the region; (2) indie labels license the recording to local labels in the territories of interest; (3) products manufactured in the United States are exported through a foreign distributor; (4) digital delivery to the consumer via the Internet. In all cases, international touring schedules are generally set up to coincide with the release of the album in that territory.

Major labels and affiliates

The major labels rely on their branch (affiliate) offices around the world to promote products deemed profitable in those markets. The majors have the lion's share of the market in most countries, with a viable recording industry as indicated in Table 16.8.

Table 16.8 Combined market share of majors	
Territory	**Combined market share of majors**
North America	81.8%
Latin America	74.0%
Europe	80.6%
Asia	62.1%
Australasia	82.5%
Africa	66.0%
WORLD	74.7%
	Source: IFPI

Marketing through an affiliate

Each regional affiliate of a major label operates independently for the most part, and each branch is responsible for its own marketing. Affiliates generally have the option to accept or decline an opportunity to release a record from an affiliate in another region. For example, if a U.S. branch wants affiliates in other territories to take on one of their releases, they would use a product presentation as an opportunity to pitch the project to affiliates in the targeted territories. If the artist is a major star with international appeal, an international release is customary. If the artist is mid-level or emerging, the U.S. branch may have to convince the affiliates that they should take on the release, even to the extent of paying a portion of the artist's travel expenses. They would also want to demonstrate the marketability of this release.

International affiliates of a major label will gather periodically for product meetings. At these meetings, product presentations by each affiliate feature the artists that they think are most appropriate for international release. The other affiliates at the meeting then take this under consideration, keeping in mind similar acts that they might be working, and consumer demand in their territory.

The potential releases are then prioritized by label, by album and by single. Having a great single can move an album up the priority list. If the label doing the pitching does not succeed with their own affiliates, they may consider contracting with an entity outside of the major label system for distribution in that region, but the affiliates get first refusal.

Touring has an impact on the decisions of whether to take on a project. An artist, their manager, and their booking agent may work with their label to coordinate an overseas release with a concert tour in that region. At times, a label's clout with one artist may help influence an affiliate to take on projects from other artists on the label. For instance, if an affiliate in Europe wanted to handle the release of a major U.S. artist, that artist's U.S. label may request or require that the affiliate handle other acts on the label as well. An artist's contract with their label may also guarantee that the record will be released internationally. If that artist is still developing, the label may scramble to find foreign territories interested in the release. For a major act, a coordinated worldwide release is normal, with all other affiliates onboard to take advantage of a global marketing campaign. On occasion, a U.S. artist will be released initially overseas to "test market" the artist before investing in a U.S. release.

For the international release of Creed's first album, Jed Hilly, who served as Vice President Marketing Services, Sony Music International, New York, had this to say (Jed Hilly, personal interview):

> "When the Creed/Wind-Up deal was announced, I worked with the Wind-Up people and we orchestrated an event. We invited marketing representatives from all over the world, about 20-something people came. Japan, U.K., France—we had guests from all over the world, to Orlando, Florida. We had an afternoon performance where we rented a suite at the local Marriott. We had Julia Darling and Stretch Princess do acoustic sets and then we went to the House of Blues later that night and Finger Eleven warmed up for Creed, for about 2,000 people. So everyone got to go experience that, meet the band members and label personnel. We did that in November and the official Wind-Up launch [internationally] was in January. We had to do this to get people on board." (Jed Hilly)

Once Creed was established in the marketplace, subsequent albums were released with an international plan in place.

Indie labels and international markets

Independent labels must work through arrangements made with labels or distributors in other territories. Arrangements with local labels usually involve licensing and arrangements with distributors involved in the transshipment of products created in the home territory. Again, touring plays a major factor in the success of indie label artists overseas.

Licensing

Licensing was discussed in an earlier section dealing with parallel imports. The label originally releasing an album licenses a label in a foreign territory to release the recording in that territory. The licensor (original label) provides the licensee with a master recording and all related artwork in exchange for a fee and a percentage of sales. The licensee is responsible for manufacturing, distribution and marketing in the designated territory. Compensation is paid to the licensor at generally 8–17% of the local retail list price, or computed as a percentage of published price to dealer (PPD), or wholesale price (Lathrop, T. and Pettigrew, J. Jr., 1999).

When licensing is done, indie labels give up much of the potential for profit, as well as control over manufacturing, sales and marketing. Generally, the licensor does not place requirements or restrictions on how a record is to be marketed. However, requirements may include the quantity of product for which royalties will be paid.

Some indie labels prefer to avoid licensing if they have the resources to engage in marketing activities on their own or with the assistance of a local distributor. It allows them to gain more control over the marketing and distribution process. However, this may vary depending on the territory. One independent label, Compass records, prefers to use licensing arrangements in most of Asia, while relying on local distributors in other territories (transshipments) (Thad Keim, personal interview).

Transshipments/exports

In this situation, finished recordings are rerouted or exported to a foreign distributor. For the importing territory, they are not considered parallel imports unless a locally licensed version is also available. Indie labels that engage in using foreign distributors must work hard to promote the recordings in those regions. They must contact local press, and generally coordinate closely with any touring arrangements. It is also wise to price the product competitively with local pricing, rather than pricing as an "import." Indie labels may use different distributors for each territory and perhaps even different distributors within a territory, dependent upon the music genres they are involved with.

The timing of the release date must be taken into consideration for each territory. An artist who does well in Australia may want a fourth-quarter release date. The same artist may not be as well known in the U.S., and may require a release date in the first quarter of the following year.

"When Compass released The Waifs, 'A Brief History ...' CD in January 2005, it followed the release date in their native Australia by almost three months. While the ideal situation would have been a simultaneous release schedule, it's very difficult for an independent label to release a CD in the U.S. during November (and remain competitive) with all the major labels, hit product being released during 4th quarter. Since among other things, marketing costs at retail average three times higher during this period, we decided to wait until January for the U.S. release. The same restrictions don't apply to The Waifs in Australia, as they are a platinum selling act down there." (Thad Keim, Vice President, Sales & Marketing, Compass Records)

Timing is also important to minimize the potential impact of parallel imports.

"When Compass set up Paul Brady's 2005 release, for the U.S. and Europe, we were in a position where our release dates in certain territories had to be within a 1–2 week window of one another or we would run the risk of exports from one territory making their way into another and upsetting the domestic release. Exports from the U.S. were of the most concern because the dollar is rather weak against the Euro and Pound right now. So in this case, the risk of exports did have some influence on setting our release dates." (Thad Keim)

Internet sales and distribution

The Internet has created opportunities for smaller record labels to compete internationally by offering products for sale through their website. Such sales fall into two categories: (1) the shipment of physical products to global consumers, and (2) the commercial sale of downloads. Target marketing is no longer limited to geographic segmentation, as the Internet has eliminated geographic barriers to marketing, sales and distribution. This is beginning to complicate the issue of territorial rights, differential pricing and staggered release dates.

Marketing strategy

The basic marketing aspects of radio, press, tour support, retail promotion, and video are utilized to market overseas, but the relative importance of each may vary from country to country. Many foreign countries have a different radio industry structure from what has been established in the United States. Government-run radio, such as the BBC, dominates many foreign countries that have only recently introduced privately-owned radio networks. MTV is popular in Europe and Asia and has been proven successful in marketing, but is beyond the reach of all but the majors. Tour support becomes a more important aspect of many of the U.S. acts promoted overseas.

Music genre classifications vary from country-to-country, and may be vastly different from what we are used to in the U.S. For example, in Europe, country music encompasses folk and Americana.

It is precisely because of these vagaries that marketing in a particular region should be left up to that local affiliate or a distributor with the resources to assist in marketing. They know best what will work in their market. The pre-release set up must be done up to two months in advance of the release in order to get press coverage such as magazine covers and media reviews at the time of the release. That can become difficult if the materials are not ready, or the label is not willing to release these materials (including an advance copy of the album) overseas in advance of the album release.

With a new artist, an overseas release usually comes after some success with sales in the United States demonstrating the profitability of the recording. It also allows the artist to wrap-up a U.S. tour before going overseas to support the release with the international tour. For an established artist, a worldwide release date is common to reduce the likelihood of parallel imports and piracy.

Conclusion

The international recording industry faces many of the same challenges that face the RIAA and the U.S. recording industry, and add to that the increased level of piracy and complications stemming from international trade. The industry is monitored by the IFPI, which provides data on recorded music sales and economic climates of the various member nations. The industry numbers in this chapter were the most current available at the time of printing. The IFPI releases annual global statistics in the fall of each year. By applying the concepts outlined in this chapter to economic indicators in subsequent years, the student of the international recording industry can continue to gain insight in the complex factors influencing international record sales.

Opportunities for growth in the recorded music industry will mainly stem from new technologies and the development of new markets as global trade becomes more commonplace. Overseas marketing can be lucrative for the home label and the artist since the expenses associated with pre-production (mastering, video production, and so forth) and marketing are picked up by the overseas affiliate or business partner, and risk is minimal for the home label.

> "This is a global marketplace and you can't approach a pop record and just be thinking about America, because music is one of America's greatest exports. It's the global perspective." (Jed Hilly)

Glossary

Censorship – The process of banning or deleting content considered unsuitable for reading or viewing.

Cultural imperialism – The extension of rule or influence by one society over another.

Emerging markets – Countries that are still growing their industrial sector, moving from predominantly agricultural products to manufacturing.

General Agreement on Tariffs and Trade (GATT) – Established in 1948 between member countries to facilitate international trade. *GATT o make trading easier*

International Federation of Phonographic Industries (IFPI) – The global trade organization for recorded music industries.

Licensing of recordings – The original record label grants a foreign record label a license to issue an artist's recording in its local or regional territory.

Market saturation – The quantity of products in use in the market place is close to or at its maximum. *CDs in America*

Mature market – A market that has reached a state of equilibrium marked by the absence of significant growth or innovation.

Oligopoly – A market dominated by a small number of participants who are able to collectively exert control over supply and market prices.

Parallel imports – The importing of copyright goods into a country that were lawfully made in a foreign country, without obtaining permission of the holder of the copyright for those goods in the recipient country.

Super audio compact disc (SACD) – A Sony formatted compact disc.

Trade-Related Aspects of Intellectual Property Rights (TRIPS) – An agreement that establishes a minimum amount of copyright protection for creative works and specifies how countries should protect and enforce these rights.

Transshipment – The act of sending an exported product through an intermediary before routing it to the country intended to be its final destination.

World Trade Organization (WTO) – An organization responsible for monitoring international trade policies, handling trade disputes, and enforcing the GATT agreements.

Bibliography

BBC Website (2004). http://www.bbc.co.uk/info/purpose/.

Bessman, J. (2002). Slow Global Music Growth Predicted, *http://www.bandname.com/ magazine/articles/display__article.asp?article = slow.*

Brenneman, E. S. (2003). China: Culture, Legislation and Censorship. Excerpts from Chinese Cultural Laws Regulations and Institutions, *http://www.freemuse.org/sw5220.asp.*

Broward Daily Business Review (2004). Annual Global Sales of Recorded Music Fall Again, *Broward Daily Business Review*, **45**, (84). http://www.web.lexis-nexis.com/.

Burnett, R. (1996). *The Global Jukebox,* New York: Routledge.

Cloonan, M. and Reebee G. (2003). In Cloonan, M and Reebee G. (ed) *Policing Pop,* Philadelphia: Temple University Press.

IFPI press release (March 22, 2005). Global music retail sales, including digital, flat in 2004, *www.ifpi.com.*

IFPI (2004a). The Industry in Numbers, *International Federation of the Phonographic Industry.*

IFPI (2004b). The Recording Industry in Numbers annual report, *www.ifpi.org.*

Index Online (March 7, 2003). Report: Morocco: Judge jails Moroccan heavy metal fans, *http://www.indexonline.org/indexindex/20030307__morocco.html.*

Lathrop, T. and Pettigrew, J. Jr. (1999). *This Business of Music Marketing and Promotion*, New York: BPI Publications.

Lathrop, T. and Pettigrew, J. Jr. (1999). *This Business of Music Marketing and Promotion*, New York: BPI Publications.

Orban, M. (2003). Censorship in Music, *http://www.bgsu.edu/departments/tcom/faculty/ ha/sp2003/gp9/docs/TCOMPAPERONCENSORSHIP.doc.*

Press Release (2005). No Growth in Global Music Sales Before 2005, *http://www.the-infoshop.com/press/fi16775_en.shtml.*

PricewaterhouseCoopers (2004). http://www.pwcglobal.com/extweb/ncpressrelease.nsf/docid/418FBBF773CAD0A4CA256ECA002F4489.

The Foreign Exchange Market (2004). http://www.worldgameofeconomics.com/exchangerates.htm.

Throsby, D. (2002). *The Music Industry in the New Millennium: Global and Local Perspectives,* Paper prepared for the Global Alliance for Cultural Diversity; Division of Arts and Cultural Enterprise, UNESCO, Paris.

Vaknin, S. (2001). The Benefit of Oligopolies. http://www/suite101.com/files/topics/6514/files/capitalism.rtf.

www.encyclopedia.com (2004). www.encyclopedia.com. World Trade Organization (2004). http://www.wto.org/english/thewto_e/whatis_e/inbrief_e/inbr00_e.htm.

17 Tour Support and Promotional Touring

Paul Allen

Tour support

Among the tools the label's marketing department uses to promote a recorded music project is tour support for the artist. Most labels view radio airplay as the most important promotional outlet for new music, but right behind radio are live performances through touring.

Often, tour support comes in the form of money advanced to the artist to cover the losses incurred from being on the road. But it can also be time and energy spent by the label to promote the artist while they are touring, performing, and promoting their new album. This chapter will look at what tour support is, and how it figures into the marketing mix as a component of the marketing plan.

The decision to support a tour

Label support for a tour is based entirely on doing "anything it takes to get the artist to perform live on the road"[i] to help promote a new album. A decision by a record label to support a tour with either money or services depends a lot on the strength of the proposed tour, whether the tour can be expected to result in significant added sales, and on the caliber of the team that supports the artist.

---

[i] Interview with John Grady, March 14, 2005.

A good agent who is planning the tour will work with the label to ensure everyone agrees that the markets/cities in the plan are prime locations to sell tickets (to pay for the tour) and benefit from the performances, which will drive sales of the album. Market considerations will include the following:

- Has the artist done well in this market before?
- Are radio stations supporting the single and the album with airplay?
- Does the local print media write about and support entertainment?
- Does the tour sponsor have a presence in the market?
- Is there a possibility for a media partner in this market for the tour?
- Is this a good retail market for the genre and the artist?

Having a strong artist manager is helpful in the label's decision to provide tour support for the new artist. An experienced manager is an assurance to the label that the tour will be organized and well directed on behalf of everyone.

And finally, the decision to support a tour for an artist may depend on whether the artist's recording contract includes a provision for the label to earn a percentage of ticket and merchandise sales. If a label has access to these income streams of an artist, there is an incentive for the label to be more forthcoming with support and reduce the risk that the recording project will not recoup.

Direct financial support for a tour

Tour support can be in the form of financial aid paid to an artist to support their own tour, or to be part of the tour of others. New artists especially have a difficult time being able to fund all of the costs associated with touring, and it becomes necessary to go to the label and ask for a subsidy to cover the losses they will face on the road. Among the touring costs that cash from a label helps to support include:

- The purchase or lease of a van or bus
- Hotel costs for the artist and support personnel
- Per diem (daily food costs) for the artist and band members
- Rental or purchase of equipment for performances
- A tour manager

When a label offers financial support for a tour, it usually requires a proposed budget for the entire tour with an explanation of anticipated losses from touring. Up to reasonable limits (typically less than $100,000) (Passman, D., 2004), the label will advance the money with the expectation that there will be an accounting for the use of the money for the tour, and that any excess will be returned to the label when the tour is complete. Since the money is provided to support the tour, which is helping to promote the new and active single and album, tour support in this form is generally recoupable, meaning that the artist must repay the advance as a deduction from royalties earned from the sale of recordings. Also, it is not unusual for an artist who has purchased equipment or a vehicle for the tour from advance money, to give the items purchased to the record label when the tour is over.

Most advances in the recording industry are immediately commissionable to the artist manager, meaning the manager takes part of it right away. The artist-manager contract usually requires that a percentage of the artist's earnings—including advances—be paid to the manager. However, in most instances, advances from labels for tour support are offered to an artist with the requirement that it is not commissionable by the artist-manager. The logic behind this is that the tour is trying to find a break-even point with the label's support, and the manager should be paid when the artist is actually making money from the tour.

Major label tour support

Major labels have considerably more financial ability to support new artists with tour support in both cash and services. Labels are willing to support recorded music projects in this way because of the large numbers of unit sales the company is expecting from the marketing plan.

Tours that help promote a major release of a new artist require considerable coordination and cooperation by the stakeholders in the project. The label is often willing to provide publicity support for the tour when it is initially announced, and then will continue the media blitz with each stop of the tour. This means setting up each performance in the press in advance, and following up tour stops with post-event press. The artist must be willing to accommodate the local media at each stop with press interviews, visit the local radio stations that play (or should be playing) the single from the album, and they must be willing to appear in local record stores to help promote the new album. The artist manager must be willing to travel with the artist or have someone support the artist on the road in the role of a surrogate for the manager, meaning a tour manager or road manager.

Tour support is also provided for new artists who open concerts for established major headliners. For appearances such as these, the artist typically is not paid by the major act. The value to the artist and the label is the opportunity to appear in front of their target audience to promote a hit single on the new album. The headliners themselves make the decisions on who opens for them. To a major act, that decision is partially political, meaning business relationships can help get the act on the bill. But the bottom line to that decision *is* the bottom line: Does the act have appeal to the audience and can the artist help sell tickets? New artists usually do not have the resources for travel to open for headliners, so there is heavy reliance on label tour support for this kind of performance.

> There are times when a release by a major label requires a high degree of tour support because radio is not playing a single from the project. Such was the case with the soundtrack from *O Brother, Where Art Thou?* Mercury Records released the soundtrack in December 2000, three weeks before the movie went into national distribution. It became one of the most successful soundtracks in history, selling seven million units with virtually no airplay on commercial radio. John Grady was vice-president of marketing for Mercury when the project was released. Among the things he credits for its success is the commitment to a tour featuring the numerous artists on the album, coupled with tour support by the label. Grady says,
>
> *Touring and tour support were major parts of the marketing of the* O Brother *project. After the first performance [on the initial tour], the next was at Carnegie Hall where it sold out in three hours at $115 per ticket. That's when you can tell there's an appetite. We took the tour out 65 times, and it became a major event when it came to town.*

▲ *Figure 17.1 Touring and the "O Brother" project (Source: Interview, March 2005)*

Independent label tour support

Artists signed to independent labels find that tour support in the form of cash is very limited, often non-existent. The economics of independent labels prevents them from having enough resources to financially support touring artists with anything but the smallest amounts of cash. Independent labels rely on grassroots elements to support tours of their artists.

Touring for the independent label is especially critical because very few receive airplay on commercial radio, which limits the exposure of their artist's music to the public. On a nationwide basis in the United States, major labels can expect as high as 76 million weekly exposures of a hit record to individual listeners to commercial radio (Airplay Monitor, 2004). Independents can't compete at that level, so they must rely on live performance and touring as the best—and sometimes the only way—to present the music of their artists.

The position of the independent label is expressed best by David Haley of Compass Records. He says, "If you're not touring and not present in the market, there's no interest ... there's no retail interest, there's no consumer interest, there's no radio interest, no print interest, no nothing." He says that without an artist continually going back to perform in markets that work, it becomes difficult for independent labels to stimulate interest in the record company's recordings.[ii] Touring keeps retail record stores energized, because without the label's projects available at retail, the label cannot thrive.

Tour support for many independent labels comes in the form of mobilizing street teams and e-teams, through the media via the label's publicity specialist, and by using many of the tools outlined in the Grassroots Marketing chapter in this book. Haley says dealing with gatekeepers causes his label's staff to become more and more creative. "Unfortunately, the way the system is setup, there are a ton of gatekeepers and our job is to deal with those people while we figure other ways around them." In some ways this is not much different from the music business in general.

Tour sponsors

Sometimes labels can assist with indirect tour support by working with artist management to find compatible sponsors for tours. Sponsors pay money to the tour in exchange for having a presence at live performances. Products and service companies often link themselves to specific tours because they can connect with the similar target markets delivered by certain touring acts.

An example is the decision of Boost Mobile and Motorola's agreement for a label-wide deal with BMG to sponsor tours. Boost Mobile was launching service in 30 of the markets of Mario's 2004–2005 tour, and with Motorola, sponsored album-release parties for Mario and had company logos imprinted on music CD samplers (Kipnis, J., 2004).

Another example was the decision of Coors to sponsor Brooks & Dunn's Neon Circus Tour in 2002. The Coors logo was integrated into the Brooks & Dunn logo for the tour, and was featured in all promotional materials much like a title sponsor. For their sponsorship, they provided 1,230 cases of Coors beer (tba Entertainment, 2002) plus $2 million in sponsor support. The sponsorship helped fund the 41-date tour that required 140 people working with the tour on the road.

[ii] Interview with David Haley, March 17, 2005.

The payoff for record labels

Touring can contribute to record sales, especially in the absence of any mass media presence. Indie artists who lack airplay and television exposure depend on touring for income, and their labels depend on touring to sell records. But major acts who tour can also reap the benefits in record sales. The following chart in Figure 17.2 shows how sales in local markets tended to rise just before and after each local performance date for artist U2 and their Elevation 2001 tour.

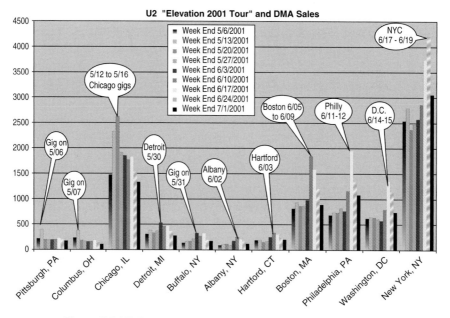

▲ Figure 17.2 U2 Elevation 2001 tour and local sales (Source: SoundScan)

Promotional touring

Ben Berkman, in an article for *StarPolish*, describes a promotional tour as "the process of taking an artist across the country to push their music to radio stations" (Berkman, B., 2002). Add to that the potential to visit press and retail accounts, which are other industry gatekeepers. The main distinctions of a promotional tour, compared to a concert tour, are: (1) it is not designed for or funded by public performances, and (2) it is intended to promote the product to the industry before promoting it to the consumer.

Obviously, in the marketing plan, the radio promotional tour would occur long before the album release and shortly before the release of the first single to radio. Promotional tours are commonly set up for new artists to introduce them to radio. Promotional touring for radio can take on two different forms: (1) the artist travels to radio stations, meeting with program directors and doing on-air interviews; (2) regional showcases are setup and area program directors are treated to a private concert. Berkman states that bringing an artist to the radio station demonstrates the label's commitment to marketing the artist and it helps radio personnel put a face with a name and song.

> "Touring is one of the most important ways that an independent artist can spread the word about their music within both the industry and the general public. It can serve as an excellent catalyst for radio airplay; articles, stories and reviews; retail placement, on-air performances; in-store performances and other promotional opportunities. Touring is also expensive and exhausting for most independent artists".
>
> - INDIEgo.com

▲ Figure 17.3 INDIEgo on promotional touring

Promotional touring aimed at retail, called *in-stores*, are usually targeted at launch date, often to boost first-week sales at SoundScan to create buzz and reorders. So a record's debut on the album charts can be a self-fulfilling prophecy, creating awareness and demand at both retail buying offices and end consumers. *you'll make $ for you + the store*

Promo tours to retail are very similar to radio, and can be the very same event. If a label goes through the expense to create an event showcasing an act, it makes sense for as many decision-makers as possible to be there, taking in the same event. Labels have artists visit retail at their buying offices prior to the initial order date. Meeting key personnel and playing music for the office staff creates good will and connectivity with the artist and his release.

Glossary

Advances – Cash given to the artist in the form of an interest-free loan that will be repaid as a recoupable expense, meaning it will be deducted from royalties.

Agent – Someone who negotiates agreements for live concert performances with promoters on behalf of an artist.

Media partner – A local newspaper, radio station, or television station that helps promote a concert in exchange for a promotional presence in advertising and at the event.

Recoupable expense – An item of cost paid by the record label that the artist agrees to pay back to the label in the form of a deduction from earned royalties.

Tour manager – An individual who coordinates an artist's live performances while on the road, and works directly for the artist manager.

Tour support – Money or services provided by a record company to offset the costs of touring.

Bibliography

Airplay Monitor (March 5, 2004). *Airplay Monitor*, pp. 22.

Berkman, B. (2002). http://www.starpolish.com. /advice/print.asp?id = 73. Promotional Tours StarPolish.com.

Kipnis, J. (Oct 2, 2004). Labels, Marketers Mix at Roadshow, *Billboard*, pp. 6.

Passman, D. (2003). *All You Need to Know About the Music Business,* New York: Simon and Schuster, pp. 144.

tba Entertainment (2002). *Neon Circus Program Book,* tba Entertainment.

18 Special Products and Special Markets

Amy Macy

To try and distinguish between special products and special markets can be difficult. It would be similar to asking the age-old question, "What came first ... the chicken or the egg?" Are there special markets because the industry has created special products, or are there now special products because special markets now exist? Many would say "YES!"

Lifestyle

With the emergence of lifestyle and specialty product retailers, a new outlet for music has evolved. When grabbing a cup of coffee from Starbucks, it's hard to ignore the "cool" music playing and the CD display that's right under your nose. A visit to Pottery Barn or Restoration Hardware elicits more than just home décor, but an ambience that is enhanced by the music playing, which you can replicate in your own home via the purchase of the "branded" CD.

More and more retailers are creating "branded" CDs as product extensions of their retail offerings. Between 1995 and 2003, Pottery Barn sold over 3 million units of music, all of which are an "easy way to put lifestyle in a home," according to Patricia Sellman, marketing vice-president at Williams-Sonoma (Strasburg, J., 2003). Starbucks promotes music at another level. Not only does the retailer sell music CDs on-site, but also through the website www.hearmusic.com; Starbucks logs the playlists of songs being aired in its stores nationwide. If a consumer hears a song he likes, he can go the website and find out the title and performer, with additional links to the performer's catalog and online sales site (Finn, T., 2002).

In an unprecedented event, Starbucks has continued its evolution as a significant music retailer. Coinciding with the national movie release of "Ray," Ray Charles' final recording "Genius Loves Company" debuted on the Billboard Top 200 charts at number two. Starbuck's sold more than 44,000 units, which was approximately 22% of total album sales. This percentage outsold both traditional music retailers and mass merchants—making Starbucks the number one seller during the CD's first week of sales. "Starbucks continues to be thoughtful and strategic as we extend our reach into the music and entertainment arena," said Ken Lombard, president, Starbucks Entertainment. "Our expertise in music through the Hear Music brand, coupled with our unparalleled distribution and consumer reach, created what we hope to be the first of many successful music collaborations. It is truly gratifying to see our customers embracing the fabulous CD and purchasing it in our stores" (Business Wire, 2004).

So what is going on here? Why are music consumers purchasing their CDs from the most unlikely brick-and-mortar stores? Starbucks looked at the tumultuous music-retailing environment as "the perfect storm." As retailers began their pricing wars with the $9.99 CD, music became devalued, and the actual purchasing experience less than sexy. Coupled with the consolidation of radio, which diminished the importance of disc jockeys and their abilities to educate listeners to new music, Starbucks has capitalized on these forces that "have fractured the traditional music industry and diminished the sense of discovery many consumers used to feel about music, giving nontraditional retailers an 'in' to capture those consumers" (Yerak, B., 2004).

About Hear Music:
Founded in 1990, and acquired by Starbucks Coffee Company in 1999, Hear Music is dedicated to helping people discover great music. Not a traditional record label or distributor, Hear Music, with a catalog of more than 100 CD compilations, handpicks songs from new and classic records to create CDs that help people discover music they might not hear otherwise. Hear Music creates CDs and music programming for Starbucks coffeehouses worldwide, as well as for the newly announced 24-hour "Starbucks Hear Music" channel launching in the fall of 2004, under an exclusive, multi-year strategic marketing alliance with XM Satellite Radio. Hear Music operates innovative retail stores in California including the newly-opened Hear Music Coffeehouse in Santa Monica, the first retail setting in America where customers can burn custom CD mixes in minutes. Hear Music CDs are featured at Hear Music and Starbucks retail locations, as well as at www.hearmusic.com.[i]

[i]Business Wire, Sept 9, 2004

▲ *Figure 18.1 Starbucks and Hear Music (Source: Starbucks press release, Sep 9, 2004)*

The benefits

Why, then, are artists aligning themselves with retailers? What is the benefit to them and their existing record label partners? Initial response would be money. The licensing of existing recordings from labels can be very lucrative. Because labels never know the value of their recordings long term, record companies find it hard to calculate the forecast of licensed product. Often, the licensing of music is considered "gravy" — monies generated beyond the "meat and potatoes" of the business. But to address the burgeoning market, many conglomerates have created "special markets" departments devoted to mining the catalog of their holdings and finding new, innovative homes beyond traditional music channels. Additionally, private marketing entities have emerged as potent messengers for products including an innovative organization called Rock River, a leader in branded CD production.

Artists also win in the exchange of private label products that carry their music and image. Hallmark Gold Crown stores have partnered with several artists, with one being James Taylor. His "A Christmas Album" sold over 1 million units during the 2004 holiday season. Taylor's last release "October Road," released in 2002, was the final recording for his Columbia Records contract and he had not re-signed with any other label. "Mr. Taylor is thrilled that so many people are getting to hear the music and that he had an opportunity to record this album," said Gary Borman, who manages Mr. Taylor and stars like Faith Hill for Borman Entertainment. "There were no financial-incentive issues to worry about, and no artistic issues, either. It was a very fair sharing of the pie. It made it simple" (Ogunnaike, L, 2004). With the purchase of three Hallmark cards, consumers could purchase James Taylor's "A Christmas Album" for $6.95, or could buy it direct for $10.95.

Hallmark released a Valentine's Day-themed CD for 2005. Using market research, "We went to our core consumers and asked what artists they're connected with. Martina [McBride] was on the top of the list. It's to drive new consumers in; Martina McBride fans that may not currently shop in the store," says Ann Herrick, integrated marketing manager for Hallmark (Ebenkamp, B. Wasserman, T., 2004).

In the case of James Taylor, Hallmark offered an after-burn market for the artist and his music, capitalizing on name value and music, as well as tapping older consumers familiar with his work who are no longer active music purchasers. Martina McBride may find new fans, since consumers with similar buying practices to current fans already frequent Hallmark, making the alignment a win-win for all three parties: Hallmark, Martina McBride, and her record label, RCA.

Enhancing careers

"Matching music to a vehicle's image is essential to developing a consistent and credible message," says Rich Stoddart, Ford Division's marketing communications manager. In recent years, car advertisers have paired Led Zeppelin with the new generation of Cadillac, Aerosmith's "Dream On" with Buick LaCrosse, and Jimi Hendrix's rendition of the "Star-Spangled Banner" with the 2005 Ford Mustang. While music "won't sell the car," says Gordan Wangers, president of AMCI Inc., an automotive consulting firm in Marina Del Rey, Calif. "... it is capable of grabbing the attention of the right audience to say, 'Here's my new car and here's my message'" (LaReau, J, 2004).

Ford uses country music to market its F-series pickups. The theme song, "Ford Truck Man," was written and performed by country superstar Toby Keith. Clearly, the attention of the right audience has been grabbed. Research validates that pickup truck owners are country music listeners, and Ford is trying to reach this broad audience where it taps its foot ... and gas pedal. Aggressive advertising, including radio, television and print, prominently display the artist's image along with the pickup, ensuring a connection between message and messenger to the consumer.

But in the case of Mitsubishi, the music can get in the way of the auto-maker's marketing message. Their TV advertising campaign featured a band called Dirty Vegas, showing young people dancing in their cars to the group's techno-music. "Mitsubishi sold a ton of records for Dirty Vegas but not a lot of cars. It was a marketing debacle," stated market analyst Gordan Wangers. The 2002 release by Dirty Vegas has scanned over 550,000 units, with the bulk of those sales during the Mitsubishi advertising push. Since then, Dirty Vegas has released several more albums, with none of them scanning over 10,000 units.

Product extensions

Under the description of "special products," many labels are attempting to capture more of their current consumers' dollars by creating product extensions of a specific artist's release. The concept is to cover the losses being generated by file-swapping and burning by creating "super-size" versions of the bigger releases, and targeting hard-core fans that are willing to dig deeper into their pockets in order to have the "special" release. According to research done by the Handleman Company, which provides music for both Wal-Mart and Kmart, "23% of music buyers account for an estimated 62% of album sales, buying an average of a CD every month" (Leeds, J., 2004). Though sales vary, a deluxe edition commonly accounts for 10–20% of an album's overall volume; and they usually carry a much higher profit margin, executives say.

"In this climate, where everyone is bemoaning the death of the CD, and we're all talking about price pressure, there is a growing market, which record companies are hoping to develop, of people that are happy to pay more money for value," said Steve Gottlieb, chairman of the independent powerhouse TVT Records. They released two versions of Lil Jon & The Eastside Boyz' rap hit "Crunk Juice," one with a single CD, and one with two CDs and a DVD. Joining the "special" release crowd is Interscope's release of U2's "How to Dismantle an Atomic Bomb." This release has three versions: The 11-track, $10 basic version; or for $32, a fan can purchase the "collector's edition," including a DVD and 50-page hardcover book. The mid-priced version has the DVD but no book. But if a consumer buys a new "U2" Apple iPod, they get a pre-loaded version of "How to Dismantle an Atomic Bomb" for free.

Other examples include Green Day's "American Idiot," whose deluxe version sells for $25 and includes a 52-page hardcover book. Eminem's "Encore" CD collector's edition sells for $27, and includes 25 photos of the rapper along with a bonus disc granting access via the Internet to Eminem cell phone ringtones.

"The concept of holding on to a fan was eroding for a lot of reasons— other interests in the digital world," said Tom Whalley, chairman of Warner Brothers Records. "We've sold the CD as if there's one type of buyer for 15 years. There are different kinds of buyers." Warner Brothers Records released three versions of last year's Josh Groban album "Closer," including a "fan edition" that was sold only through the singer's website and fan club. The basic CD included thirteen songs, the limited edition included two extra songs and a DVD, and the fan version, listed for $29.98, included two more songs and the DVD. Sales ranged from more than 4.1 million copies for the regular CD to about 15,000 of the top-of-the-line version (Leeds, J., 2004).

Labels have continued to expand the limit of their CD offerings, and they "found that there's not a ceiling," said Jennifer Schaidler, vice president of music for Best Buy, the electronics retailer based in Richfield, Minnesota. "If you put the right value to the consumer, you can charge more."

Compilations

Since moving pictures became "talkies," music has been an integral element to the success of the movie business. Early in the evolution of movies, the industry would hire actual singers to play a part, integrating music and story on-screen. Often, the music from these movies would become nationwide hits and the popularity of the singing stars would explode, making many of them classic voices of the genre. Movie soundtracks have continued to evolve, with the 1960s delivering *Sergeant Pepper's Lonely Hearts Club Band*, and *The Graduate* with its hits "Mrs. Robinson" and "Scarborough Fair." The 70s included *Urban Cowboy*, *Grease*, and *Saturday Night Fever*, all containing a common thread of one very hot actor, John Travolta. The 80s brought about *Footloose*, *Dirty Dancing*, and Roy Orbison-famed *Pretty Woman*. The 90s elevated Whitney Houston to a new level of superstardom with her film debut in *The Bodyguard*. And Celine Dion reached new heights with her mega-hit "My Heart Will Go On" from the epic *Titanic*. All of these soundtracks contained songs that would eventually become huge pop hits, selling millions of units and igniting a new kind of consumer to purchase music.

Soundtracks have learned the lessons of the past, while keeping up with the purchasing practices of today's music consumers. With file-sharing, many music purchasers have become their own disc-jockeys, collecting and burning their "personal soundtracks" containing the individual songs that they want to hear. By compiling songs that capture this spirit while representing the theme of a movie, soundtrack managers have fashioned a "genre" of music that has produced sales.

But sales for the genre have slipped in recent years. In the last two years, 2003 produced four soundtracks that sold more than 1 million units, including *Chicago*, *Bad Boys II*, *The Lizzie McGuire Movie*, and *8 Mile*. For the same year, soundtracks comprised about 5.1% of total album sales. For 2004, soundtrack sales sold 4.2% of total album sales, with no single title passing the 1 million unit sales mark.

Movie popularity can carry soundtrack sales, even without radio airplay. In 2004, *The Lord of the Rings: The Return of the King*, and *The Passion of the Christ* were two best-selling soundtracks that clearly illustrate sales power without radio. Sony Music Soundtrax president Glen Brunman says consumers believe a great soundtrack is "not only a souvenir of the movie but also something that you can experience as a good collection" (Hay, C., 2004). Other soundtracks released in late 2004 that had substantial sales without radio airplay include *Ray*, *Shall We Dance?*, *The Polar Express*, *Mile*, *Dora the Explorer*, *The O.C.: Music From the O.C.: Mix 2*, *Garden State*, and *Team America: World Police*.

There was a time when soundtracks needed superstar performers and top hits to drive sales. But a new consumer is recognizing the value of lesser-known, mid-level acts. "You don't need a star act," Universal Pictures president of film music Kathy Nelson says. "So if that means [retaining] an artist who is relatively unknown but people see the movie and say, 'Wow, that's fantastic,' that's great." Targeting an audience and cross-promoting seems to be a powerful combination in connecting soundtracks with consumers. Disney continues its bond with a younger audience by synergistically focusing its marketing strategies where its consumers exist—the Disney Channel and Radio Disney. "The Cheetah Girls," a movie and soundtrack, has scanned over 750,000 units due to its repeat airings on TV and promotional activities on radio. Plus, the ongoing web presence of Disney-branded products validates to power of targeting an audience (Hay, C., 2004). *Camp Rock, Hannah, Jonas (show, movie, albums, tour, merch)*

Record label A&R reps, along with music publishers, mine film and television opportunities aggressively. Placing a recording and/or song on a movie soundtrack or television score can only positively affect the bottom line of a company. Developing relationships with music supervisors for film and television is the key to securing these coveted positions. The *Hollywood Reporter*, along with *Variety* magazine, keeps industry watchers in the know about upcoming productions, cueing label and publishing reps to the opportunity to place a song in a soundtrack. In addition to in-house representation, both labels and publishers are using the knowledge and connection of specialty brokers. Similar to that of independent radio promoters, these brokers will represent a catalog in the marketplace and are rewarded in the success of the movie or television placement by receiving a percentage of the royalties.

There is a new concept in which movie makers are contracting prominent musicians and artists to conceive the music and create an original score. In this case, the movie production company has created a "work for hire" scenario, where the film company then owns the music. And of course, if the movie and/or soundtrack gain popularity, it's the movie production company that reaps all the benefits (David Haley, Compass Records, personal interview).

Taking a page out of the soundtrack playbook, non-movie compilations have emerged as powerful sales items, as well as marketing tools. The Now! Series music soundtrack is considered the most successful compilation collection in the history of music sales. The U.S. participating collaborators include EMI Recorded Music, Universal Music Group, Sony Music Entertainment, and the Zomba Group, with the albums rotating among Sony, Universal, and EMI for marketing and distribution. The U.S. series has generated about $325 million in billing since its inception in 1998. The key to its success has been the hits, since the brand relies on the top singles to drive sales.

Intangible is the impact on sales of participating artists. Someone purchasing a Now! CD might know three or four of the artists on the package, but the consumer most likely is introduced to a new act that they might also enjoy. The residual purchase of the full-length product of that new act is the result of this cross-promotional item.

Products in the music

Busta Rhymes single "Pass the Courvoisier" was a hit at both radio and with the consumers of Allied Domecq brand of cognac. Although the rapper was not paid to write the top-shelf liquor into his song, the artist wrote the hit after trying the drink, compliments of hip-hop entrepreneur Russell Simmons, who instigated the relationship between the brand and the artist. Simmons continues to brand extensions, including a clothing line he started called *Phat Farm*. Motorola has partnered with the attire division to position its phones as a fashion accessory. Rappers 8Ball and MJG's song about Grey Goose vodka increased the sales of the liquor by 600%, prompting a relationship that included a two-year Grey Goose Music Tour featuring these artists, plus Bone Thugs-n-Harmony, L'il Jon & The East Side Boyz, Musiq Soulchild, and other artists.

Maven Strategies, a marketing agency that specializes in placing corporate brands into songs, believes that more artists will consider product placement deals since hip-hop specifically sells product. To minimize risk for the corporate partners, Maven Strategies created compensation programs that reward the hits that receive heavy airplay, while songs that do not do well on the radio, receive less. "We value what the hip-hop community brings to the table, but we deal in an environment of unknowns. It's important to structure agreements that fairly compensate the artists for their capabilities while minimizing the investment risk for our clients," says Tony Rome, president of Maven Strategies (Banerjee, S., 2004).

Retail exclusives

To lure consumers into a specific store, many music retailers are negotiating exclusive product for their customers. This "value-added" product is called by the industry "superior" versions of an album, and according to the

Coalition of Independent Music Stores (CIMS), creates an unleveled playing field. For the 2004 holiday selling season, Best Buy secured "superior" albums for Atreyu and Queen, and scored the exclusive rights to a four-disc DVD set from Elton John. Mass-merchant store Target arranged for an extra cut on the new "Simple Plan" album from Lava, and negotiated three extra tracks on Twista's "Kamikaze" release. Who wins but the consumer that chooses the "right" store in which to purchase their new music? But CIMS looks at the "superior" product practice to be anti-competitive and hostile, and has decided to apply sanctions within their store environments. Pricing and positioning advertising along with awareness programs will not be enforced, and the stores will refuse to report sales or chart positions for any artist and product released by the affected label. Although these stores can be instrumental in launching the careers of developing acts, the overriding power of the larger merchants usually undermines these types of policies, and the practice continues.

Nontraditional music retailers are getting into the act as well. Through the 2003 holiday season, Amazon.com offered daily exclusives as a promotion to drive holiday sales, while increasing pre-orders of upcoming releases. Concert footage from a fall Springsteen concert at Fenway Park was given to those ordering the upcoming release, "The Essential Bruce Springsteen." Other exclusives included a full-length Counting Crows concert, filmed exclusively for Amazon, and video footage of R.E.M. at a spring 2003 rehearsal in Vancouver (Garrity, B., 2003).

The trade-off for giving away exclusive tracks is gaining valuable marketing real estate and promotional efforts put forth by the retailer. Specifically in the holiday season, marketing programs such as front window displays, product dumps, and in-store positioning can double in price. But at any time of the year, co-op advertising and marketing efforts with national retailers is expensive, so leveraging the assets of the record company, the addition of a single track of music, which may have never otherwise seen the light of day, can be smart business and a profitable venture.

Tour sponsorship and more ...

Not to be confused with label-funded tour support, tour sponsorships are considered those ever-important financial relationships between product and service providers with artists who can emote their message. Beyond tour sponsorship and sync licensing, many record labels wanting to tap the deeper pockets of consumer brands, have positioned their acts for lucrative partnerships. Activities such as product placement, DVD underwriting, and digital tie-ins are being explored at the L.A. Office RoadShow's Music Day. BMG scored a label-wide deal with Boost Mobile and Motorola. Boost and Motorola will sponsor R&B artist Mario's promotional tour in 30 major markets where the Boost service is launching. The companies will also sponsor Mario's release parties with their logos placed prominently on his new CD release.

Traditionally, tour sponsorship was monopolized by "sin" products, since there were so many restrictions of ad placement and targeted audiences. But new entries in the tour sponsorship business reflect star power and the need to sell ... anything and everything. In the summer of 2004, Jessica Simpson's 41-stop tour was sponsored by Proactiv® Solution for $225,000. The skin care brand wanted to build awareness using the positive association with her celebrity status. In exchange for the sponsorship, Proactiv received on-site branding, placement in programs, radio station ticket promotions, and a behind-the-scenes look at Jessica's skin care regiment.

The Destiny's Child tour for 2005 included a U.S. sweep, plus a nine-country international tour and McDonald's sponsored these concerts for $5 million dollars. Building on the pop music relationship of the past, including Justin Timberlake and Alejah Fernandez, Destiny's Child will help brand McDonald's internationally by appearing in TV ads and in-store marketing elements. They will also serve as worldwide ambassadors for World Children's Day at McDonalds, funding Ronald McDonald House and other children's causes.

Not just artists, but venues are catching some of the action as well. Clear Channel Communications secured a $750,000 deal where 16 of its venues will be sponsored by True, an online dating company. Many of the venues' second stages carry the True logo, live dating games occurred between acts, and radio promotions through Clear Channel's website were offered.

These deals are secured in various ways. Sometimes the record label and its conglomerate will help broker the agreement by being the intermediary of the artist and product manager. Additionally, artist managers can be approached directly by product representatives. And, on occasion, a third-party music marketing company will help with a hook-up between a product and an artist, assisting in the navigation of very different corporate cultures and agendas.

As the discretionary income of consumers continues to fragment to the many entertainment choices, music makers are becoming very creative in connecting with buyers. As the "perfect storm" continues to rage, with file-sharing, price wars, radio consolidation, and conglomerate mergers, record labels and their artists will need to continually mine the marketplace for willing partners and entertainment synergy. No longer are "special" markets and products considered incremental business to the bottom line, but this evolving revenue source is essential to the survival of the industry.

Glossary

Branded CDs – A CD sponsored by and sporting the brand of a company not normally associated with the release of recorded music. Examples include product designed for and sold at Pottery Barn, Pier 1 Imports and Victoria's Secret.

Brick-and-mortar stores – Businesses that have physical (rather than virtual or online) presences, in other words, stores (built of physical material such as bricks and mortar) that you can drive to and enter physically to see, touch, and purchase merchandise.

CIMS – The Coalition of Independent Music Stores. *Don't like the big places selling stuff*

Compilations – A collection of previously released songs sold as a one album unit, or a collection of new material, either by single or multiple performers, sold as a collaborative effort on one musical recording. *Soundtracks, greatest hits*

Cross-promotion – Using one product to sell another product, or to reach the market of the other product. *Disney &*

Exclusives – Retail exclusive marketing programs that are not offered to other retailers, but arranged specifically through one retail chain. *Wal-Mart + the Eagles*

Private label – A label unique to a specific retailer. *Wal-Mart*

Tour Sponsorship – A brand or company "sponsors" a concert tour by providing some of the tour expenses in exchange for product exposure at the events.

Superior products – A value-added version of a product that is sold in the "regular" version elsewhere. *Demi's CD @ Target*

Value-adds – To lure consumers into a specific store, music retailers offer exclusive product for their customers.

Bibliography

Brandweek, Sept 20, 2004 v45 i33 p10; Martina McBride, Hallmark, Fall in Love This Valentine's Day With My Heart; Grammy Award®-winner's CD to be available exclusively through Hallmark.

Business Wire (August 31, 2004). Starbucks Coffee Company and Concord Records Release Landmark Recording.

Ebenkamp, B. Wasserman, T. (2004). http://pressroom.hallmark.com/music__val__martina.html.

Finn, T. (September 3, 2002). Stores from Starbucks to Pottery Barn are marketing recordings, *The Kansas City Star.*

LaReau, J. (Dec 20, 2004). Music is key to carmakers' marketing. *Automotive News,* http://goliath.ecnext.com/coms2/summary__0199-3523938__ITM&referid=2090. v79 i6126 p22

Ogunnaike, L. (December 17, 2004). James Taylor's Got a Friend at Hallmark Cards, *New York Times,* http://www.nytimes.com/2004/12/17/business/media/17adco.html.

Strasburg, J. (July 25, 2003). Bands to fit the brand; SF company's CDs sell for likes of Pottery Barn, *The San Francisco Chronicle.*

Yerak, B. (Nov 11, 2004). Retailers branch out for holiday, *Chicago Tribune.*

Leeds, J. (2004). New York Times, Dec 27, 2004; $10 for a Plain CD or $32 with the Extras.

Hay, C. (2004). Movies & Music: Director Hackford on the Genius of 'Ray', *Billboard,* **116,** (47), pp. 10. Nov 20, 2004.

Banerjee, S. (2004). Digital Entertainment: New Ideas, New Outlets, *Billboard,* **116,** (45), pp. 47. Nov 6, 2004.

Garrity, B. (2003). Amazon offering exclusives through holidays. *Billboard Bulletin,* Nov 4, 2003.

19 Marketing Research

Thomas Hutchison

Marketing research is important in every aspect of marketing recorded music. Research is used to find out about the marketplace, about who is buying what products, and what they like and dislike about existing products in the marketplace. Basically, marketing research is conducted to learn more about the market and to learn how to improve the product. Chapter 1 outlined how diffusion theory can be applied to improve upon products and make them more appealing in the marketplace. Marketers learn from research what the most popular product attributes are for each market segment, and then they retool the product and the promotional messages to focus on those attributes. For example, toothpaste users can be segmented into those whose choices are driven by ability to: (1) whiten teeth, (2) freshen breath, (3) fight cavities, (4) fight gum disease, and so forth. It is research that gives marketers the insight into who wants what.

What is marketing research?

Marketing research is defined as the systematic design, collection, analysis, and reporting of data and findings relevant to a specific situation facing a company (Blankenship, A.B. and Breen, G.E., 1993). The American Marketing Association describes it as ''marketing research links the consumer, customer and public through information—information used to identify and define marketing opportunities and problems; generate, refine, and evaluate marketing actions; monitor marketing performance; and improve understanding of marketing as a process.''

The information gathered with marketing research helps companies make marketing decisions; decisions about where to advertise, how to allocate the marketing budget, how to craft marketing messages, and how to present the product in the marketplace.

Much of what is done for research is designed to predict future behavior of consumers. Research is commonly conducted using *samples*, a small representative subsection of the market segment. Then, the findings are generalized to the entire market segment. Sampling is done because of the physical impossibility of interviewing, measuring, or monitoring every member of a market segment. Therefore, much of the information reported in research findings, being predictive in nature, has a preset level of *margin of error*: the level of uncertainty associated with the research. Margin of error is usually expressed as a plus or minus factor, for example 75% ± 2% (see section on research validity).

> "Research is conducted for so many reasons:
> - To check out a new product or concept idea,
> - To ask consumers what they think of an artist's new singles, videos, image changes,
> - To touch base with a particular consumer group, either genre or demographic-based to see what they are thinking about artists, radio, video play,
> - To ask questions about how music is fitting into people's lifestyles as compared to other entertainment items like videogames, movies, DVD, cell phones, etc.,
> - To understand better the relationship between an artist and other brands."
>
> - Linda Ury Greenberg
> VP, Sony BMG Consumer Research

▲ *Figure 19.1 Why marketing research is conducted*

Another aspect of marketing research involves collecting data on previous sales patterns and then projecting that to future sales, whether it's the same products or similar products. SoundScan data is used to examine sales of one release to help guide and shape the marketing decisions of other releases. Researchers also use test marketing: introducing the product into a small portion of the marketplace and monitoring sales before financing a large scale release.

Syndicated vs. custom research

The two basic categories of marketing research are *syndicated* and *custom*. Syndicated research is the gathering of continuous or periodic information that is sold in standardized form to all companies involved in the industry. Examples include the work of Arbitron, Nielsen, BDS and SoundScan. Custom research is usually tailored, proprietary, initiated by a company, and often specific to a particular product developed by that company. Examples include product surveys (found inside the product or attached to the warranty card), focus groups, customer satisfaction surveys and personal interviews.

The research process

There are six stages to the research process, starting with identifying the problem that must be researched. Through the subsequent stages, the research project is shaped and designed to discover ways to overcome the marketing challenge or problem.

1. **Problem identification**—Before a survey is developed, or a research project is designed, the questions to be addressed by the project must be identified. Stage one begins with a definition of the marketing problem. This leads to a preliminary statement of research objectives. The goals of the research may include a description of the market, a prediction of how the product will perform in the marketplace, or an evaluation of the product or a portion of the marketing plan. For example, before marketing the release of a new album, the marketing department may wish to learn more about the target market: what they read, where they hang out, how they find out about new music, and so on.

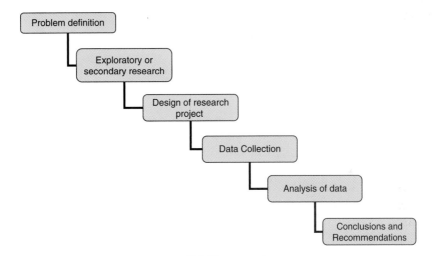

▲ *Figure 19.2 The research process*

2. **Exploratory/secondary research**—Stage two involves finding out what has already been learned on the topic, and what information is already available. This is known as secondary research, defined as going over data that has been previously collected for a project other than this one at hand (Zikmund, W. G, 1991). Examining previous research is necessary to avoid repeating mistakes and to gain an underlying knowledge of the area of study. This helps guide the researcher toward the type of project that will most likely address the research questions, so that in stage three, the research project is designed.

Exploratory research is defined as initial research conducted to clarify and define the nature of the problem. It is done to help diagnose the situation, to allow for screening of alternatives, and to discover new ideas. Concept testing is frequently done in exploratory research. Concept testing is a means of evaluating ideas by providing an early indicator of the merits of an idea prior to the commitment of resources to development and manufacturing—in other words, a prototype. Some record labels will concept test music earmarked for an upcoming album before deciding which songs should be included and which should not.

3. **Design of research project**—The three most popular research designs are experiments, descriptive, and qualitative (focus groups).

Experiments are designed primarily to determine cause and effect. For example, experiments were conducted in the 1960s and 70s to determine if smoking cigarettes caused lung cancer. Ongoing research experiments seek to determine if observing violence on television increases the likelihood of aggressive behavior. Experiments usually involve an experimental group and a control group. The experimental group gets the treatment (cause), or is subjected to the marketing program, and the control group is not. In marketing, experiments may be conducted by placing product or promotional materials in particular locations in some stores, and other locations in similar stores, and then monitoring the differences in sales. Or, finding similar geographic units or markets and implementing advertising plan A in one market and plan B in another.

Descriptive research is designed to describe something, and is often used to describe the market. Surveys are the most common instrument used in descriptive studies. Consumers are asked questions about who they are, what they like and what they do. Record labels commonly use consumer reply cards embedded in CD jewel cases to survey consumers. Questions are designed to provide information of value to marketers. For example, the following survey questions are designed to guide marketers or identify consumers in particular ways.

Table 19.1 Survey questions and research goals	
The survey question	**The research question or goal**
1 What is your age? Gender? Race? Address?	To know more about the demographic and geographic makeup of our market.
2 How did you hear about this CD? What influenced you to buy this CD? (e.g., radio, TV, in-store, concert, magazine, etc.)	Which of our marketing efforts are most successful in reaching our customers?
3 How many songs on this disc were you familiar with before purchasing? 1 2 3 4 5+	Perceived risk: How much risk was the customer willing to take on whether they would like the rest of this CD?

Table 19.1 Continued

	The survey question	The research question or goal
4	What are your three favorite tracks on this CD?	Which songs should we be looking at for future release as singles? Which songs are most likely driving sales of the album?
5	How many other albums do you own by this same artist?	Loyalty: How loyal are you to this artist?
6	What do you read? Watch? Listen to?	How can we best reach you with future advertising and media placement?
7	Do you buy online? Mail order? Catalog? Toll-free phone numbers?	To what extent should we be using alternative distribution methods?
8	What is your favorite beverage? Sneaker company? Car or truck? Music store?	What companies should we team up with for promotional tie-ins?
9	How much time do you spend online? Shopping? Out to dinner? At the mall?	Where can we find you to target you with our marketing efforts?
10	Do you own a DVD or video game player, computer, cell phone? Do you go to the movies, clubs?	What is competing with music for your entertainment dollar?
11	Do you own or use any of the following consumer electronic devices: DVD, MP3 player, CD-R burner, computer music software, multiuse cell phone, etc.?	What configurations should we use to deliver our recorded music and how is the consumer then reformatting our music for consumption?

Thank you for buying this album. We'd love to hear your feedback. After you've had time to listen to the album, please complete this card and return it for a chance to win valuable prizes. Thanks!

(1)
First Name_____
Last Name_____
Date of Birth_____ / _____ / _____   Gender: ☐ male ☐ female
Address _____
City: _____   State:_____ Zip _____
Email: _____

Where did you purchase this album?
☐ Store: _____
☐ Ordered from TV ad
☐ Internet: _____
☐ Received as gift
☐ Won in Contest

(2)
How did you find out about this album? (check all that apply)
☐ Print
☐ Radio
☐ Internet
☐ Listening stations
☐ Friends
☐ TV/cable
☐ Fan club
☐ Store browsing
☐ At Artist's performance
☐ Ad on radio
☐ Ad on TV
☐ Ad in magazine/newspaper
☐ Other: _____

(3)
How many songs on this disc were you familiar with before purchasing?
1   2   3   4   5+

(4)
What are your three favorite songs on this disc?
1. _____
2. _____
3. _____

(5)
Not including this album, how many CDs, cassettes, or LPs do you own by this artist?
0   1   2   3   4   5   6   7   8+

What TV shows that feature music do you watch regularly?
☐ Saturday Night Live
☐ David Letterman
☐ Tonight Show
☐ Conan O'Brien
(6)
☐ 120 Minutes
☐ Total Request Live
☐ Showtime at the Apollo
☐ Yo! MTV Raps
☐ Soul Train
☐ Jimmy Kimmel
☐ Other:

What are your favorite types of music? (1- favorite, 2 - next, 3 - third)
☐ Pop   ☐ Jazz
☐ Rock   ☐ Club/Dance
☐ Rap   ☐ Classical
☐ Country   ☐ Gospel
☐ Blues   ☐ Alternative
☐ Metal   ☐ Soul/R&B
☐ Latin   ☐ Folk
☐ Reggae   ☐ Oldies
☐ Other

Which of the following music video channels/shows do you watch?
☐ MTV   ☐ Video Soul
☐ BET   ☐ Rap City
☐ VH-1   ☐ Video LP
☐ TNN   ☐ Tht
☐ CMT   ☐ Jukebox Network
☐ GAC   ☐ Other

Which of the following daytime TV shows do you watch?
☐ CBS This Morning   ☐ Phil Donahue
☐ Montel Williams   ☐ Oprah Winfrey
☐ Geraldo Rivera   ☐ Regis and Kelly
☐ Today Show   ☐ Dr. Phil
☐ Sally Jesse Rafael ☐ Other
☐ Ellen Degeneres Show
☐ Good Morning America

What are your three favorite TV shows?
1. _____
2. **(6)**
3. _____

Which of these magazines do you read regularly?
☐ Spin   ☐ US
☐ Rolling Stone ☐ People
☐ Details   ☐ TV Guide
☐ N.M.E.   ☐ Melody Maker
☐ Time   ☐ Newsweek
☐ Metal Edge   ☐ Jet
☐ Ebony   ☐ Essence
☐ Down Beat   ☐ Entertainment Weekly
☐ Elle   ☐ Other

List three magazines that you read regularly:
1. _____
2. _____
3. _____

Aside from this artist, whose albums have you purchased recently?
1. _____
2. _____
3. _____

Have you purchased a music download in the past year?
☐ No   ☐ Yes   **(11)**
If yes, how many?
What sites? _____

Do you order from any of the following mail order catalogs?
☐ Crutchfield   ☐ Victoria's Secret
☐ Spiegel   ☐ L.L. Bean
☐ Tweeds   ☐ Eddie Bauer
☐ J. Crew   ☐ J&R Music World
☐ Lands End   ☐ Bose Music Express
☐ Hear   ☐ Record Roundup
Others:
1. _____
2. **(7)**
3. _____

Do you purchase items from any of the following Internet sites?
☐ Amazon   ☐ Target
☐ CD Baby   ☐ J.C. Penney
☐ CD Now   ☐ Buy.com
☐ Wal-Mart.com ☐ Ebay
☐ Half.com   ☐ Record Roundup
☐ Yahoo shopping ☐ Bizrate.com
Others:
1. _____
2. _____
3. _____

What is your favorite...?
Soft drink:_____
Fast food chain:_____
Sneakers:_____
Jeans:_____   **(8)**
Clothing store:_____
Book store:_____
Car/Truck:_____
Software:_____
Videogame:_____
Sports event:_____

Do you own a DVD player?
☐ No   ☐ Yes
If yes, how long?_____

How many DVDs have you purchased in the past 12 months?_____

Do you own a home computer?
☐ No   ☐ Yes
If yes, how long?_____

How many hours per week do you spend online?_____

Do you own a videogame player?
☐ No   ☐ Yes
If yes, how long?_____

How many hours per week do you spend playing videogames?_____

Do you own a cell phone?
☐ No   ☐ Yes
If yes, how long?_____

How many hours per week do you spend on your cell phone?_____

What are your favorite radio stations?
1. _____
2. **(10)**
3. _____

How many hours per week do you spend listening to the radio?_____

How often to you go to the movies? _____ per _____

What are your three favorite movies?
1. _____
2. _____
3. _____

Do you go to clubs to hear live music?
☐ No   ☐ Yes
If yes, how often?_____per_____

How many concerts have you attended in the past 6 months?_____

How many CDs have you purchased in the past 6 months?_____

Would you like to receive information and special music offers via email?
☐ No   ☐ Yes
Email address:_____

▲ Figure 19.3 Sample consumer survey

Focus groups, or other qualitative methods such as personal interviews, are used when marketers want more in-depth information from the market. Focus groups are unstructured, free-flowing group interviews designed to gain a deeper understanding of consumer feelings, beliefs, and attitudes. Responses to questions can be *probed* with further questioning to find out why consumers hold these attitudes. Focus groups are good for test marketing new music and finding out why consumers like or dislike certain songs. They are not good for generalizing and providing statistical data about a market.

Table 19.2 Research design		
Research design	**Uses**	**Weaknesses**
Experiments	To determine cause and effect. To find out which of our marketing efforts are most effective. Good predictive validity.	You must have full control over all aspects of the experiment to isolate the cause and be certain the desired effects are not being caused by something else.
Description	Surveys are used to describe the market. Easy and cost-effective. Can be generalized.	Less predictive value than experiments.
Qualitative	Good for in-depth information and understanding why. Good for test marketing new products.	Not generalizable to the general market. Relies on just a few people to provide information about a much larger population.

For greater accuracy, sometimes *triangulation* is used where two methods are employed and if findings are consistent, researchers are more confident in the results. Triangulation is the application and combination of several research methodologies in the study of the same phenomenon (Zulkardi, 2004). By looking at something from two different angles, more accurate conclusions can be drawn.

▲ *Figure 19.4 Triangulation*

4. **Data collection**—Data are collected by conducting the experiment, survey or focus groups. Once the design has been determined, a sample must be selected. Rarely does a research project rely on a census (a measure of everyone in the population or market) because of the costs involved in collecting the data. Therefore, a sample is selected to represent the market, and data are collected from this sample.

Sampling must be scientific and follow certain guidelines, known as *sample design*. This includes the method of selection, the sample structure and plans for analyzing and interpreting the results (Tutor2U, 2005). Common folklore has people believing that population size is very important in determining sample size. In other words, if I want to generalize to the entire U.S. population, I would need a larger sample than if I wanted to generalize to all college students. This is not true. Population size does not normally affect sample size. But the sample does have to be proportionally representative of the population, known as *sampling frame*. For example, if I wanted to know what high school males and females thought about algebra, it would not be wise to conduct the survey of female shoppers in the mall during regular school hours. Instead, I would need to sample a certain number of high school males and females; the proportion of males to females would be dependent upon the proportion in the population. The sampling frame must be representative of the population, and is defined as a list of elements from which a sample may be drawn. Often these elements can be identified by existing information. For presidential election polls, the sampling frame would be all registered voters, or likely voters.

There are two phases in the process of gathering data: the *pre-testing phase* and the *main study*. Pre-testing is done to ensure the validity of the instrument (i.e., a survey) used to measure the research issues. After pre-testing, the revised survey is administered, or the experiment performed, or the focus groups are conducted. The data must then be converted into a form that allows for systematic analysis, whether that be numerical data that is input into a program, or recordings and transcripts of interviews and focus groups for analysis.

5. **Analysis of data**—Data analysis begins with editing and coding. *Editing* involves checking the data collection forms for omissions, legibility and consistency in classification, especially for fill-in-the-blank questions. The data are then coded. *Coding* refers to the systematic process of interpreting, categorizing, recording and transferring the data to the data processing program. *Data analysis* is the application of logic to the data, looking for systematic patterns and details that will answer the research questions.

The appropriate analytical technique is determined by the survey design. Surveys are generally analyzed using a statistical software program. Interviews and focus groups are often transcribed, and sometimes software is used to look for key terms in the text and identify patterns of responses. In experiments, the experimental variable (i.e., the presence or absence of cancer in smokers, presence or absence of aggressive behavior after watching violence on television) is measured with the appropriate instruments.

6. **Conclusions and recommendations**—The research report is the main tool for communicating the findings and recommendations from the research company to the client. A research report is an oral presentation and/or written report, and the purpose is to communicate the results, findings, and recommendations to the marketing client. The research report states the conclusions from the data analysis, first as straightforward responses to the measurements. These are then compared to the original research questions and goals to make sweeping conclusions and recommendations. A typical written report contains an executive summary, and introduction (including objectives), methodology, the literal results, and conclusions and recommendations. The executive summary is a one or two page bulleted-list summary of the highlights of the findings. It is designed to be read by those people who need to know the main points, but do not have time to read through the report to find them.

Research validity

Caution must be taken to ensure any research project is valid. Validity is a term that describes the ability of a research project to measure what was intended to be measured. If the goal is to measure *purchase behavior*, asking questions about *attitude* toward various products may not provide a valid measurement of their likelihood to actually purchase the product. Consumers may have a more positive attitude toward a product that they have no intention of buying because of the price or some other factor. Instead, questions should be asked about actual purchase behavior and purchase intention.

There are three approaches to establishing validity: face or content validity, criterion validity, and construct validity. Face validity refers to a subjective understanding among experts that the research design and measurement logically appear to actually measure what it intends to measure, (i.e., Does it logically make sense?). Criterion validity relies on what has been done in other studies and whether it meets common standards, (i.e., Is this the way similar research has been successfully conducted in the past?). Construct validity is the ability of a measure to provide empirical evidence consistent with a theory based on the concepts. Often in marketing research, the marketing client will have expectations of the possible outcomes of a research project, based upon theory.

Confidence level and margin of error

Because almost all research involves taking a sample of the market rather than measuring everyone, there is the possibility of some error—that is, the findings from the sample are not exactly what occur in nature. The margin of error, which is the plus or minus three percentage points seen after percentages are listed in reporting results, has to do with the probability that the findings represented by the sample will be true for the population at large. In other words, if a result is presented as 36% ±3%, it means that while the sample found a result of 36%, there is a strong likelihood that in the population, the true result is between 33% and 39%. The margin of error decreases as the size of the sample increases, but only to a point. Researchers generally weigh the risk of error against the additional costs required to fund research for a larger, more accurate sample. The level of confidence is a measure of how confident we are in a given margin of error, and the general acceptable standard is a 95% level of confidence with a margin of error of ±2 or 3%. (For example, if we ran this experiment 100 times, it would produce findings within the margin of error 95 times.)

Table 19.3 Margin of error

Survey sample size	Margin of error percent*
2,000	2
1,500	3
1,000	3
900	3
800	3
700	4
600	4
500	4
400	5
300	6
200	7
100	10
50	14

*Assumes a 95% level of confidence

The application of research principles

Marketing researchers must have an understanding of when to use various research designs and measurements, and also when they may not be appropriate.

When focus groups are appropriate

Focus groups are considered exploratory research, but can be conducted either before a full-scale survey is launched, in conjunction with a survey, subsequent to a survey, or in rare cases, as a stand-alone research project. Focus groups are usually conducted before a major survey in order to gain an understanding of the target market for the survey. The group interviews help the researcher understand context, vernacular, and how various survey questions will be interpreted.

When triangulation is advised, focus groups may be run in conjunction with a survey, with the survey providing the information to be generalized to the population and the focus groups providing depth of understanding to the responses. Occasionally, focus groups are conducted after a survey is completed, especially if there are lingering questions about the findings that need clarification from members of the market. Stand-alone focus groups are conducted when researchers want some insight into the minds of the consumers, but do not need quantitative data and do not intend to generalize the findings to the larger population.

Focus groups are good for concept testing. Record labels use them to judge consumer opinions about artists and their music, their appearance, the quality of the recordings, the lyrics, and the artist image. In these situations, focus group attendees are exposed to elements of the marketing plan (the music, video, photos, and so on) in a controlled setting, and reactions to these elements are noted. The groups can be conducted early in the project, before the music has been finalized for the release and before imaging materials have been selected. The attendees may be presented with photos of the artist with different looks, and then asked to comment on each one. They may also be presented with recordings and asked to judge and comment on them. The results may not guarantee acceptance in the marketplace, but at least the labels will be aware of the various "groups of thought" regarding their products.

Focus groups are not appropriate when attempting to analyze the market for an artist. The market must be identified prior to selecting participants for focus groups. Focus groups are not appropriate for generating quantitative information such as ratio of males to females, a comparison of market segments on an issue, or to identify traits in the market.

When surveys are appropriate

Surveys are best suited to situations where the research questions can be answered in a straightforward manner, when more information about aggregate consumer groups is needed, and when that information will need to be generalized to a larger population. Surveys are good for identifying characteristics of target markets, describing consumer purchasing patterns, and measuring consumer attitudes. Surveys provide an inexpensive, efficient, and accurate means of evaluating information about a market by using a small sample and extrapolating the results to the total population or market.

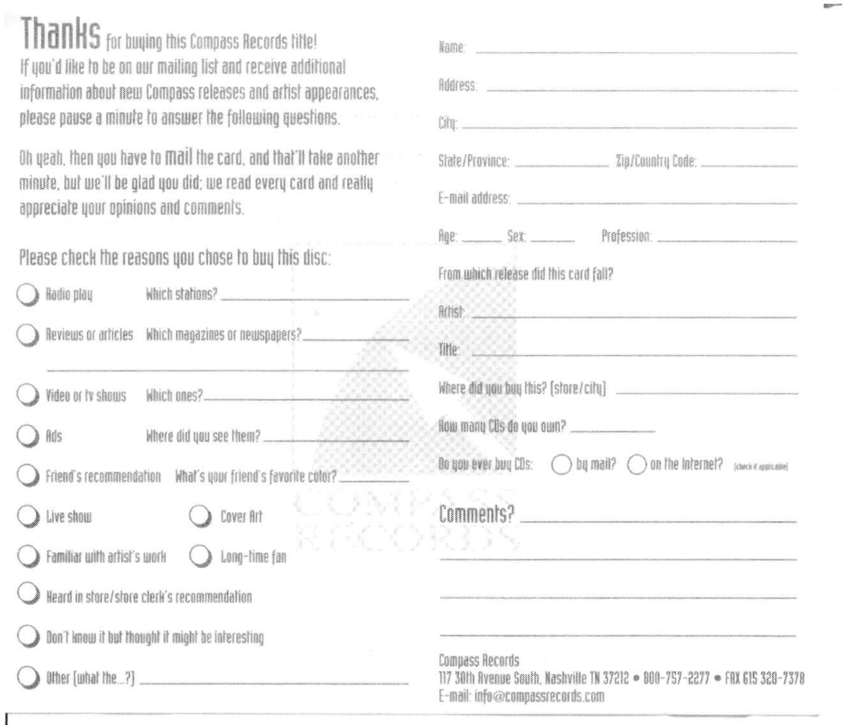

Thanks for buying this Compass Records title!
If you'd like to be on our mailing list and receive additional information about new Compass releases and artist appearances, please pause a minute to answer the following questions.

Oh yeah, then you have to mail the card, and that'll take another minute, but we'll be glad you did; we read every card and really appreciate your opinions and comments.

Please check the reasons you chose to buy this disc:

- Radio play — Which stations? ______
- Reviews or articles — Which magazines or newspapers? ______
- Video or tv shows — Which ones? ______
- Ads — Where did you see them? ______
- Friend's recommendation — What's your friend's favorite color? ______
- Live show
- Cover Art
- Familiar with artist's work
- Long-time fan
- Heard in store/store clerk's recommendation
- Don't know it but thought it might be interesting
- Other (what the...?) ______

Name: ______

Address: ______

City: ______

State/Province: ______ Zip/Country Code: ______

E-mail address: ______

Age: ______ Sex: ______ Profession: ______

From which release did this card fall?

Artist: ______

Title: ______

Where did you buy this? (store/city) ______

How many CDs do you own? ______

Do you ever buy CDs: by mail? on the Internet? (check if applicable)

Comments? ______

Compass Records
117 30th Avenue South, Nashville TN 37212 • 800-757-2277 • FAX 615 320-7378
E-mail: info@compassrecords.com

Source: Compass Records

▲ *Figure 19.5 Customer reply card for Compass Records (Source: Compass Records)*

Entire textbooks are devoted to survey design. The questions must be worded and presented in a clear, concise and unbiased manner to ensure validity. Measurement scales are developed to measure more abstract concepts such as personality and attitudes. These scales are banks of questions that have been carefully tested and found valid when applied appropriately to measure the concept they were designed to measure. If the survey has been judged to be valid and the sampling procedures are followed correctly, the results of the survey should convey a degree of confidence that the entire market or population exhibit the opinions and characteristics similar to those found in the sample.

Surveys are used extensively inside products to gather information on customers. Often the survey is part of the warranty card, but in the recording industry the cards are inserted in jewel boxes. The cards have different names depending upon the label; at some they are *tossback* cards, at others, *bounceback* cards, and still others may refer to them as customer response cards or business reply cards. The sample base for these cards should not be considered random or representative of the market overall. Sample bias may occur because consumers who send in the cards may be systematically different from customers who purchase the product but do not send in the cards. Nonetheless, the cards are beneficial in gaining an understanding of the market.

In the early to mid-1990s, Geffen Records was interested in learning more about young male music fans who were buying the metal and grunge recordings released by the label. Bounceback cards were placed inside specific releases and the results were analyzed to gain an understanding of the market. Label personnel were surprised that many of the fans read *Rolling Stone* and *Spin* magazines, rather than the heavy metal genre-specific magazines targeted to those consumers. A comparison of magazine preferences with geographic location (urban, rural, and suburban) led to the conclusion that these more specific magazines were not available on news racks in many locations and therefore not as popular. It was also learned for the first time that many of the grunge fans enjoyed skateboarding, and that tattoo parlors were a good place to promote heavy metal music.[i] (In today's business climate, these findings are intuitive, but at the time they were not as obvious.) How did the label learn of the connection between their music and skateboarding? The skateboard magazine, *Thrasher*, kept showing up in the "top ten favorite magazines" listed by those who returned the bounceback cards. The label was also able to find out about the interest in vinyl recordings and usage of the Internet at that time. These types of findings help a record label make better decisions about where and how to market their products, and which media outlets to use.

---

[i] Hutchison, Thomas. (1992-1997) primary research data.

Bounceback card for Geffen Sampler *Buy Product*

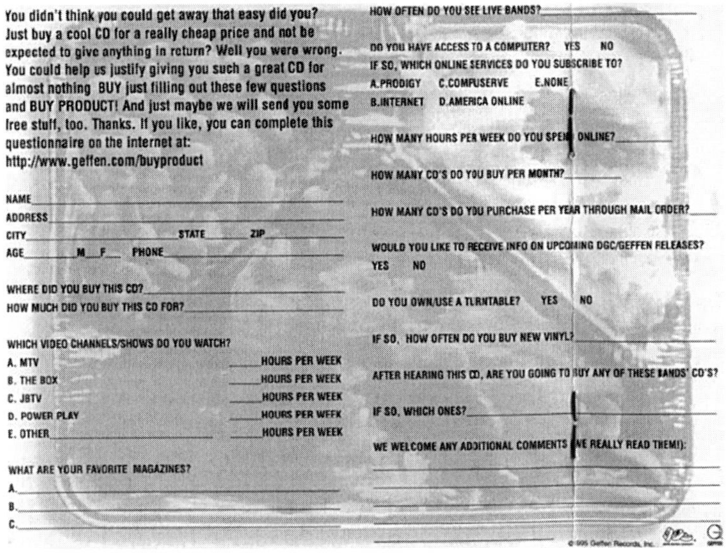

You didn't think you could get away that easy did you? Just buy a cool CD for a really cheap price and not be expected to give anything in return? Well you were wrong. You could help us justify giving you such a great CD for almost nothing BUY just filling out these few questions and BUY PRODUCT! And just maybe we will send you some free stuff, too. Thanks. If you like, you can complete this questionnaire on the internet at: http://www.geffen.com/buyproduct

NAME
ADDRESS
CITY _____ STATE ___ ZIP
AGE ___ M __ F __ PHONE

WHERE DID YOU BUY THIS CD?
HOW MUCH DID YOU BUY THIS CD FOR?

WHICH VIDEO CHANNELS/SHOWS DO YOU WATCH?
A. MTV _____ HOURS PER WEEK
B. THE BOX _____ HOURS PER WEEK
C. JBTV _____ HOURS PER WEEK
D. POWER PLAY _____ HOURS PER WEEK
E. OTHER _____ HOURS PER WEEK

WHAT ARE YOUR FAVORITE MAGAZINES?
A.
B.
C.

HOW OFTEN DO YOU SEE LIVE BANDS?

DO YOU HAVE ACCESS TO A COMPUTER? YES   NO
IF SO, WHICH ONLINE SERVICES DO YOU SUBSCRIBE TO?
A.PRODIGY C.COMPUSERVE   E.NONE
B.INTERNET D.AMERICA ONLINE

HOW MANY HOURS PER WEEK DO YOU SPEND ONLINE?

HOW MANY CD'S DO YOU BUY PER MONTH?

HOW MANY CD'S DO YOU PURCHASE PER YEAR THROUGH MAIL ORDER?

WOULD YOU LIKE TO RECEIVE INFO ON UPCOMING DGC/GEFFEN RELEASES?
YES   NO

DO YOU OWN/USE A TURNTABLE?   YES   NO

IF SO, HOW OFTEN DO YOU BUY NEW VINYL?

AFTER HEARING THIS CD, ARE YOU GOING TO BUY ANY OF THESE BANDS' CD'S?

IF SO, WHICH ONES?

WE WELCOME ANY ADDITIONAL COMMENTS (WE REALLY READ THEM!):

▲ *Figure 19.6 Bounceback card for Geffen Records (Source: Geffen Records, reprinted with permission)*

When experiments are appropriate

Scientific experiments require that all variables be controlled. *Variables* are any factors that might influence the outcome. In the case of marketing, the outcome is sales of the recording. The various factors or variables that might influence this include any and all marketing efforts—radio airplay, TV performance, touring, in-store activities, local and national press coverage, word of mouth, and so forth. In a normal experimental setting, one or more of the variables would be *manipulated*, and the other variables would be held constant or controlled. For example, experimental research on the impact of television advertising on sales would involve at least two groups in the market that are identical in all aspects, except exposure to the advertising variable: a flight of commercials. Other factors that might influence sales would have to be kept identical for both groups. This is much easier to accomplish in a laboratory setting, but not practical for marketing research, since people do not shop in a vacuum.

Field experiments are conducted in a natural setting, despite the fact that complete control of extraneous variables is not possible. Attempts are made to find two groups of consumers that are alike in as many aspects as possible. Then one group may be exposed to a particular marketing strategy, while the other is not. This is not popular in the recording industry because it involves withholding some marketing efforts to a portion of the market during the crucial window for generating sales. Controlled store tests (or test marketing) may be conducted by providing products to some stores to gauge sales before investing in a nationwide product rollout. Results from this type of testing do not automatically generalize to the entire market and guarantee national success, due to other factors that may not have been accounted for during the controlled store tests.

The use of syndicated research

Since the introduction of SoundScan and BDS, the use of syndicated research has become a valuable tool for making marketing decisions in the record business. Chapter 6 illustrates how SoundScan data can be used as a basis for more in-depth research to detect sales trends and the impact of marketing strategies. Data from BDS can be merged with SoundScan to determine a more precise impact of radio airplay on record sales than was possible 15 years ago. SoundScan and primary research are combined for tour analyses, market profiles and to persuade radio stations to increase airplay. Data from SoundScan helps determine who (what geographic locations) to select for focus groups.[ii] The use of SoundScan, coupled with the business reply cards, syndicated research from other sources, and occasional focus groups, give marketing departments a better idea of the marketplace and allow for better prediction of marketplace performance of products.

Arbitron provides information on radio listening audiences, and much of that information is valuable to the record business. NARM and the RIAA conduct research projects and provide the results to its members. The U.S. Census Bureau provides basic demographic and geographic information that is used to develop marketing strategies. Other syndicated research services provide information about advertising effectiveness in general, and help marketers determine which media outlets to use to reach a particular market segment.

Research sources for record labels

Tracking services

Nielsen SoundScan is arguably the most important company providing research data and information to record labels. They are self-described as ''an information system that tracks sales of music and music video products throughout the

---

[ii]Ury Greenberg, Linda. (February 21, 2005). Personal Interview.

United States and Canada. Sales data from point-of-sale cash registers is collected weekly from over 14,000 retail, mass-merchant and nontraditional (online stores, venues, etc.) outlets." Sales results can be compared with marketing efforts for evaluation of the impact of those marketing events.

Weekly data from sales are compiled by SoundScan and made available every Wednesday. "Nielsen SoundScan is the sales source for the *Billboard* music charts" (SoundScan, 2005). Nielsen also owns Broadcast Data Systems (BDS). BDS provides airplay tracking for the entertainment industry using a digital pattern recognition technology. "Nielsen BDS captures in excess of 100 million song detections annually on more than 1,200 radio stations in over 130 markets in the U.S." (BDS, 2005). MediaBase also tracks radio airplay (see Chapter 8). In 2001, MediaBase joined efforts with *Radio & Records* to provide data for compiling their charts.

In the following example, the effects of artist appearances on Prairie Home Companion and the impact of the artist winning the Shure Vocal Competition at the Montreux Jazz Festival are evident in the sales patterns for jazz artist Inga Swearingen.

Inga Swearingen Sales

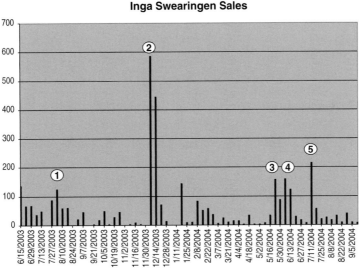

1. The artist won the 2003 Shure Montreux Jazz Voice Competition in July of 2003.
2. The artist appeared on the National Public Radio show Prairie Home Companion on November 29, 2003.
3. A second appearance on Prairie Home Companion on May 15, 2004, followed by
4. an appearance on May 29, created some sustained sales over several weeks.
5. Another PHC appearance on July 3, 2004 created this final spike in sales.
• And no other marketing or promotional activities were being conducted during this time.

▲ *Figure 19.7 Inga Swearingen Sales Spikes (Source: SoundScan)*

Trade association research services

All the major trade organizations provide research for their members. NARM publishes its monthly *Research Briefs,* where they have worked with NPD and Ipsos for industry/technology information. NARM provides research findings at its annual convention on a variety of current and ever-changing industry topics. The RIAA provides annual data on shipments, and contracts with an outside research firm to conduct its annual consumer profile. For this project, nearly 3,000 music buyers are surveyed. The RIAA has worked with Taylor Research on a large-scale consumer research survey, with Pricewater-houseCoopers (PWC) on a study of shipments and returns, and NPD Digital for data on what consumers are doing in the digital area. The RIAA also provides research information on illegal downloading. The IFPI collects data from member countries and publishes an annual report called *The Recording Industry in Numbers*. The International Federation of Phonographic Industries (IFPI) also releases periodic reports on digital music and global piracy. *Billboard* publishes *MarketWatch* in its weekly magazine. The associations tend to conduct issue-oriented research of benefit to all members of the record industry.

Commercial research firms

Forrester Research is one of the leading market research firms focused on the Internet and technology, and conducts research for the recording industry on all aspects of music and the Internet. Forrester also offers custom research and consulting services to its clients. *Jupitermedia Corporation* is a leading provider of original information, images, research, and events for information technology, business and creative professionals. They are often hired by the associations to conduct and report on online music consumers. *Edison Media Research* is a leader in political, radio and music industry research with clients that include major labels and broadcast groups. *Music Forecasting* does custom research projects on artist imaging and positioning.

The NPD Group, which was recently acquired by Ipsos, provides marketing research services through a combination of point-of-sales data and information derived from a consumer panel. Their research covers music, movies, software, technologies, video games, and many other product groups. *ComScore* offers consulting and research services to clients in the entertainment and technology industries, and conducts audience measurements on website usage through its *Media Metrix* division. *Taylor Nelson Sofres*, a UK firm, provides both syndicated and custom research of media usage and consumer behavior. Based in France, *IPSOS* is a global group of researchers providing survey-based research on consumer behavior. *BigChampagne* (owned by Clear Channel), tracks online P2P usage and reports, among other things, most popular songs on P2P networks. *ResearchMusic* is an online music service that also offers free music market research to registered customers. *OTX* is an online consumer research and consulting firm.

Conclusion

Marketing research is a scientific process designed to ascertain information about consumers and their attitudes toward products. It follows a process of problem identification, exploratory research, the research design, data collection, analysis, and conclusions or recommendations. Since markets are always a moving target, and the recording industry is constantly releasing new and unproven products, marketing research has become a valuable tool enabling record labels to more accurately plan marketing strategies. The recording industry has embraced marketing research to a greater extent in the past 20 years as marketing has grown more competitive and sophisticated.

Glossary

Marketing research – The systematic design, collection, analysis and reporting of data and findings relevant to a specific situation facing a company.

Sample – A small representative subsection of the market segment used as a basis for analysis in research.

Syndicated research – The gathering of continuous or periodic information which is sold in standardized form to all companies involved in the industry. Examples include Arbitron, Nielsen, BDS and SoundScan.

Custom research – Tailored, proprietary, initiated by a company, and often specific to a particular product developed by that company. Examples include product surveys, focus groups, customer satisfaction surveys and personal interviews.

Secondary research – Going over data that has been previously collected for a project other than this one at hand.

Exploratory research – Initial research conducted to clarify and define the nature of the problem. *Explore then form*

Concept testing – A means of evaluating ideas by providing an early indicator of the merits of an idea prior to the commitment of resources to development and manufacturing.

Experiments – A form of research in which conditions are controlled so that (an) independent variable(s) can be manipulated to test a hypothesis about a dependent variable.

Descriptive research – Research designed to describe characteristics of a population or phenomenon.

Focus groups – Unstructured, free-flowing group interviews designed to gain a deeper understanding of consumer feelings, beliefs and attitudes.

Triangulation – The application and combination of several research methodologies in the study of the same phenomenon.

Instrument – A data collection form, such as a questionnaire, or other measuring device.

Coding – The process of identifying and assigning a numerical score or other character symbol to previously edited data.

Data analysis – The application of logic to the data, looking for systematic patterns and details that will answer the research questions.

Executive summary – A one or two page bulleted-list summary of the highlights of the findings, designed to be read by those people who need to know the main points but do not have time to read through the report to find them.

Research report – An oral presentation or written statement of research results, strategic recommendations, and/or other conclusions given to the marketing client.

Validity – The ability of a scale to measure what was intended to be measured.

Measurement scales – Any series of items on a survey that have been scientifically designed to measure or test a concept.

Bibliography

BDS (2005). http://www.bdsonline.com/about.html.

Blankenship, A. B. and Breen, G. E. (1993). *State of the Art Marketing Research*, Chicago: NTC Business Books.

SoundScan (2005). http://www.soundscan.com/about.html.

Tutor2U (2005). Market Research – Sampling. http://www.tutor2u.net/business/marketing/research__sampling.asp.

Zikmund, W. G. (1991). *Exploring Marketing Research*, Orlando: The Dryden Press.

Zulkardi (2004). Triangulation in Educational Research. http://www.geocities.com/Zulkardi/submit3.html.

20 The Recording Industry of the Future

Thomas Hutchison

The record label of the future is expected to be very different from what is in place today. The current business model is structured around music as a physical product, sold through brick-and-mortar retail outlets. The record label of the future may be shifting away from that of product provider, to more of a service provider, much like television networks are today. Their role may be limited to developing *content*, and then marketing that content to consumers. This represents a drastic paradigm shift for record labels that have resisted moving into the digital age.

Before unraveling the mystery of the future of recorded music, we must first examine the basic principle behind recorded music and how it is consumed. Then we must examine how, through the course of time, new innovative methods replace old methods of accomplishing this task. Recorded music allows the consumer to possess a captured musical performance, and replay that performance whenever and wherever desired. The music business for the past 100 years has been supplying that to consumers in the form of a physical product. But many things have changed, especially in the fields of entertainment and communication. To understand the future, we must move away from the concept of supplying consumers with a physical product. After all, the goal is for the consumer to replay the recording at will, and modern technology does not mandate that it must be in a physical form. Instead, we should look at the record business as *music delivery systems*, since the goal is to get the consumer to pay for the convenience of listening to their favorite music on demand.

The new model for record labels as service providers involves a drastic overhaul of the label system. The following chart illustrates how within the past 100 years, with the exception of radio, all new music delivery systems have followed the same basic format of creating a physical product. The one exception, radio, is not an on-demand system capable of being user-controlled; and at this point, radio does not provide direct income for the labels.

Table 20.1 Communication flow				
Source		**Medium – Channel**		**Receiver**
Labels – product	→	Cylinder	→	Phonograph, Graphophone
provider	→	Vinyl disc	→	Gramophone, Victrola
Service provider	→	Radio broadcasts	→	Radio
Labels – product	→	8-track/Cassette	→	Tape player
provider	→	CD	→	CD player
Service provider	→	MP3 via Internet	→	Computer
	→	Digital wireless	→	Universal mobile device

Adapted from C. E. Shannon and W. Weaver, The Mathematical Theory of Communication (1949);

Label partnerships

Changes in music delivery systems have at times driven changes in record label ownership and partnerships. When radio was first developed in the 1930s, radio networks sought to own record labels. RCA, with its NBC Radio Network, purchased Victor records in 1929, and CBS purchased Columbia and Okeh Records in 1938 (Sony website, 2005). In the 1970s through the 1980s, entertainment hardware companies sought alignment with labels. Philips electronics created PolyGram Records in 1972 (Hoover's, 2005), and Thorn Electrical Industries purchased EMI in 1979. Sony purchased CBS records in 1988, and Matsushita bought MCA Records in 1990. This alignment started to change in the 1990s when telecom and utility companies took an interest in record labels. Vivendi bought Universal in 2000, and AOL merged with Time Warner (parent to the Warner Label Group) in 2001. Bertelsmann AG, parent company to RCA and BMG, ventured into the broadcast media industry (PBS, 2002). These mergers and relationships were designed to create *synergy*, defined as a mutually advantageous conjunction or compatibility of distinct business participants or elements (as resources or efforts) (Merriam-Webster Online, 2005). Synergy allows for a company to use one branch to support the activities of another branch. Disney is an example of a company that has used synergy to create a "superbrand" of some of their assets using a collaborative effort among their film, animation, music, theme park and product merchandising branches to push a brand into the marketplace. In the 1980s, MCA President Irving Azoff managed to use Universal's film division to propel his artists' careers (Dannen, F., 1991). BMG has the potential to use their magazine and television affiliates to promote an artist.

Under the old label/receiver relationship model, electronics companies sought to promote new consumer entertainment hardware through the acquisition of music and movie catalogues. After Sony lost the format debate on VCRs[i], they pursued the purchase of CBS Records, and Columbia and Tri-Star Pictures. The idea was that they could then provide pre-recorded software in a format that supported their new hardware. The Sony MiniDisc is an example. Sony attempted to make the MiniDisc into the next recorded music platform by releasing pre-recorded titles from the CBS catalog. At one time, five of the six major labels were affiliated with electronic hardware companies.

Table 20.2 The label/receiver relationship model

The label/receiver relationship model		
Source	**Channel**	**Receiver**
Sony/CBS Records		Sony Electronics
MCA Records		Matsushita
RCA		General Electric
WEA	Warner Cable and TV	
EMI		Thorne Electronics
PolyGram		Philips Electronics

As the promise of digital delivery of movies and music loomed, companies began to realign themselves in partnerships involving content (source) and service providers (channels). The expectation was that the synergy created by aligning record companies with online and cable service providers would attract customers and increase sales. America Online merged with Time Warner, gaining access to vast amounts of content to fill the AOL "channel." Bertelsmann AG, parent company to RCA and Arista, began to increase its presence in the European media landscape (Frontline: Merchants of Cool, 2001). Vivendi purchased Universal in 2000 combining "Vivendi's telecommunications assets with Seagram's film, television and music holdings (including Universal Studios) and Canal+'s programming and broadcast capacity" (Frontline: Merchants of Cool, 2001). Universal had previously purchased Polygram under the leadership of Edgar Bronfman, Jr.

---

[i]Sony introduced the Betamax format for VCRs but ultimately lost out to the more popular VHS format of its competitors.

Table 20.3 The label/channel relationship model

The label/channel relationship model		
Source	Channel	Receiver
Sony/CBS Records		Sony Electronics
MCA/Universal Records	Vivendi	
RCA/BMG	Bertelsmann AG	
WEA	AOL	
EMI		
PolyGram	Part of Vivendi/Universal	

But as of 2005, these companies have yet to capitalize on the synergy they hoped to create. The labels and other divisions of these media giants have been criticized as being too entrenched in their business practices to open up to innovative synergistic opportunities. AOL has divested itself of Warner Music Group, rumors have been flying about Vivendi parting with Universal, and BMG and Sony Music Entertainment have further consolidated the industry by creating a joint venture involving just the label operations and distribution.

New media concepts

The study of new media, including the recording industry, requires an understanding of what constitutes *new media* and how it affects the future of music delivery. To understand how media is evolving, it is necessary to look back at earlier technology waves and apply them to the context of diffusion theory. We have previously discussed the evolution of fixed-music formats, noting that radio is the only historical innovation to deliver music in a format other than physical product. But the future holds more promise in the area of providing music as a service rather than a product, especially one that generates revenue for the record labels.

The development of online services

The telephone can be considered the first streaming communication device to proliferate into U.S. households throughout the twentieth century. The public utility business structure made it possible to supply most households with a telephone line at a reasonable cost, and adoption became widespread. The basic limitation of the twentieth century telephone line was limited *bandwidth*. Bandwidth can be described as the capacity, or the maximum data transfer rate of an electronic communications system. The telephone line is limited and restrictive. When community antenna television (CATV) became popular in the 1970s and 80s, it became necessary to install an additional cable in the home— a *coaxial cable* (a transmission line of high bandwidth television signals that consists of a tube of electrically conducting material surrounding a central conductor held in place by insulators). The problem with initial coaxial cables from the CATV providers was that the cable only allowed for one-way communication; it was not interactive. Interactivity is described as a two-way interaction between sender and receiver. While the standard telephone line is interactive, it is not capable of rapid multimedia transmission.

By the 1980s, homes were wired with one cable and one telephone line, each for separate tasks. However, recent advances in both areas have caused some convergence of media in the home: interactive cable and broadband telephone lines. This convergence brings a higher level of service into the home, but telecom and cable companies have been wrangling over who will be allowed to provide what services. This may have actually delayed the onset of comprehensive media and communication services.

Table 20.4 Phone lines vs. Cable		
	Telephone line	**Cable TV**
Advantage	Interactive, two-way communication	Wide bandwidth, fast transmission rate
Disadvantage	Slow data transfer speed	Not originally capable of interactivity

Online and offline content

We receive our entertainment and information through two types of channels: *offline* and *online*. Online is considered any type of transmission that is streaming in real time, and includes airwaves and wireless communication transmission (radio and cell phones). We also receive information and entertainment through fixed-media such as CDs, videocassettes, DVDs, and newspapers. Under those circumstances, we are considered offline. Reception of offline content often requires some physical product, and the inconvenience of driving to the video rental store or music store, or ordering through the mail. The fixed-media content business has been allowed to thrive mainly because the streaming media infrastructure was not in place to provide the same media products at a reasonable cost and convenience (of both hardware and service) to the consumer.

Table 20.5 Fixed vs. streaming media		
Content	**Offline-fixed**	**Online-streaming**
Music	CDs	Radio, Internet
Movies	DVDs, videocassettes	Cable channels, video-on-demand
News	Newspapers, magazines	Broadcast and cable TV, radio, Internet
Personal Communication	Letters	Email, telephone, fax

Technology diffusion and convergence

The new media model accounts for a proliferation of new technology devices at a reasonable cost to the consumer. Under this model, a variety of services are offered to consumers as various submarkets begin to embrace the new delivery methods. Predicting *who* will adopt *what* is a challenge for marketers, but diffusion theory can offer some guidance. The rate of diffusion is positively related to the standardization and simplicity of use. The new technological devices that offer consumers the opportunity to listen to music more conveniently and with less cost may prevail. Some of the issues that may affect adoption are shown in Table 20.6.

Table 20.6 Attributes of new media		
Attribute	**Industry challenge**	**Present situation**
Affordability	The industry must determine a pricing structure that will attract and sustain consumers.	Pricing at this point is experimental. Pricing models are not yet in place that works for both suppliers and consumers. Subscription or a la carte?
Simplicity	New technology must be easy to use, without the constraints of digital rights management systems that would impede consumer's normal fair use of music purchased.	Digital rights management (DRM) systems create limitations on fair use that are leading to frustration and confusion among customers.
Compatibility	A plethora of noninterchangeable platforms is guaranteed to confuse and discourage consumers.	Downloads from different services do not all work on the same platforms. Formats are proprietary. MP3 is the most popular software format, but the iPod dominates the hardware market.
Value	The hardware and software must be sustainable enough for the consumer to consider the initial investment in the hardware.	With no competition in the iPod format, hardware pricing is high. Subscriptions are not attractive until offerings become more all-inclusive.
Convenience	One article describes it as "smaller, faster, portable"[i], devices and music must be continuously available.	Access is currently limited to computers and the Internet. Portable devices must be loaded from computers.
Availability	Consumers must have one-stop shopping access to everything available.	Not everything is available on all downloading services. Consumers must check several services to fill their needs.

[i] Keaten, Jamey. (2005) *The Future of Music: Smaller, Faster, Portable*. Associated Press. http://www.examiner.com/article/index.cfm/i/012705bu__music

Digital rights management

Digital rights management (DRM) is a systematic approach to copyright protection for digital media, designed to prevent illegal distribution of paid content over the Internet. This was developed in response to the rapid increase in online piracy of commercially marketed material, which proliferated through the widespread use of peer-to-peer file exchange programs. Most DRM programs require the user to register or verify online before transferring the protected music to a portable device. This two-edged sword has offered some security to record labels, but has frustrated consumers and discouraged the adoption of digital downloads. Rob Enderle of TechNewsWorld comments: ''I, like many others, see DRM as a pain. I just don't want to deal with it. So I've gone back to ordering CDs from Amazon.com, ripping the CDs into MusicMatch, and then putting the music anyplace I want—in my home, on my person or in my car—without violating the Digital Millennium Copyright Act (DMCA)'' (Enderle, R., 2004).

Technology convergence: the universal mobile device

Convergence of technology is appearing in many forms today. Cable TV has incorporated high-speed Internet access; computers have become music management devices; the television and stereo system have now become an integrated media system; the laptop computer doubles as a DVD player. Convergence exists when two or more separate entities combine to form one.

At some point, consumers draw the line at carrying around an MP3 player, cell phone, portable video game *and* a digital camera. Enter the 3G handset technology, developed by cell phone manufacturers. This third generation (3G) cell phone technology incorporates text and video messaging, music downloads, games, and mobile TV into one handset and offers a new outlet for the music industry. This *music-to-mobile* (M2M) market was first developed through the offering of ''ringtones.'' Consumers can purchase and download these pieces of music to their cell phones to ''personalize'' their ring. Original ringtones were a series of beeps or sounds, rather than an actual recording. Record labels were not the beneficiary of licensing for these early ringtones, only music publishers. The more recent ''ringtunes,'' or master recording ringtones use an actual piece of the recording, and thus generate a royalty for the label and the artist. Similarly, ring back services allow telephone owners to customize the sound a caller hears before the owner answers the call by replacing the traditional ''ring-ring'' sound with actual recordings or other sounds. According to the IFPI, music-to-mobile is already big business in Japan, South Korea and Taiwan. In South Korea, music-to-mobile services generated $29 million for record companies in 2003, a full 18% of the total music market.

This represents a potential new market for the sale of recordings. In March 2004, *Variety* reported that more than 1.3 billion people have mobile phones— exceeding the combined number of PCs and televisions (McCartney, N., 2003). This is predicted to reach 2 billion by 2008; yet more recent predictions from Deloitte and Touche place the 2 billion milestone at the end of 2005 (Reuters News Service, 2005). Kusek and Leonhard, in their book, *The Future of Music: A Manifesto for the Digital Music Revolution*, state that if all cell phone users around the world paid just $1 per month for access to music, it would amount to half of the current value of annual worldwide recorded music sales. And the mobile phone market is growing faster than any previous technological innovation. The potential of M2M offers enormous opportunities to grow the market for recorded music. China and the U.S. are regarded by the IFPI as the best markets for potential growth of M2M; China because of its large population, increasing disposable income and vast young consumer base; the U.S. because it is the largest market for recorded music sales, and 3G networks are currently being implemented.

The 3G networks have the capacity to deliver a multitude of content to mobile devices, but music delivery may become the driving force in the adoption of these systems by younger users. BBC commentator Bill Thompson remarks, "just as the World Wide Web was the 'killer application' that drove Internet adoption, music videos are going to drive 3G adoption ... Why should I want to carry 60 GB of music and pictures around with me in my pocket when I can simply listen to anything I want, whenever I want, streamed to my phone?" (Thompson, B., 2004).

New markets: what does the consumer want?

Kusek and Leonhard state: "Music wants to be mobile." They point to the overwhelming success of the Sony Walkman. They also point out that today's teenagers have grown up in an era of technology access with the cell phone and the Internet. Today's kids, whom they call *screenagers,* are active users of video games, cell phones, instant messaging, email, surfing the web, and controlling and manipulating their own music collection. Much like the generations before them, the *Net Generation* is interested in sharing their music interests with peers—only with today's technology, it has become much easier and at the expense of the record labels. But these consumers don't just pass around free and illegal music files. They also visit artist websites, listen to online radio, attend concerts and discuss music with their friends. They are active music consumers, if not paying customers of the record business. The potential to actually sell music to these consumers is great, but only on their terms. Young consumers do not like being required to purchase a complete album of songs when they just want one or two singles. When asked in focus groups, they will often comment that $15–$20 is too much to pay just to get the one or two songs they want.

The success of Napster in 2000, and of subsequent peer-to-peer (P2P) file-sharing services, proved that the consumer was ready and eager to embrace downloading digital music. The IFPI states that the key drivers of the M2M market are consumer demand and technology. Of course pricing factors in also. Young consumers like to personalize their handset and are willing to pay premium prices for ringtones. Content owners, network operators and handset manufacturers all consider music as a motivating force in attracting new customers and driving repeat purchases. Consumer research by Nokia found that music surpassed other forms of entertainment and content in popularity.

▲ *Figure 20.1 3G Handsets*
(Source: Sony Ericsson)

Table 20.7 Consumer interest in 3G uses	
Content	**Consumer interest**
Music	65%
Film	54%
Games	44%
Sports	40%

Increased penetration of 3G devices is necessary for the growth of M2M. Increased storage capacity of portable devices is necessary for music collections. In 2005, Sony-Ericsson announced plans to introduce a music-player mobile handset marketed under the Walkman brand, and Nokia setup a partnership with Microsoft to allow users to download tracks to their handsets and then transfer them to a computer for CD burning. Improved visual screen technology or head-mounted displays may drive the film and music video adoption, but it is anticipated that users will only be interested in short video features, not 30 or 60 minute programs. This makes the music video an ideal format to drive sales of 3G technology.

The desire to own physical product may still exist, with consumers continuing to collect and possess the music that they value most. Consumers may begin to prioritize music based on their subjective value of the music: (1) music that they love and wish to own, and add to their collection, (2) music they like and are willing to pay to listen to, and (3) music that they are willing to listen to at no cost. This could create a tiered system of music consumption among customers and could sustain sales of physical units. The market for Internet-based digital music services will continue to grow, but subscription-based systems will only take off when the consumer can access music remotely through portable devices.

The new music model

As delivery systems change, the record industry may need to reinvent itself as a service provider, rather than a product provider. Music is moving from an off-line product-based commodity to an online commodity, and ultimately to an online service. Record labels may have to find income sources from licensing for streaming and music downloads. Licensing by record labels to third-party subscription-based digital music services is necessary for the label industry to survive in the digital economy. Digital rights management will need to evolve to a transparent system that does not impede on the consumer's right to create personal copies. Starbucks has already initiated a music downloading system that provides adequate copyright protection for providers without DRM systems. Starbuck's customers can use their laptops and the venue's wi-fi system to create custom CDs devoid of DRM.

Presently, the limitation is in the offerings. Many labels have yet to license their entire catalog for digital downloads and are still reluctant to do so. Kusek and Leonhard believe that statutory compulsory licensing will be necessary to provide what they describe as a ubiquitous presence of recorded music. What would bring ultimate copyright protection and satisfy consumer desires to copy and share music would be widespread adoption of subscription-based services. But until more content is licensed and made available, consumers may remain reluctant to embrace subscription-based services with limited offerings. With a broad consumer base all subscribing to the same service, copy protection would become unnecessary as all consumers would have access to—and be paying for—their favorite music. Consumers could share music with one another, provided they are all "paying members." Personal music collections could become unnecessary under the celestial jukebox model. Music downloading service *Rhapsody* describes the celestial jukebox as "immediate access to more music than you ever imagined. Thousands of albums by top artists, bands, and composers—all in CD-quality sound and delivered to you effortlessly via the Internet [or digital mobile devices]" (listen.com, 2005).

Marketing under the new model

It is expected that music markets will continue to fragment, especially with new forms of music distribution. This means that promoting music through mass media will become only one of many ways to market and promote new music. Because there is so much music in the marketplace, consumers have traditionally relied on gatekeepers and tastemakers to guide them in their music choices. Radio playlists serve to reduce the plethora of musical offerings to just the most popular songs. Kusek and Leonhard state:

> "Most consumers want and need tastemakers or taste-agents . . . that package programs and expose us to new music. This is why word-of-mouth marketing works so well . . . A friend is a true tastemaker."

Music placement in videos, movies, video games, and commercials is on the rise and will become even more popular as a way to introduce consumers to music. Direct marketing is becoming more prevalent with the use of the Internet. And marketing to fans, through online fan clubs, is also on the rise, producing a word-of-mouth phenomenon. As consumers begin to rely less on mass appeal products such as Top 40 radio and MTV, they will need guidance to wade through the overabundance of recorded music to find what they like. But new marketing models may be developed that rely more on a pull strategy (the consumer seeking information) than a push strategy (the consumer being bombarded with information). As consumers seek out new music to their liking, collaborative filtering software will help consumers find what they like by developing individual taste profiles.

Collaborative filtering software examines a user's past preferences and compares them with other users who have similar interests. When that user's interests are found to match another group of users, the system starts making suggestions of other things this group likes (Weekly, D. E., 2005). Suppose you normally never listened to jazz music, but you liked artists A, B, and C a lot. If numerous other people who don't normally listen to jazz also like A, B, and C, but also like band D, the system might suggest D to you and be relatively confident that you'll like it. Amazon.com uses the technology to recommend other products with their "people who bought this product also purchased these other products." Such marketing is less intrusive and most importantly, more intuitive. Consumers are reluctant to pay attention to messages that are not tailored to them—advertising that is general in nature and doesn't understand their individual preferences. However, the more intuitive recommendation engines are highly likely to suggest products that the consumer actually likes, therefore, becoming the ultimate in niche marketing, where each niche consists of one person and their individual taste profile. Customers are less likely to ignore marketing that is precisely targeted to their interests.

We are likely to see more alignment of music labels with digital wireless providers in the future, much like the previous relationships with hardware companies. New music delivery methods have the potential to reinvigorate the struggling recording industry; but in order for the new music model to work, these systems must be simple, affordable, available, compatible, convenient, and valuable.

Glossary

3G – Third-generation cell phone handsets and systems capable of transmitting, storing and playing multimedia content.

Bandwidth – The capacity or the maximum data transfer rate of an electronic communications system.

Coaxial cable – A transmission line of high bandwidth television signals that consists of a tube of electrically conducting material surrounding a central conductor held in place by insulators.

Collaborative filtering software – Examines a user's past preferences and compares them with other users who have similar interests.

Convergence – Moving toward union or uniformity.

Digital rights management (DRM) – A systematic approach to copyright protection for digital media, designed to prevent illegal distribution of paid content over the Internet.

IM – Instant text messaging by the user of a computer or wireless device.

Music-to-mobile (M2M) – Offering the ability to download music to wireless portable devices using cell phone, wi-fi or satellite technology.

Online – Any type of transmission that is streaming in real time, including airwaves and wireless communication transmission (radio and cell phones).

Ringtones – A tune composed of individual sequential or polyphonic beeps based on an existing song.

Ringtunes – or master recording ringtone. A ringtone that uses an actual sound recording by the original artist.

Tastemakers – Opinion leaders whose personal tastes in music, movies or other popular culture items influences others.

Wi-Fi (short for "wireless fidelity") – A term for certain types of wireless local area networks.

Bibliography

Dannen, F. (1991). *Hit Men,* Vintage. (0679730613).

Enderle, R. (2004). Starbucks and HP: The Future of Digital Music, *Ecommerce Times.* http://www.ecommercetimes.com/story/33164.html.

Frontline: Merchants of Cool (2001). http://www.pbs.org. /wgbh/pages/frontline/shows/cool/giants/bertelsmann.html.

Frontline: Merchants of Cool (2001). http://www.pbs.org/wgbh/pages/frontline/shows/cool/giants/vivendi.html.

Hoover's (2005). Philips Electronics. Hoover's Company Profiles.

listen.com (2005). http://www.listen.com/rhap__about.jsp?sect=juke.

McCartney, N. (03/29/03). Cellphone Media Making Noise, *Variety.com.*

Merriam-Webster Online (2005). http://www.m-w.com/cgi-bin/dictionary?book=Dictionary&va=synergy&x=23&y=17.

PBS (2002). http://www.pbs.org /wgbh/pages/frontline/shows/cool/giants/bertelsmann.html.

Reuters News Service (01/18/05). Cell Phone Market Seen Soaring in '05. *Reuters News Service.*

Sony website (2005). http://www.sonymusic.co.uk/uk/history.php.

Thompson, B. (11/22/04). *Music Future for Phones.* http://news.bbc.co.ukBBC News/go/pr/fr/-/1/hi/technology/4032615.stm.

Weekly, D. E. (02/08/05). http://david.weekly.org/writings/collab.php3.

21 The Marketing Plan

Thomas Hutchison

The marketing plan

For a record label, the marketing plan is the cornerstone of each release, and every record released by the label has its own unique marketing plan tailored to the target market and is based on expectations of performance in the marketplace. It is the marketing plan that helps coordinate the various aspects of promoting the record. There are two basic target markets for the marketing plan: The *consumer* and the *trade*. The trade consists of those people within the industry for which business-to-business marketing is done. This includes radio program directors, journalists, editors, retail buyers, distributors, and others involved in the "push" strategy.

Who gets the plan?

The plan is distributed to everyone involved in marketing the record, including people both inside and outside the label. It is important for everyone involved to understand what is being done in other departments, and how synergy is created when all of the elements come together. The following chart in Table 21.1 gives some examples of who may be involved in the execution of the plan.

There are other departments that benefit from and are included in the plan, including grassroots marketing (street teams), new media (Internet), and at

Table 21.1 Departments involved in the marketing plan

Function	Internal	External	Comments
Publicity	Label publicist	PR firms, press and media outlets	Goals may include getting reviews, features, photos, interviews, appearances
Radio	Promotion department	Indie promoters, radio programmers	Both work with radio to get airplay
Sales	Sales staff	Distribution, account buyers	The label sales staff works with distributors and retailers on retail promotions
Advertising	Developed by creative services, coordinated by marketing department	Outside firms sometimes hired to develop and implement programs	Much of advertising is coordinated with retail accounts
New media/ Grassroots	Web gurus, street team coordinators	Web design firms, Internet promotions	See chapter on Internet promotions
Video	Promotions	Production	Production is external to the label, while promotions may be either
Touring	Publicity, retail	Concert promoter	Label ensures there is product at retail, press coverage

some labels, tour support. Coordination is necessary for synergy to occur. The publicity department may develop materials necessary to send to retail accounts and radio (such as press kits). Advertising needs to be coordinated with retail accounts for sales in the marketplace and with the promotions department for trade advertising aimed at radio. The sales department works with the promotion department to ensure that product is available in all markets where airplay is prevalent. Advertising and publicity go hand-in-hand, sometimes combined in a "media plan," because in many instances, the same media outlets are targeted for both. Outside of the company, the marketing plans may go to account buyers in retail, program directors in radio, the artist's booking agency, and they are always presented formally to the artist/manager. Often, the manager will help develop the plan if corporate sponsors are involved or if there are special events that could be mined for extra marketing punch.

What's in the plan?

The plan provides basic information about the release, and a description of what is occurring in each department or area. A timetable is often included to ensure that timing and coordination are optimal. Financial specifics are often eliminated from the copies circulated at the marketing meetings, but are an integral part of the plan. Sales goals and shipment quantities, however, are often included in the plan. The following is a list of some basic information included in a marketing plan:

- Project title and artist name
- Street date
- Price, dating and discounts
- Selections or cuts
- Overview or mini-bio
- Configurations and initial shipment quantities
- Selection number and bar code
- Target market
- Promotion (radio)
- Names and release dates of singles
- Promotion schedule
- Promotional activities
- Publicity
- Materials
- Target media outlets
- Specific goals (interviews, appearances, reviews, etc.)
- Advertising
- Materials
- Target media outlets and coordination with retail accounts
- Ad schedule
- Sales
- Sales forecast
- Account-specific promotions
- POP materials
- Video
- Production information
- Distribution and promotion
- New media/Internet promotions
- Street teams/grassroots marketing
- Promotional tie-in with other companies/products
- Tour support (usually tied in with retail and publicity)
- Tour dates – help with spread of product at retail and tie-ins with radio
- International marketing
- Artist contact information (manager, publicist, and so on)

Timing

Timing is crucial to the success of an album, including the time of year the album should be released. The album release date is called the *street date*. Albums are strategically released on Tuesdays to take advantage of a full week of SoundScan sales numbers. The time of year varies depending on the stage of the artist's career. Established artists may often release an album in the fourth quarter to take advantage of holiday sales. New artists are advised to release during other times of the year, when competition is less fierce. An album may be strategically timed to coincide with a concert tour or a major media event for the artist.

In the following illustration, some retailers must have violated the street date by selling early. Missy Elliot sold 1,484 units the week before the album was officially released. Then on the week of the release, the album sold 143,644 units—enough to place it at number thirteen on the chart. But if the previous week's sales not been made before street date and instead sold during the first week, the additional 1,484 units would have been enough to rank the album at number twelve for the week, above Nelly, except for the fact that he also had some early sales.

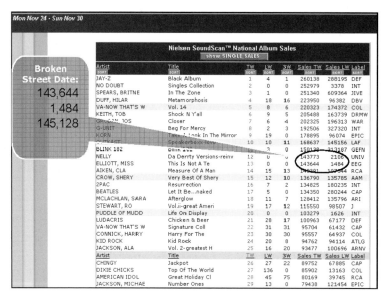

▲ *Figure 21.1 Effect of breaking street date on chart position*

Timing of the individual elements is also crucial. The first single is usually released several weeks before the album, to create an initial demand for the album. Then a second single is released to further support the album once it has been in the marketplace for a short while. Some new artists on major labels may release a single to radio in the fall to introduce the artist into the marketplace, but wait until early the following year to launch the album and release the second single. The actual timeline varies for different genres. Rap/hip-hop albums have a shorter life cycle and the marketing activities may be more compressed. Country music has a longer life cycle and the release of singles may come at longer intervals. It takes a popular rap single only 10.6 weeks to rise to a "Top 5" chart position, whereas it takes 14.8 weeks for R&B, 13 weeks for Top 40, 26 weeks for Adult Contemporary (AC), and 18.6 weeks for rock. The country chart is larger and it takes 19.5 weeks for a popular single to enter the Top 5. In 2000, top pop singles spent the most time on the *Billboard* chart, at 29 weeks, with Rap singles spending the least amount of time.

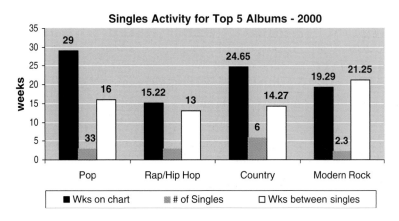

▲ *Figure 21.2 Singles activity for top 5 albums*

Rock and metal acts that depend more on touring may plan events around Ozzfest, The Warped Tour, or other summer tours and festivals. Smaller indie labels may depend more upon touring to sell records, since radio is often too expensive.

Some publicity materials are sent to print vehicles in advance of the release date so that album reviews and features will appear during the week or month of the release. Trade advertising and publicity to radio occurs prior to the release of each single; and to retail, advertising is timed with the release of the album.

All the elements are timed and coordinated to create a comprehensive snowball effect, generating a buzz in the marketplace and convincing gatekeepers and consumers that this project is the "hottest thing" in popular music.

▲ *Figure 21.3 One sheet for Olivia*

MARKETING PLAN

DeMand
De World Over

The following marketing plan is for a fictitious band, DeMand, and is not intended to represent any artist or band, living or dead. The plan is a simulation of the elements that an actual marketing plan would contain.

DeMand
De World Over

Malign Records
Phase 1 Marketing Plan
(updated as of November 26, 2002)

Prepare for *DeMand* Domination! First, there was the platinum debut album *DeMand* in 1996 that featured the smash singles "Talk To Me," "On Your Mark" and "Never Break A Promise." Next, came the multi-platinum, follow-up release <u>Enter DeMand</u> in 1998 that took the group to even greater heights with chart topping hits "What Is Your Love" and "There Are Some Tunes." On November 26th, witness history in the making with the third installment in the *DeMand* saga. Last seen together in 1999, Jacqo, Sharp, Marco, Nada and newest member Kino return with <u>De World Over</u>, the long-awaited new album featuring the runaway hit single "I Could Have Been."

Project facts	
Artist:	DeMand
Members:	Bobby "Kino" Keller, Jack "Jacqo" Brown, Leroy "Sharp" Baskins, Timothy "Nada" Bordeaux and Mark "Marco" Adams
Residence:	Baltimore, MD
Album Title:	De World Over
Release Date:	November 26, 2002
Producers:	Kino, Maxie, Sharp, Bobby Campbell, Michael Gaines, Shapia, Dr. Spearson,

Configurations: CD   0 440 052 476-2 $_3$
Cass 0 440 052 476-4 $_7$
LP   0 440 052 477-1 $_5$

Price Code: C18/T12/L15

Order Due Date: 11/8/02

File Under: DeMand

Genre: R&B

Box Lot: 30

Track listing

1. Love Happens
2. Rapturous
3. I Could Have Been ...
4. Should I
5. Get It Done
6. W.A.R.
7. Best Love
8. What She Tells Me
9. All The Time
10. My True Love
11. Always In Love With You
12. Seem To Say I'm Sorry
13. You Did What?

First Single: "I Could Have Been ..."

Servicing Date: September 9, 2002 to Urban and Urban AC radio

Impact Date: October 7, 2002

Project contacts

Management: Kxxxxxx Cxxxx   Assistant: Sxxxxxn Hxxxxx
CMG
Cell: (310) 555-2020

Mxxx ''Hxxxx'' Cxxxxx
Mobile: (818) 555-4040
2way: 7750XXX@skytel.com
Email: XXXXX@XXX.com

Label: Align/Malign
XXX XXXXXX Avenue
New York, NY 10019
Phone: (212) 555-5680

Publicists: Jxxxxxxx Xxxx
Phone: (212) 555-2481
Fax: (212) 555-8264
Email: jXXX.fXXXXXXXX@umusic.com

Cxxxxxxx Bxxxxx
Phone: (323) 555-7409
Fax: (323) 555-0564
Email: sXXXX@aol.com

A&R: Jxxx Jxxxxxx
Phone: (212) 555-5460
Fax: (212) 555-5591
Email: jxxx.jxxxxxx@umusic.com

Marketing: Dexter Story
Phone: (212) 555-4845
Fax: (212) 555-4615
Email: dexter.story@umusic.com
Cell: (917) 555-7419

Assistant: Txxxxxx Gxxxe
Phone: (212) 555-6648
Email:
txxxxxx.gxxxx@umusic.com
Cell: (917) 555-7549

Background information/marketing

Album History:

- ''DeMand'' 1996 – 1.62 million scanned
- ''Enter DeMand'' 1998 – 2.24 million scanned
- Jacqo ''Unleash The Dog'' 1999 – 4.19 million scanned
- Jacqo ''Return of Dog'' 2001 – 446k scanned
- ''De World Over'' – Now

Single History:

- "On Your Mark" 1996 – 1.3 million scanned
- "Talk to Me" 1996 – 756k scanned
- "Never Break A Promise" 1997 – 756k scanned
- "We're Not Into Love" 1997 – 726k scanned
- "There Are Some Tunes" 1997 – 276k scanned
- Jacqo "Uncomparable" 2002 – 1.02 million scanned

Top Soundscan Markets:

- New York
- Los Angeles
- Chicago
- Philadelphia
- Washington DC
- Atlanta
- Dallas
- Boston
- Detroit
- Baltimore
- SF-Oakland-San Jose

Target Audience:

- Core DeMand fan
- Dedicated Jacqo fan
- 18-34 age
- Female 70/30
- Urban, crossover, R&B-savvy

Marketing Tools:

- Bios
- Glossies
- :15 Access Granted World Premier TV spot
- :15 album TV spot
- :30 album TV spot
- :60 radio
- Personality posters
- 2,000 teaser flats
- 10,000 teaser flyers
- 3×1 flats

- Die cut clings
- T-shirts
- Tour posters
- Baseball cards

Marketing objective

Malign Records considers it a privilege to carry on the successful DeMand brand. Designed to reach a broad demographic (e.g. core fan base, Jacqo enthusiasts and beyond), this plan utilizes aggressive and timely marketing. The De World Over project must explode visually and facilitate Malign's radio, video, sales, publicity and new media departments in bringing the BEST maximum visibility to the group. The first "looks" are:

- Online: (9/3) Yahoo.com World Premier download
- Radio: (9/4) single servicing and (10/18) interview at a Baltimore station
- Print: (week of 10/7) Airplay Monitor tip sheet cover (teaser using logo only)
- Group appearance: (10/16) LL Cool J performance at Hammerstein Ballroom
- Television: (10/23) BET Access Granted ("I Could Have Been ..." video shoot – Los Angeles)
- Group performance: (10/29) BET Walk of Fame: Stevie Wonder tribute ("Living For The City")
- Live performance: (11/20) Red Star/Heineken Music Initiative presents ...

Key dates

10/23 – BET Access Granted airing
10/24 – Online premier of "I Could Have Been ..." video on Yahoo.com
10/29 – BET 106th and Park airing
10/29 – BET Walk of Fame: Stevie Wonder tribute airing
11/4 (week of) – MTV2 Soul host appearance
11/6 – MTV DFX video premiere
11/8 – Multi-city promo tour launch through 11/27
11/20 – Red Star/Heineken Music Initiative presents ... DeMand Live at The World (New York)
11/25 – Midnight sale
11/26 – Album release date, BET 106th and Park appearance airing, release party (Baltimore tentative)

11/27 – BET Blueprint Special airing, retail in-store (Baltimore)
12/7 – Dale Tans promotion through 12/11 (St. Croix)
12/9 – 13th Annual Billboard Awards (Las Vegas)

A&R/Artist development/setup

Every effort has been made to transition DeMand from a 4-year hiatus to a full-time endeavor. Back from recording the album in a remote area on the island of St. Croix, the group is willingly engaged in taking themselves to the next level. Artist development and set up include:

- Management by veterans Kelly Smith and Mike "Diggy" Alden
- Media training with the highly-effective Dana Wilson
- Photo shoot with photographer Matthew Miller
- Choreography with talented Rodney Wallace
- Art direction by the awesome Roger Sanjen

Following is our working tag line:

First, there was <u>DeMand</u>!
Next, there was <u>Enter DeMand</u>!
Now, get ready for De World Domination!
DeMand (logo)
"I Could Have Been ..."
The New Single From The Long-Awaited Album
<u>De World Over</u>.

Publicity

Malign will implement a press/promotional plan that targets maximum exposure through print, radio and television. Album reviews and feature stories will be timed to the release of the De World Over album. Some publications and/or television shows will depend heavily on chart positioning and first week sales success. Long and short lead press strategies are as follows:

- Independent publicity coordinated by Cathy Bailey
- Fanzine shoot in September with Anthony Catello in September
- Bio written by Abby Edgar in October
- Extensive EPK shoot in conjunction with the new media department in September/October

- Special press kit servicing in October
- Intimate listening session for key press in November

The biography (subject to revision):

DeMand have shattered sales records and racked up numerous awards. Their singles have soared to the top of the charts, their albums have sold in the millions and they've set the gold standard for R&B. So, after all the number one hits, the platinum CDs, the awards and the love, what could DeMand possibly do to top themselves?

Do it again.

De World Over is the eagerly anticipated third album from one of R&B's most popular and successful acts. Featuring the emotional new single "I Could Have Been", and production and songwriting contributions from Kelly, TJ, Allday and DeMand, De World Over is the album fans have been waiting for and DeMand has always wanted to create. "I really feel that this album is our best effort to date," "declares Jacqo. "We all came to the project really focused and with more energy, intensity and desire and you can feel it." "In many ways this record is a new start," chimes in Nada, who handled the majority of the production. "It wasn't a record we were obligated to make; it was something we really wanted to do."

Wanted to do but waited to do as well. It has been four years since Enter DeMand, the group's multi platinum sophomore CD. The reason for the long hiatus? Jacqo offers,"People had things to deal with in their own lives. We all needed time to do that and then be able to come back fresh. We had solo projects (Jacqo's Soul Music album and Marco's Unleash the Dog and Return of Dog), but finally we had the opportunity to come together."

"It was just time," says Nada, "but more than that it was the call from the street and from the fans. Everyone who had supported us over the past four years and from the very beginning. Making another record was one of those things that you couldn't prevent. All roads were leading to it."

DeMand started down the road towards De World Over with renewed enthusiasm, wisdom, confidence and a new member. Joining the foursome is long time friend and musical collaborator, Kino. Although De World Over marks Kino's DM debut he has been down with the fellows from the very beginning, but due to contractual issues, was prohibited from recording with his friends. "

We were always tight, "Kino offers. "In many ways I was always there, I was the ghost fifth member." Kino officially joined DeMand in 2001, right about the time the group decided to enter the studio. Says Nada: "He came over and started singing and that was pretty much it. He'd written songs on Jacqo's album and done production work with me, so he already had the DeMand Sound."

The new DeMand was undaunted by industry expectations or the pressures of a comeback. "We weren't nervous,"Nada explains, "because we left off at a high point and it never went anywhere. Nobody ever came and knocked down that bar we set. There's not much pressure, it's just making sure we retain and maintain what makes DeMand." "I love it!" laughs Jacqo."I like the pressure. We've always had that driving force, that energy. That constant movement."

And as for any fears as to how DeMand might fit into the current R&B trends Nada shrugs,"All of our music is emotions. It's real life and that doesn't ever go out of style."

Those sort of timeless emotions are evident on the lead off single "I Could Have Been." The passionate mid tempo love song was written by newcomers Allday, who DeMand discovered at a Baltimore talent show. "We saw these guys on stage," Marco recalls, "and they were good. They kinda reminded me of us and after they performed they handed us a CD of their material. I played it and one of the songs was 'I Could Have Been.' As soon as I heard it, I knew it was a hit." In an age of big name producers and starmaking songwriters, handing the creative reigns over to unknowns was a gutsy move. Jacqo dismisses the concerns. "A lot of times people go for the most expensive producers, the hottest guys, but by having Nada, who produced 80 percent of the album, we didn't have to look elsewhere. We are self-contained because we have writers and producers in the group. It was cool that these young guys could come up with a track that embodied that youthful feel and energy, and for us to have a song that wasn't by well-known guys. Just four kids from Baltimore. They captured the DeMand spirit."

More of that spirit is felt on "Rapturous," which Jacqo exclaims is "Pheno-menal!" Written by Nada, TJ and Clive, and produced by Nada, "Rapturous" is a sexy love ballad that deals with the mind games played between lovers. Another sensual ballad is "What She Tells Me" featuring Kinky of EndLess. Offers Nada, who wrote and produced the song, "What She Tells Me" is a different vibe than you might get from an ordinary ballad because it talks about a man talking to another woman about his woman and the comforting feelings that arise from that situation. Everyone who has heard the song is really touched by the depth of the emotions." Emotions of another variety are explored on the rugged "W.A.R." which was written and produced by Nada, Jacqo and Kino. Featuring a sizzling cameo from Gavaera, "W.A.R." is a straight up, gangsta club banger.

The versatility to rock the house *and* whisper sweet nothings has always been at the core of DeMand. DeMand started singing together while they were barely in their teens. In 1992 the group's soaring harmonies and musicality (each of the guys plays several instruments), caught the attention of a local manager and by '95 word of their talent had made its way to then President of Bonk Records Black Music, Che Hendricks. Hendricks asked the teens to record "Talk to Me," which was included on the Chuckie soundtrack. The single went platinum, became a Top 20 pop hit and hit the top of the R&B charts, becoming the first of a string of DeMand singles to do so. DeMand released their debut album DeMand in 1996.The album was an immediate smash, selling over a million copies and yielding hits "On Your Mark (Billboard's #1 single for 1997) "Never Break A Promise" (the group's third consecutive #1 R&B chart topper), and the gold single "You're Not In Love," which was a Top 5 pop and R&B smash.

The group toured extensively around the world and in 1998 was invited to perform at a special celebration for then President Nelson Mandela's 80th birthday in South Africa.

That same year DeMand dropped their sophomore CD Enter DeMand, which debuted at #2 on the Billboard 200 album chart. The CD yielded the hits "What Is Your Love" and "There Are Some Tunes," both of which hit the top of the charts. In 1999 DeMand appeared on Will Smith's hit "Wild Wild West."

Later in 1999 Jacqo released his multi platinum solo debut Unleash The Dogs that contained the international smash "Sarong Song." In 2001 Jacqo followed up that success with the best selling Return Of Dogs. "Even though we did solo stuff, the plan always was for us to get back together, "Jacqo says "I'd always bring it in up in interviews, it was just a matter of waiting for the right moment."

Asked what the difference is this time around, the guys are thoughtful. "We've learned how to deal with and respect each other," Sharp says. "We each had our chance to shine and now coming back together it's made us that much stronger." "I'm really impressed by the versatility of the vocal arrangements this time," Jacqo offers. "In the past most of our songs featured either Sharp or myself on lead but now this is the biggest group effort we've ever made. Everybody is singing and to me it's comparable to what the Temptations did."

Revitalized, renewed and blessed with a collection of honest, emotional songs, DeMand is poised to regain their crown by doing what they do best. "It's like what the intro says," Nada smiles. "'Love me or hate me. I'm gonna take you back to the beginning.'"

THE END

Lifestyle trade and news publication targets include, but are not limited to:

- Vibe
- Jet
- Honey
- Essence
- Ebony
- Upscale
- Savoy
- USA Today

Television targets include, but are not limited to:

- Wayne Brady
- Regis and Kelly
- Jay Leno
- Soul Train
- BET NYLA
- Showtime at the Apollo
- Carson Daly
- Billboard Music Awards

Radio

Phase 1:

First Single: "I Could Have Been ..."

Servicing Date: September 9, 2002

Tip Sheet Advertising: R&B Airplay Monitor (cover), R&R, HITS and SIN

Impact Date: October 7, 2002

Goal: #2 most added at Urban mainstream and top 5 most added at Urban AC on 10/7

Previous early airplay markets (+15 spins):

Hartford, CT
Buffalo, NY
Baltimore, MD
Washington DC
Raleigh, NC
Greenville, SC
Memphis, TN
Augusta, GA
Miami, FL
Orlando, FL
Columbus, OH
Flint, MI
Toledo, OH
Cincinatti, OH
Chicago, IL
St. Louis, MO
Tulsa, OK
Shreveport, LA

National promotion concepts include:

- "I Could Have Been ..."– themed national sweepstakes involving local band competition for an opening slot on upcoming DeMand tour in 2003
- "I Could Have Been ..."– themed national sweepstakes involving flyaway for two to St. Croix for exclusive concert attendance and vacation with DeMand at live performance
- "I Could Have Been ... Your Sweetie" sweepstakes involving female radio call-ins to enter to win flyaway with to undisclosed location
- Win-It-Before-You-Can-Buy-It promotions at major airplay market stations.

Phase 2 (subject to revisions):

On deck is the rhythmic second single "Get It Done" with its anthemic shouts. The single's remix version features multi-Platinum Align South artist Luscious and Shannon (of DRT fame).

Ship Date: 12/16 (vinyl)
                 1/6 (Cdpro)

Video

Production:

Known for their trend-setting visuals, DeMand enlisted prolific and talented director San Johnson to bring "I Could Have Been ..." to life. Shot in an abandoned Masonic Temple in San Francisco, California, the video captures the essence of the strong individual personalities as well as the group synergy. Fans will not be disappointed as they view the signature slick performance in addition to renewed camaraderie among the augmented DeMand.

- Date: September 15–16
- Location: San Francisco, CA
- Theme/treatment: Performance video in avant garde venue
- Director: San Johnson

Promotion: "I Could Have Been ..."

Plans include partnering with the networks to time video-play, appearances, contests and charting with the album release and single phases.

BET

- Video servicing: (week of October 14th)
- Video add date: (week of October 21st)
- Access Granted: taping (September 15th and 16th) and airing (October 23rd)
- 106th and Park walk-on: taping (October 28th) and airing (October 29th)
- Walk of Fame - Stevie Wonder Tribute: taping (October 19th) and airing (October 29th)
- Peak spins and impressions surrounding (November 26th) release
- 106th and Park walk-on: taping (November 19th) and airing (November 26th)
- Blueprint Special: taping (November 6th) and airing (November 27th)
- Turnstyle taping (November 14th) and airing (January 2003)

MTV/MTV2

- Video servicing: November 2002
- DFX: taping (October 29th) and airing (November 6th)

- <u>TRL</u>: taping (TBD) and airing (TBD)
- <u>M2 Soul</u>: taping (November 1st) and airing (week of November 4th)

Regional

- Video servicing: (week of October 21st)

<u>Phase 2 (subject to revisions)</u>:

Second single video: "Get It Done"

- SHOOT – Middle to end of January 2003
- SERVICING – February 2003

New media

The new media campaign—led by Thelma Olivia-Hendon and staff—will complement the overall strategy with innovative website imaging and aggressive viral marketing. The following items are based loosely on the comprehensive plan that is forthcoming:

- 9/3 www.yahoo.com World Premiere of "I Could Have Been..."
- Zen-themed, fan-friendly www.demand.net launch on **November 18th** powered by OKtoSay
- House of Blues Artist of the Month Fall 2002 during upcoming HOB tour
- 10/25 Online World Premiere of DeMand's debut video, "I Could Have Been", on the day after BET's Access Granted premier (7 day exclusive event)
- Win A Date With DeMand/Match promotion on BET.com (Match is the leading online dating service, and this contest will essentially be an online personals ad, where women can match their interests with those of the group. Five lucky winners will win a date with DeMand)
- DeMand EPK "Making The <u>De</u> World Over" posted on website including footage from the video, photo shoot, Quad and Hit Factory recording studios and exclusive interviews
- Online premiere of the DeMand album on BET.com during the week of release as a part of their "Get It First, Get It Fast" promotion
- Building online street team with FanScape
- "REMIX" contest featuring an a cappella download and a final judge (phase 2)
- BET.com "Video Girl" contest (proposed)

Advertising

Television:

- A 1-week :15 teaser flight world that premiers the BET Access Granted airing on October 23rd
- A 5-week :15 spot flight at BET targeting the highly-rated 106th and Park, Hits from the Street, Comic View and NYLA/BET.com programming (11/18 – 12/9)

Radio:

Urban and cross-over radio time buys planned surrounding album release

- Spot: :60
- Flight: 11/22 – 11/29
- Markets (Stations):
  - New York (3 stations: WQHT, WWPR, WBLS)
  - Philadelphia (2 stations: WUSL, WPHI)
  - Boston (WJMN)
  - Baltimore (2 stations: WERQ, WXYV)
  - Washington, DC (2 stations: WPGC, WKYS)
  - Norfolk (WOWI)
  - Charlotte (WPEG)
  - Raleigh (WQOK)
  - Atlanta (WVEE)
  - Detroit (WJLB)
  - Chicago (WGCI)
  - Los Angeles (KKBT)

Print: full-page album release advertising with Sam Goody retail tag in the following trades:

Vibe	(street date: December 3rd – circulation: 777k)
King	(street date: November 19th – circulation: 150k)
One World	(street date: December 2nd – circulation: 215k)
Word Up!	(street date: December 10th – circulation: 85k)
Right On!	(street date: December 10th – circulation: 76k)
Sister2Sister	(street date: December 13th – circulation: 154k)
Black Beat	(street date: December 27th – circulation: 76k)
Honey	(street date: January 7, 2003 – circulation: 350k)

Out of home: MovieTunes December run pending approval

UMVD books ship: 10/23

Album solicitation period: 10/28 - 11/8

Initial ship: 300,000

- National Account Visibility Co-op programs in development
- Mid-Level Chain and Urban Indie campaigns in development
- Beginning November 1st, DeMand will be part of the monthly Sam Goody Hot 97 mixshow spots and each of their stores in the Northeast region will participate in a pre-order campaign

Phase I: Creating the buzz at retail

Local market retail representatives tied the DeMand album release into the Heineken Music Initiative: Red Star parties in order to create a buzz with the key retail consumers. Major markets where events took place included:

Dallas
Atlanta
Philadelphia
Los Angeles
Chicago
Baltimore
New York

Phase 2: Visibility

Prime visibility was secured for the DeMand release at all key Urban Independent Coalition and Urban Leaning Local Chains. Front windows were purchased at accounts in the following top 20 markets:

New York
Philadelphia
Los Angeles
Chicago
Washington, DC
Baltimore
Virginia
Raleigh

Greensboro
Charlotte
Atlanta
Detroit
Dallas
Houston
Jackson, MS
Cleveland
San Francisco
Seattle

Visibility Programs are running from November 20th through January 1st

Belly Bandit shrink wrap program with remix CD including Target campaign (pending)

Phase 3 : Added Value Promotions

Each key Urban Independent Retailer and Mid-level Chain in key airplay markets will execute an added value promotion in which each account will receive a quantity of customized 18×24 DeMand Personality Posters to reward their customers with upon each purchase of the album.

International

Initial Press Interviews: October 25th

Proposed travel: January 2003 (Europe), March 2003 (Europe/Japan)

Plan: De World Over will have a 'soft' release overseas simultaneously with the U.S. to avoid imports. However, we will go after the commercial single and a major marketing push in 2003.

Therefore, we are concentrating on the major markets in the next 2 months, Canada being most important, along with long lead press in the UK and Japan, and Germany, with France and Holland following as second priority.

Promotion plans are coming in from the above listed territories and priority interviews and phoners will be slotted into the 12pm–3pm interview time from Oct. 22–Nov. 3, where time is available (will be submitted via AIS).

Red Star Sound presents ... DeMand Performance: Using this event for media to get their first look at the new DeMand. Have asked for international promotion time the day before, so we can invite media to fly over for an interview and to cover the gig.

Will service bio, some publicity photos, EPK, and other tools when they become available, to the territories. Will do major mail-out in 2003 with highlights from the U.S. story, and will set up overseas promotion trips around single release in 2003.

Tour

It's business as usual with De World Over as DeMand will support the album release with a 2-week national promo tour in November featuring radio, local video, local press, retail and new media elements. Further, club promotions and charitable activities will also take place in select markets.

In 2003, DeMand will embark on their signature live performance tour. The group will headline a theatre run in the 2,000-4,000-seat range with 2 to 3 opening acts.

- 11/26 – Promo tour launch in the following cities:
  - Baltimore, MD
  - Atlanta, GA
  - Birmingham, AL
  - Memphis, TN
  - Chicago, IL
  - Detroit, MI
  - Philadelphia, PA
  - Washington DC
  - Richmond, VA
  - Raleigh, NC
  - Charlotte, SC
  - New York, NY
  - Los Angeles, CA
- 12/09 – Promo tour ends (in Baltimore)
- 12/11 – Dale Tans promotion in St. Croix
- January 2003 – Head lining small venue or theatre performance tour

Featuring:

- Tour name (TBD)
- Ticket giveaway at Urban and cross-over radio with national contest

- On-site retailer with DeMand catalog
- Merchandising

Agency: TBD

Set list (subject to revision):

- "Talk to Me"
- "On Your Mark"
- "There Are Some Tunes"
- "Five in Time"
- "What Is Your Love"
- "Splendor"
- "You're Not In Love No More"
- "I Could Have Been"
- "My True Love"

| **Street team** |

Malign's world renowned street team will national target lifestyle events and locations frequented by our targeted Urban and cross-over demographic. The online street team will also be enlisted targeting Align's email lists and Urban lifestyle websites. An initial street team assault is as follows:

- <u>10/19</u> Howard Homecoming weekend in Washington DC with 1×1 flats and 6×8 flyers advertising the album in store date and key BET information
- <u>11/26</u> Release party at a Baltimore location TBD
- <u>12/09</u> Los Angeles-based assault in early December with a 2-day party and live performance (tentative)

| **Lifestyle/Cross-marketing** |

Malign is in pursuit of corporate partners and lifestyle companies that are committed to tapping into DeMand's and the label's own Urban demographic.

Red Star Sounds Presents . . .

- "W.A.R." song inclusion on the Heineken beverage compilation CD in stores November 5th
- Event sponsorship in New York in November

- Mention on all merchandise including 1×1 flats, clings, snipes and flyers for distribution in October
- Contest tie-in's and entry at malign.com, demand.net and bet.com (11/5 – 12/10)

Others TBD

Timeline

AUGUST

8/21 Creative Conference Call (Stylist, photographer and groomer selected)

8/22 Placeholder launch for www.demand.net

8/26-27 EPK shoot (Los Angeles)

8/28 Album delivered

SEPTEMBER

9/3 "I Could Have Been" World Premiere on Yahoo.com

9/4 "I Could Have Been" ships

9/8-14 Rehearsal (Baltimore/Los Angeles)

9/13 Fitting (video)

9/15-16 Video shoot (San Johnson - Los Angeles)

9/18 Fitting (photo)

9/19 Photo shoot (Matthew Miller - Los Angeles)

9/20 Fanzine shoot (Anthony Catello – Los Angeles)

9/22-24 Media training begins (Dana Wilson)

9/30 Video servicing; album

OCTOBER

10/1 "I Could Have Been" commercial vinyl ships

10/7 Advance Retail Sampler ships, "I Could Have Been" radio impact date

10/7 (week of) Media Training/EPK interviews (Baltimore, MD)

10/9 Album readers circulate, media enhancing, biography interviewing, LL Cool J party attendance (New York, NY)

10/10 EPK shoot/taping (New York, NY)

10/16-17 Stevie Wonder tribute rehearsal (Baltimore, MD)

10/17 Vibe interview (part 2)

10/18 Stevie Wonder dress rehearsal (Washington, DC); radio visit (DC)

10/19 BET Walk of Fame Stevie Wonder tribute performance taping (Washington, DC), Vibe feature interview (Washington, DC)

10/21 Advance Retail Sampler ships, video servicing

10/22 Album film ships; masters ship; P.O.P. allocation; marketing plan due

10/23 BET Access Granted airs

10/24 Press day; Band rehearsals through 11/4 (New York, NY)

10/25 Press day

10/28 BET New Joint walk-on taping

10/29 BET New Joint walk-on airing, MTV DFX taping, BET Walk of Fame Stevie Wonder tribute airing

10/31 Vibe photo shoot (New York)

NOVEMBER

11/1 MTV2 Soul taping (New York)

11/4 MTV DFX airing on MTV2

11/5 Advanced Edited LP ships

11/6 Universal Music Distribution books ship, BET Blueprint special taping (New York)

11/8 Advance edited vinyl ships; begin album solicitation; promo tour launches

11/18 BET National buy begins (:15), www.demand.net launches

11/18 House of Blues promo tour starts through December

11/19 BET 106th and Park walk-on performance taping

11/20 Red Star/Heineken Music Initiative presents ... DeMand Live at The World (New York)

11/22 – 11/29 Radio buys begin in 12 major markets

11/26 Album release (Worldwide), MTV DFX airing, BET 106th and Park walk-on performance airing, release party (tentative)

11/27 End album solicitation, BET Blueprint special airing, retail in-store (Baltimore)

DECEMBER

TBD ''I Could Have Been ... Opening For DeMand'' radio contest launch through December

12/9 2-day party and live performance (Los Angeles)(tentative)

12/10 Right On!/Black Beat magazine street date (full-page ad)

12/11 Dale Tans promotion (St. Croix)

12/9 BET National buy ends (:15), 13th Annual Billboard Music Awards appearance (MGM Grand Hotel – Fox)

12/16 "Get It Done" vinyl ships

12/11 "I Could Have Been..in St. Croix" performance/flyaway (tentative)

JANUARY 2003

1/6 "Get It Done" CD ships

TBD Theatre performance tour launch

TBD 2nd single launch

FEBRUARY 2003

TBD "Get It Done" impact date

2/2 – 2/12 International promo tour

Album in Stores November 26th
2nd Phase Marketing Posts

Tools:

- T-shirts
- Information booklets (ex. Musiq)
- Get It Done postcards
- Baseball cards
- Glossies
- Tour posters

Advertising:

- Radio:
  - Flight: Surrounding peak airplay period
  - Markets: Top 20 airplay markets
  - Creative: Spot consists of girls singing the chorus that leads into the real music and a voice-over that says "Get It Done (about getting) the new single from DeMand's De World Over in stores now"
- Television
  - MTV – :30 local flight surrounding peak videoplay for "Get It Done"
  - BET – :15 national flight on 106th and Park, HITS, BET.com and Comic View
- Print
  - Full page ads and contest incentives in fanzines such Word Up!, Right On!, Black Beat and J-14

Tour:

- HOB tour and spot radio dates in January 2003 with live band and teaser 30 minute performance
- International promo tour in February 2003 hitting European territories and press

- Live theatre tour in March 2003 featuring opening new artist and established act

Other live tour elements include full service merchandising, radio visits, BET NYLA/MTV News coverage and film taping for DVD or 3rd single video content

Press/Publicity:

Print:

Targeting 2nd phase covers and features in major Urban and lifestyle trades like Vibe, Ebony, Savoy and Upscale as well as fanzines such as Word Up!, Right On! and Black Beat.

Television:

Targeting the following shows for 2nd phase visual injection

- TRL
- Soul Train
- Jay Leno
- David Letterman
- Carson Daly
- Conan O'Brien
- BET NYLA

Fanzine shoot #2

Radio:

"Get It Done" featuring Luscious and Shannon

Ship Date: TBD
Impact Date: TBD

Contest Idea: "Get It Done ... The Music of DeMand" contest where station plays a quick montage of DeMand hits (or songs from the new album) and then callers guess the correct songs to be entered to win a trip to Shaolin

Contest Idea: "Get It Done ... We're Making A Difference" contest where competing schools collect the most winter coats for the homeless and enter to win a free assembly concert from DeMand

<u>Contest Idea</u>: "Get It Done" sweepstakes where female callers at radio sing the chorus or verse on air to win the album

Video:

Fun and moving performance video concept with crisp imaging of the group is a must for "Get It Done." The video must be a signature DeMand visual, but communicate a rare sense of humor and choreography. San Johnson is a strong candidate, but talents like Erik White and Benny Boom also come to mind to direct.

Shoot Date: January 2000

Servicing Date: February 2003

<u>MTV Making of the Video</u> would be a welcome element to the campaign.

New Media:

- Regular <u>www.demand.net</u> updates
- Contest pop-up with gadget or overseas trip giveaway (with tie-in ads at fanzines)

Cross-marketing:

Targeting lifestyle companies for tour and campaign funding including but not limited to the following:

- Coca Cola/Sprite
- MGD
- Target
- Jeep
- Sears
- Gap

Index